NON-EQUILIBRIUM STATISTICAL MECHANICS

ILYA PRIGOGINE

Professor of Physical Chemistry and of Theoretical Physics
Université Libre, Brussels, Belgium

DOVER PUBLICATIONS, INC.
MINEOLA, NEW YORK

Bibliographical Note

This Dover edition, first published in 2017, is an unabridged republication of the work originally published in 1962 by Interscience Publishers, New York, as Volume I of the series "Monographs in Statistical Physics and Thermodynamics."

Library of Congress Cataloging-in-Publication Data

Names: Prigogine, I. (Ilya), author.
Title: Non-equilibrium statistical mechanics / Ilya Prigogine.
Description: Mineola, New York : Dover Publications, Inc., 2017. | "This Dover edition, first published in 2017, is an unabridged republication of the work originally published in 1962 by Interscience Publishers, New York, as Volume I of the series "Monographs in Statistical Physics and Thermodynamics"--Title page verso. | Includes bibliographical references and index.
Identifiers: LCCN 2016042731| ISBN 9780486815558 | ISBN 0486815552
Subjects: LCSH: Statistical mechanics.
Classification: LCC QC175 .P762 2017 | DDC 530.13--dc23 LC record available at https://lccn.loc.gov/2016042731

Manufactured in the United States by LSC Communications
81555202 2017
www.doverpublications.com

Contents

Introduction

1. The earliest quantitative definition of irreversibility occurs in the formulation of the second law of thermodynamics. Here the introduction of the entropy concept permits the classification of natural events into reversible processes, in which entropy remains constant, and irreversible processes, in which entropy increases with time. The classification is purely phenomenological, however, and is clearly insufficient to establish the connection between irreversibility and the laws of mechanics.

The link with mechanics first appeared through the kinetic theory of gases, whose foundation was laid down more than a hundred years ago in a rapid succession of papers by Kronig (1856), Clausius (1857), and Maxwell (1860). These researches culminated in the famous H-theorem of Boltzmann (1872). Its fundamental importance arises from the introduction of the quantity H, defined in terms of the molecular velocity distribution function, which behaves exactly like the thermodynamic entropy.

A characteristic feature of this stage of the theory was the free mixing of mechanical and probabilistic concepts. In Kronig's own words (1856): "The path of each molecule must be so irregular that it will defy all calculations. However, according to the laws of probability theory, one can assume a completely regular motion instead of this completely irregular one." Such considerations also appear in Boltzmann's integro-differential equation for the rate of change of the velocity distribution. In the absence of external forces this well-known equation takes the form

$$\partial f/\partial t + \boldsymbol{v} \cdot \partial f/\partial \boldsymbol{x} = (\partial f/\partial t)_{\text{coll}}$$

Now the "flow" term $\boldsymbol{v} \cdot \partial f/\partial \boldsymbol{x}$ for a system of non-interacting particles can be derived from mechanics (Liouville's theorem). On the other hand, the collision term is not deduced from mechanics alone, but contains a probabilistic assumption about the number of collisions (the so-called "Stosszahlansatz"). Moreover, the collision term is estimated as though there were no flow, and the forces responsible for collisions are neglected in the flow term.

There is certainly a profound physical meaning in Boltzmann's equation, as is borne out by its remarkable agreement with experiment in the calculation of transport coefficients for dilute gases. What remains unclear in Boltzmann's derivation is the range of validity of his equation. This question, as do the controversies which were raised about Boltzmann's work by Loschmidt (1876), Zermelo (1896), and others, originates in the somewhat uncritical use of probabilistic concepts.

The task of formulating a general theory of irreversible processes has acquired new urgency in recent years. The reason is not only that time, so closely related to irreversibility, remains one of the basic problems of physics, but also that there is an enormous range of experimental conditions in which transport or relaxation phenomena are now being studied. Starting from low temperature transport processes in liquid helium or in superconductors and mounting to high temperature processes in fully ionized plasmas, the range of energies covers ten powers of ten! And yet until recently all attempts to extend Boltzmann's original derivation to situations different from those for which it was derived have failed.

To us the only hope for obtaining a general theory of non-equilibrium processes seems to be to reformulate the entire problem in a more systematic way on a purely mechanical basis. This will be our main goal here.

2. One of the main purposes of such a theory is to achieve a generality comparable to that of equilibrium statistical mechanics.

While the formal structure of equilibrium statistical mechanics already appears clearly in the fundamental investigations by Gibbs (see his *Collected Works*, 1928), its power was only recognized around 1930 when among many other advances Ursell (1928) and Mayer (1937) applied it successfully to the problem of the equation of state (see Fowler and Guggenheim, 1939). It was therefore natural to try to adopt a similar, general point of view for non-equilibrium processes and to investigate the relation between transport equations like the Boltzmann equation and the Liouville equation which is the basic equation of ensemble theory. The pioneering work in this direction is due to Yvon (1935), Born and Green (1946 and 1947), and Kirkwood (1946). These first attempts are discussed in many papers (see for example de Boer, 1948 and 1949; Prigogine, 1958; for a recent discussion of the theory of Kirkwood and his co-workers see specially Rice and Frisch, 1960), and we shall therefore not go into details here. The importance of these contributions lies in the generality of their starting point as well as, in the case of Kirkwood's theory, in the possibility of applications to dense media. However many aspects of these theories remain obscure. Supplementary assumptions have to be introduced and no *systematic* way of going beyond the classical Boltzmann equation is indicated.

At the time of the work of Born and Green and Kirkwood, Bogoliubov (1946) proposed a different and very original approach. An important feature of his theory is a clear distinction between the time scales involved. We have at least two characteristic times involved: the duration of an interaction t_{int} ($10^{-12} - 10^{-13}$ sec) and the relaxation time, which for a dilute gas is of the order of the time between collisions ($10^{-8} - 10^{-9}$ sec). Now Bogoliubov assumes that after a time of the order of t_{int} there occurs a great simplification in the description of the system: the one-particle distribution function f_1 satisfies a separate equation and the many-particle distribution functions become functionals of f_1. Using these assumptions, Bogoliubov was able to rederive Boltzmann's equation in an elegant way and indicated how at least in principle corrections due to higher densities could be obtained. It is rather remarkable that the general theory we shall develop in this monograph permits us indeed to justify Bogoliubov's assumptions for well-defined classes of initial conditions.

Perhaps the first case in which the possibility of a purely mechanical theory of irreversible processes clearly appeared was the case of interacting particles in a harmonic solid. Because of their linear character, the equations of motion can then be solved exactly and it can be shown that the system approaches equilibrium as closely as is "permitted" by the existence of the invariants of motion. (Klein and Prigogine, 1953; a few recent papers about this subject are Hemmer, Maximon and Wergeland 1958; Mazur and Montroll, 1960).

However in a harmonic system the energy of each normal mode is an invariant and an approach to thermodynamic equilibrium in the usual sense is only possible if these invariants are destroyed by anharmonic forces. Here, essential progress was made by Van Hove (1955) in the quantum mechanical case. Starting with a well-defined assumption about the wave function at the initial time and using only Schrödinger's equation, Van Hove was able to derive a transport equation valid for large t in the case in which the coupling between the degrees of freedom was weak.

The subsequent development of our ideas about non-equilibrium statistical mechanics has led us rather far from the methods used by Van Hove. However, we want to stress the deep influence that Van Hove's work has exerted on our theory, especially in its initial stage.

3. Basically, all problems involving the approach to equilibrium are examples of what is generally called the N-body problem. This includes all rate or relaxation processes, the study of all transport phenomena, the formulation of hydrodynamics — in other words, a great deal of physics and of chemistry. In all these problems the systems have many degrees of freedom, and it is the interactions among these degrees of freedom that permit the system to approach equilibrium. Equilibrium statistical mechanics is essentially much simpler; in such standard problems as the perfect gas or the harmonic solid the Hamiltonian H can be expressed as the sum of non-interacting contributions H_j

$$H = \sum H_j$$

(The H_j represent the molecular kinetic energies in the case of a gas, the normal modes for the solid.) The partition function thereby factorizes, and it becomes easy to calculate all relevant thermodynamic properties. But such systems are only of marginal interest in the study of irreversible processes, because, as we have already mentioned, without interactions the system cannot evolve towards equilibrium. Therefore we must focus our attention on the interactions among the degrees of freedom. The resultant complicated N-body problem can only be approached at present by the methods of *perturbation theory.*

But even the use of perturbation theory involves grave difficulties. Let us enumerate the most striking among them:

(a) The number of degrees of freedom is enormous for all systems of interest, and we are therefore led to investigate the limit $N \to \infty$, volume $\to \infty$, with the ratio of N to volume (i.e., the concentration) remaining constant. Such a limiting procedure also has to be used in equilibrium problems, and its meaning in quantum statistics and nuclear structure problems has been discussed in detail (see, e.g., Brueckner, 1958; Hugenholtz, 1958). Because of the dynamic nature of our problems, we must use this limiting procedure with even greater care.

(b) In many problems one is interested in times long with respect to the duration of an interaction process. This will be the case for example when we shall prove that for sufficiently large t the interactions drive the system to thermodynamic equilibrium (H-theorem, see Chapter 12). In this sense we need a long-time theory.

It is thus easy to see why only a few years ago it seemed hopeless to attempt to solve these formidable problems of N interacting bodies. The situation has improved greatly in recent years because of the development of the so-called renormalization methods of quantum field theory. For this purpose more powerful perturbation techniques were developed to treat interacting fields. Now a field is in effect a system having an infinite number of degrees of freedom, and hence field-theoretic perturbation problems have many features in common with the many-body problem in the limit $N \to \infty$. The perturbation technique described in this book was inspired by the methods of quantum field theory and corresponds to a kind of feedback from quantum mechanics to classical mechanics.

4. We will start from the Liouville equation describing the evolution of the density ρ of the system in phase space. As is shown in Chapter I, the Liouville equation may be written in the form

$$i\partial\rho/\partial t = L\rho$$

where L is a linear Hermitian operator in phase space. Starting from this equation, the development of non-equilibrium statistical mechanics proceeds without the introduction of any principles not already included in classical or quantum mechanics. However, one is led to adopt new points of view and, one may say, to a new conception of classical (or quantum) mechanics.

In the usual presentation of mechanics the essential quantities are the coordinates and momenta; their rates of change are given by Hamilton's canonical equations (or their equivalent). Here, however, the basic quantity is the statistical distribution function ρ, from which the average values of all functions of coordinates and momenta may be computed. Thus we may say that a knowledge of ρ implies complete knowledge of the "*state*" of the system. When ρ is developed in eigenfunctions of the Liouville operator L (a procedure which turns out to be equivalent to a Fourier expansion in the coordinates), the coefficients in the expansion express the ***inhomogeneities*** and ***inter-particle correlations*** of the system. In this development, the "state" of the system is given by the correlations and inhomogeneities, and the evolution of the system becomes a ***dynamics of correlations***, governed by the Liouville operator L.

We may say that the "objects" of our mechanics are the correlations, and not the coordinates or momenta of the individual particles. (see Chapter 7).

Thus an essential step in our method is the expansion of the phase distribution function in a Fourier series in the coordinates (or more generally in the angle variables, see Chapter 1). This amounts to a change of representation (in the quantum mechanical sense), in which the coordinates are replaced as independent variables by the Fourier indices or "wave vectors." In the new representation the part of the Liouville operator corresponding to the unperturbed Hamiltonian is diagonal, while the part corresponding to the interactions is off-diagonal. It is the off-diagonal character of this operator that allows the description of the time evolution in terms of changes in correlations arising from molecular interactions.

The theory developed in this way is an ensemble theory in the sense that the fundamental role is played by the phase distribution function ρ. As we already mentioned in § 3 of this introduction we shall be interested in "large systems" that is in the limit $N \to \infty$, volume $V \to \infty$, the concentration remaining constant and finite. About the distribution function ρ of such systems *at the initial time* $t = 0$, we shall make the following two fundamental assumptions:

(a) the correlation between two particles vanishes when the distance between these particles tends to infinity;

(b) all intensive properties (pressure, reduced distribution functions, . . .) exist in the limit $N \to \infty$, $V \to \infty$, N/V finite. These restrictions exclude situations that are too "abnormal" in which for example the pressure would in the limiting process $N \to \infty$, $V \to \infty$, become infinite in some regions and vanish in others.

As we shall show, the evolution equations maintain these conditions in time. If they are satisfied for $t = 0$ they remain satisfied for $t > 0$ (see Chapter 7, 11).

These conditions introduce a great simplification into the statistical description of the system. They permit us to separate, in the Fourier expansion of the phase distribution ρ, the space-independent part ρ_0 corresponding to the distribution of the velocities from the space-dependent part; somewhat as in a degenerate Bose gas we may separate the ground state from the excited states. In fact, in our theory the homogeneous, space-independent state plays the role of the ground state and the correlations and inhomogeneities the role of the excitations in quantum theory. This is emphasized by the diagram technique we use in which correlations are represented by directed lines.

A standard question that arises in connection with ensemble theories is their applicability to single systems. This question is discussed in Chapter 7 where it is shown that our theory is also valid for single systems considered in a coarse grained sense. This problem appears at present however as being of academic interest. For example in the quantum mechanical case it can be shown that the theory is valid both for ensembles and for classes of pure states (states characterized by well-defined wave functions) (see also Philippot, 1961).

5. The dynamics of correlations we shall develop in this monograph leads to a clear and simple physical picture of the mechanism of irreversibility, which would be much more difficult to obtain in terms of the rates of change of coordinates and the momenta of individual particles.

This mechanism may be briefly described as a "cascade mechanism" (see Chapter 12). After a time of the order of the duration of an interaction there appears a directed flow of correlations involving a larger and larger number of degrees of freedom which finally disappears in the "sea" of highly multiple, incoherent correlations.

The very existence of irreversibility is closely related to a continuous spectrum of wave vectors, that is, to the limit $V \to \infty$. This connection is studied in detail in the scattering theory discussed in Chapter 6 and the resolvent formalism developed in Chapter 8. The situation here is the same as in other fields of physics, like electromagnetism or quantum scattering theory. It is only in the limit of a large system that we may make a clear distinction between advanced and retarded solutions and therefore make full use of causality conditions. One can even say that irreversibility and the existence of transport equations appears as the consequence of causality conditions applied to N-body systems. However, our theory provides us not only with a qualitative understanding of the meaning of irreversibility, it also gives us the kinetic equations of evolution which are valid for all orders in the interaction constant or the concentration (Chapter 11).

A remarkable feature of these equations is their non-Markowian character. The change in the distribution function at a given time depends on the values of the distribution function over a time interval in the past. However this time interval which measures the "memory" of the system is of the order of the duration of a mechanical interaction. Therefore for all processes in which only the long-time behavior is important, and for which the collisions may be treated as instantaneous, the non-Markowian equations of evolution can be replaced by Markowian ones of a more usual Boltzmann kind. In this way the relation between the mechanics of many-body systems and the random processes that describe their time evolution may be studied in great detail.

For long times the kinetic equations drive the system to equilibrium in complete agreement with equilibrium statistical mechanics. In this way we obtain a dynamic derivation of equilibrium statistical mechanics as the asymptotic form of non-equilibrium statistical mechanics. This involves not only the velocity distribution function, but all properties, such as correlations or the virial expansion of the pressure, which can be expressed in terms of a finite number of particles (see Chapter 12).

This derivation of equilibrium statistical mechanics, including effects of interactions to all orders, may be considered as a wide generalization of Boltzmann's classical H-theorem, in which interactions were only taken account of as the mechanism necessary to obtain equilibrium, but neglected in the asymptotic equilibrium state itself.

The aim of this book has been to present the classical rather than the quantum mechanical form of non-equilibrium statistical mechanics. It is remarkable however that almost the whole formalism of the theory may be directly transcribed into quantum mechanical language (see Chapter 13 and for more details, Prigogine and Ono, 1959; Prigogine, Balescu, Henin, and Résibois, 1960; Résibois, 1960 and 1961).

The formalism studied in this monograph leads therefore to a great unification in statistical physics. It is applicable both to non-equilibrium and equilibrium situations, in the frame of classical or quantum mechanics.

6. This book has been written using an inductive approach in preference to a deductive one. The simplest situations are discussed first and, as the reader gains familiarity with the basic techniques, more general and complex problems are introduced. This method is occasionally repetitious, but will, we hope, help the reader to develop a real physical insight.

The basic chapters developing the general theory are Chapters 1, 7, 8, 11, and 12. Chapters 2 through 6 and 9, 10, and 14 deal with applications.

However the applications we discuss are treated more as illustrations of the theory than for their own sake. Professor R. Balescu is at present engaged in the preparation of a monograph on the statistical mechanics of charged particles in which further applications of the methods studied here will be found. We also intend to devote a separate monograph to the quantum mechanical theory of non-equilibrium statistical mechanics.

My co-workers have played an essential role in the elaboration and the development of the ideas presented in this book. I should like to stress especially the important contributions due to Professor R. Brout, Professor R. Balescu, Dr. F. Henin, Professor J. Philippot and Dr. P. Résibois.

Professor R. Brout (now at Cornell University, Ithaca, New York), was associated with me at the early stage of this work. The general methods of Chapter I as well as the theory of weakly coupled systems (Chapter 2) were developed in collaboration with him.

Professor R. Balescu has worked with me on the theory for weakly coupled gases (Chapter 4). He is associated with the development of the diagram technique which plays an important role in the whole theory (see especially Chapter 7), as well as with the statistical formulation of hydrodynamics (Chapter 10). He also worked out the application to plasmas summarized in Chapter 9.

Dr. F. Henin has contributed essentially to the theory of anharmonic oscillators (Chapter 2) and to the theory of scattering (Chapter 6). Also the general theory of the approach to equilibrium to an arbitrary order (Chapter 12) has been worked out in collaboration with her.

Professor Jean Philippot's contribution is mostly in the field of Brownian motion (Chapter 3). Other contributions of Professor Philippot to the theory of irreversible processes refer to spin relaxation and are not included here.

Dr. P. Résibois' main contribution is in his development of the resolvent formalism which allows one to handle the time dependence of diagrams (Chapter 8) in an especially convenient way. This was an essential step for the derivation of the general kinetic equation (Chapter 11) and the general H-theorem (Chapter 12). He is also largely responsible for the final version of the scattering theory in the Liouville formalism (Chapter 6). This scattering formalism is the main tool used in the discussion of the problem of the existence of analytic invariants in Chapter 14.

Moreover Professor Balescu has written Chapter 10 of this book; and Dr. Henin, Appendix I.

In addition I should also like to mention my other co-workers who have in different ways contributed to the rapid growth of the theory. They are Dr. F. Andrews, Harvard University; Dr. A. Babloyantz, University of Brussels; Professor Thor A. Bak, University of Copenhagen; Dr. B. Baranowski, Warsaw; Dr. R. Bingen, University of Brussels; Dr. L. Blum, Buenos-Aires; Mr. J. Brocas, University of Brussels; Mr. P. de Gottal, University of Brussels; Professor S. Fujita; Mr. Cl. George, University of Brussels; Miss M. Goche, University of Brussels; Mr. C. Hauboldt, University of Berlin; Dr. J. Henry, University of Brussels; Dr. J. Higgins, University of Pennsylvania; Dr. G. Klein, Professor I. Krieger, Case Institute, Ohio; Professor B. Leaf, University of Kansas; Dr. J. Light, Harvard University; Professor S. Ono, University of Tokyo; Dr. J. O'Toole, Princeton University; Professor S. Prager, University of Minnesota; Mr. George Severne, University of Brussels; Dr. H. Taylor, University of California; Professor M. Toda, University of Tokyo.

During the preparation of this manuscript I had many stimulating discussions with friends and colleagues. I want especially to express my appreciation to Professor M. Kac, Ithaca, New York; Professor L. Van Hove, Utrecht; Dr. P. Auer, General Electric, Schenectady and Dr. St. Tamor, General Electric, Schenectady.

Dr. Agnessa Babloyantz has been most helpful in the preparation of the manuscript and in proof reading.

Finally, I should like to express my gratitude to the organizations whose support has proved most helpful in the course of the development of these researches: The Air Research and Development Command, United States Air Force, which, through its European Office has sponsored this work from its very start, and the General Electric Research Laboratories of Schenectady, New York. Thanks to the generous support of these organizations, our investigations have taken a much broader scope than would otherwise have been possible.

I should also like to mention the Francqui Foundation, Brussels, Belgium and its director, J. Willems. Thanks to a special grant, I had the opportunity to work on the manuscript of this book under the best possible conditions. The hospitality of Northwestern University, Evanston, Illinois, where part of the final version was written, is also gratefully acknowledged.

CHAPTER 1

The Liouville Equation

1. The Phase Space of a Mechanical System

In this chapter we shall first briefly summarize some of the basic tools of classical statistical mechanics, and derive the Liouville equation. The rest of the chapter is devoted to a discussion of some fundamental properties of the Liouville equation, which we shall constantly use in this book.

In classical mechanics it is convenient to describe the state of a system by the values of the Hamiltonian variables, the coordinates $q_1, \ldots, q_s$ and the corresponding momenta $p_1, \ldots, p_s$. The equations of motion then take the canonical form

$$dq_i/dt = \partial H/\partial p_i \qquad dp_i/dt = -(\partial H/\partial q_i) \qquad (i = 1, \ldots, s) \quad (1)$$

where $H(q_1, \ldots, p_s)$ is the Hamiltonian. We shall always consider conservative systems in which H does not depend explicitly on time.

We see that once the Hamiltonian H is known, the motion of the system is determined.

Let us now imagine a space of $2s$ dimensions whose points are determined by the coordinates $q_1, \ldots, p_s$. To each mechanical state corresponds a point P_t of this space. Equations (1.1.1) are of first order in time. Therefore the position of the initial point P at the time t_0 together with the Hamiltonian completely determines the evolution of the system. This space is called the *phase space.*

Let us consider an arbitrary function of $q_1, \ldots, p_s$. Its change with time will be given by [see (1.1.1)]

$$\begin{aligned} df/dt &= \sum_{i=1}^{s} [(\partial f/\partial q_i)(dq_i/dt) + (\partial f/\partial p_i)(dp_i/dt)] \\ &= \sum_{i=1}^{s} [(\partial f/\partial q_i)(\partial H/\partial p_i) - (\partial f/\partial p_i)(\partial H/\partial q_i)] \\ &= [f, H] \end{aligned} \quad (2)$$

where $[f, H]$ is the so-called "*Poisson bracket*" of f with H. The condition for the invariance of f is, therefore,

$$[f, H] = 0 \tag{3}$$

Clearly

$$[H, H] = \sum_{i=1}^{s} [(\partial H/\partial p_i)(\partial H/\partial q_i) - (\partial H/\partial q_i)(\partial H/\partial p_i)] = 0 \tag{4}$$

This relation expresses the conservation of energy.

2. Representative Ensembles

We shall constantly use the idea of a representative ensemble, which was introduced by Gibbs (1903) as follows:

"*We may imagine a great number of systems of the same nature, but differing in the configurations and velocities which they have at a given instant, and differing not merely infinitesimally, but it may be so as to embrace every conceivable combination of configuration and velocities . . .*"

A Gibbsian ensemble of systems will be represented by a "cloud" of points in the phase space. To this ensemble corresponds the density in phase space

$$\rho(q_1 \cdots q_s, p_1 \cdots p_s, t) \tag{1}$$

As the number of points in the ensemble is an arbitrary constant we shall always normalize ρ.

$$\int \rho(q_1 \cdots p_s, t) dq_1 \cdots dp_s = 1 \tag{2}$$

Therefore

$$\rho dq_1 \cdots dq_s dp_1 \cdots dp_s \tag{3}$$

represents the *probability* of finding at time t a representative point in the volume element $dq_1 \cdots dp_s$ of phase space.

The change of density in every volume element of phase space is due to the difference of the flows across its boundaries. This gives us the continuity equation

$$\partial \rho/\partial t + \sum_{i=1}^{s} [\partial/\partial q_i(dq_i/dt\ \rho) + \partial/\partial p_i(dp_i/dt\ \rho)] = 0 \tag{4}$$

Using the canonical equations (1.1.1), we have

$$\partial\rho/\partial t + \sum_{i=1}^{s} [(\partial H/\partial p_i)(\partial\rho/\partial q_i) - (\partial H/\partial q_i)(\partial\rho/\partial p_i)] = 0 \quad (5)$$

This equation is known as the *Liouville equation.* It is a linear partial differential equation which determines the evolution of the phase density ρ.

Let us introduce the Poisson bracket defined by (1.1.2). We may then write (1.2.5) in the form

$$\partial\rho/\partial t = [H, \rho] \quad (6)$$

Let us now consider the properties of the Liouville equation in more detail.

3. The Liouville Equation

It is proved in all textbooks on partial differential equations that the solution of the Liouville equation and the integration of the canonical equations of motion are equivalent problems. For this reason the Liouville equation has not been studied in great detail until now. However, for systems with many degrees of freedom the Liouville equation has a distinct advantage over the canonical equations because it summarizes the behavior of the whole system in a *single equation.*

Let us multiply (1.2.6) by $i = \sqrt{-1}$ and write

$$i(\partial\rho/\partial t) = L\rho \quad (1)$$

where L is the linear operator:

$$L = -i(\partial H/\partial p)(\partial/\partial q) + i(\partial H/\partial q)(\partial/\partial p) \quad (2)$$

(To simplify the notation we consider a single degree of freedom.)

We do this multiplication for convenience, in order to obtain an operator which, as we shall prove below, is *Hermitian* as are the operators in quantum mechanics.[1]

In terms of the operator L, equation (1.1.3) becomes

$$Lf = 0 \quad (3)$$

[1] For the use of operator techniques in classical mechanics see E. Hopf (1948), where references to the original literature may be found.

L is simply the Poisson bracket considered as an operator which acts on f.

The Liouville equation (1.3.1) is of first order in time. Therefore once ρ is known at time $t = 0$, its evolution is uniquely determined by the laws of mechanics.

The formal similarity between (1.3.1) and the Schrödinger equation

$$i\hbar(\partial\psi/\partial t) = H_{\text{op}}\psi$$

is striking and will permit us to use methods for the study of classical systems which had been originally developed in quantum mechanics.

The distribution function ρ gives us the average, $\bar{F}$, of an arbitrary function, $F(q_1 \cdots q_s, p_1 \cdots p_s, t)$ of the coordinates and the momenta.[1]

$$\bar{F} = \int \prod dq \prod dp \; F(q_1 \cdots q_s, p_1 \cdots p_s, t)\rho$$

We may therefore say briefly that ρ determines the *state* of the ensemble.

Equation (1.3.1) then shows that the evolution of this state is determined by the operator L, which we shall call the *Liouville operator*. As we shall see considering the laws of mechanics in this way will lead us to what we shall call a dynamics of correlations and will play an essential role in the whole theory.

We shall now study some important properties of the operator L. This study will be quite similar to that of the usual quantum mechanical operators. We shall therefore go into a minimum of detail. More developments may be found in textbooks on quantum mechanics.

It is important to observe that because of the linear character of equation (1.3.1), we have a *superposition principle* in phase space. If ρ_1 and ρ_2 are special solutions of (1.3.1) every combination $a_1\rho_1 + a_2\rho_2$ will also be a solution. This, of course, enormously simplifies the study of Liouville's equation.

Let us now verify that L is indeed a Hermitian operator.

[1] We shall often use the notations $\prod dq$, $(dq)^N$, or even simply dq instead of $dq_1 \cdots dq_N$.

We shall use the notation

$$l_{mn} = \int f_m^*(Lf_n)\,dp\,dq \tag{4}$$

where f_m and f_n are arbitrary functions of p, q which are integrable over phase space.

The condition of Hermiticity is

$$l_{mn} = l_{nm}^* \tag{5}$$

where the asterisk indicates the complex conjugate. This implies that the diagonal elements l_{mm} are all real. The proof is straightforward. By partial integration, recalling that f_m and f_n vanish at the limits, we obtain in succession

$$\begin{aligned} l_{mn} &= \int f_m^*(Lf_n)\,dp\,dq \\ &= i\int f_m^*[-(\partial H/\partial p)(\partial f_n/\partial q)+(\partial H/\partial q)(\partial f_n/\partial p)]\,dp\,dq \\ &= -i\int f_n[-(\partial H/\partial p)(\partial f_m^*/\partial q)+(\partial H/\partial q)(\partial f_m^*/\partial p)]\,dp\,dq \\ &= l_{nm}^* \end{aligned} \tag{6}$$

We see that the multiplication by i in (1.3.1) is necessary to obtain the Hermitian operator L. Of course all the required results could be obtained without introducing this factor, but we prefer to make use of the well-known properties of Hermitian operators. For example, a classical theorem states that the eigenvalues corresponding to Hermitian operators are real.

As usual, the eigenfunctions $\varphi_k(p, q)$ and eigenvalues λ_k are defined by the relation

$$L\varphi_k = \lambda_k\varphi_k \tag{7}$$

Multiplication by φ_k^* and subsequent integration gives

$$\int \varphi_k^*(L\varphi_k)\,dp\,dq = \lambda_k\int |\varphi_k|^2\,dp\,dq \tag{8}$$

The left-hand member is precisely the diagonal element l_{kk} [see (1.3.4)] which as we have seen is real; hence λ_k is also real.

As in the case of quantum mechanical operators, we may have either a discrete spectrum, a continuous spectrum, or both. In the case of a discrete spectrum, only discrete values of λ are eigenvalues. We shall see examples of both a discrete and a continuous spectrum later in this chapter. Here we shall use a notation corresponding to a discrete spectrum.

Two eigenfunctions φ_i, φ_j, which correspond to different eigenvalues λ_i, λ_j are orthogonal:

$$\int \varphi_i^* \varphi_j \, dp \, dq = 0 \tag{9}$$

Indeed,

$$L\varphi_i = \lambda_i \varphi_i, \qquad L^* \varphi_j^* = \lambda_j \varphi_j^* \tag{10}$$

Let us multiply the first equation by φ_j^*, the second by φ_i, subtract the second from the first and integrate over p, q. We obtain

$$\begin{aligned} \int \varphi_j^* (L\varphi_i) \, dp \, dq &- \int \varphi_i (L^* \varphi_j^*) \, dp \, dq \\ &= (\lambda_i - \lambda_j) \int \varphi_i \varphi_j^* \, dp \, dq \end{aligned} \tag{11}$$

Taking into account the Hermitian character of L as well as the fact that $\lambda_i \neq \lambda_j$ we obtain the orthogonality condition (1.3.9).

Moreover, since the eigenfunctions as defined in (1.3.7) contain an arbitrary multiplicative constant, we may choose it to satisfy the normalization conditions

$$\int |\varphi_i|^2 \, dp \, dq = 1 \tag{12}$$

Thus, as in quantum mechanics, we have obtained an orthonormal set of eigenfunctions. We shall assume that they also form a complete set.[1] This means that an arbitrary function $f(p, q)$ may be expressed in the form of a superposition of eigenfunctions

$$f(p, q) = \sum_k a_k \varphi_k(p, q) \tag{13}$$

where the a_k are constant coefficients (which may depend on time if f depends explicitly on time), given by

$$a_k = \int f(p, q) \varphi_k^*(p, q) \, dp \, dq \tag{14}$$

[1] This can be verified in examples, but as in quantum mechanics a general proof is lacking.

In the subsequent paragraphs we shall give some examples of the eigenvalues and the eigenfunctions of the Liouville operator.

Let us go back to the Liouville equation (1.3.1), and expand the distribution function ρ as a linear combination of the eigenfunctions φ_k of L according to (1.3.13),

$$\rho(p, q, t) = \sum_k a_k(t)\, \varphi_k(p, q) \tag{15}$$

Using (1.3.1) we obtain

$$i \sum_k (da_k/dt)\, \varphi_k = \sum_k \lambda_k a_k \varphi_k \tag{16}$$

The orthonormality of the functions φ_k permits the reduction of (1.3.16) to

$$i(da_k/dt) = \lambda_k a_k \qquad \text{or} \qquad a_k(t) = c_k e^{-i\lambda_k t} \tag{17}$$

Therefore the formal solution of the Liouville equation has the simple form

$$\rho(p, q, t) = \sum_k c_k e^{-i\lambda_k t} \varphi_k(p, q) \tag{18}$$

This solution is of course the analogue of the well-known solution of the Schrödinger equation (see, for example, Bohm, 1951)

$$\psi(q, t) = \sum_k c_k e^{-iE_k t/\hbar} u_k(q) \tag{19}$$

where the u_k and the E_k are the eigenfunctions and the eigenvalues of the Hamiltonian operator H_{op}.

4. Formal Solution of the Liouville Equation

In order to obtain some familiarity with the formal solution (1.3.18) of the Liouville equation, let us make a few remarks. If our formalism is to make sense, the normalization condition (1.2.2) has to remain valid for all times if it is valid at $t = 0$. This is indeed so. Using (1.3.18) we have

$$\int \rho\, dp\, dq = \sum_k c_k e^{-i\lambda_k t} \int \varphi_k(p, q)\, dp\, dq \tag{1}$$

But

$$\int \varphi_k\, dp\, dq = 0 \qquad \text{for} \qquad \lambda_k \neq 0 \tag{2}$$

This results from

$$\lambda_k \int \varphi_k dp\, dq = \int L\varphi_k dp\, dq$$
$$= i \int [-(\partial H/\partial p)(\partial \varphi_k/\partial q) + (\partial H/\partial q)(\partial \varphi_k/\partial p)]\, dp\, dq$$
$$= 0 \tag{3}$$

by partial integration. Therefore the right-hand side of (1.4.1) is indeed time-independent, and normalization is preserved.

We have

$$\int \rho\, dp\, dq = c_0 \int \varphi_0\, dp\, dq = 1 \tag{4}$$

where c_0 corresponds to $\lambda_k = 0$ and similarly [see (1.3.9), (1.3.12), (1.3.18)]:

$$\int \rho^2\, dp\, dq = \sum_k |c_k|^2 \tag{5}$$

Therefore for all functions ρ whose square is integrable in phase space, the series (1.3.18) converges.

An interesting point is that the eigenfunctions of the Liouville operator are complex. Because of (1.3.2) we see immediately that the complex conjugate L^* is related to L by

$$L^* = -L \tag{6}$$

Therefore we have, taking the complex conjugate of (1.3.7),

$$\begin{aligned} L^* \varphi_k^* &= \lambda_k \varphi_k^* \\ L \varphi_k^* &= -\lambda_k \varphi_k^* \end{aligned} \tag{7}$$

If φ_k is an eigenfunction of L corresponding to λ_k, φ_k^* is an eigenfunction corresponding to $-\lambda_k$. The fact that both λ_k and $-\lambda_k$ are eigenvalues of L may be considered as a direct consequence of the dynamical reversibility of classical mechanics. Indeed, as shown by (1.3.18), the transformation $\lambda_k \to -\lambda_k$ is equivalent to the time inversion $t \to -t$.

Let us write

$$\varphi_k = \Phi_k + i\Xi_k \tag{8}$$

then

$$L^2 \varphi_k = L^2 \Phi_k + iL^2 \Xi_k$$

but

$$L^2\varphi_k = L(\lambda_k\varphi_k) = \lambda_k^2\varphi_k = \lambda_k^2\Phi_k + i\lambda_k^2\Xi_k$$

since the operator L^2 is real, we can equate real and imaginary parts. Therefore Φ_k and Θ_k satisfy the *second-order* differential equations

$$L^2\Phi_k = \lambda_k^2\Phi_k, \qquad L^2\Xi_k = \lambda_k^2\Xi_k \tag{9}$$

The expansion (1.3.18) therefore has some similarity to an expansion in running waves ($e^{i\boldsymbol{k}\cdot\boldsymbol{x}}$ satisfies a first-order equation in $\boldsymbol{x}$, see § 5) while (1.4.8) is similar to the decomposition of this running wave into standing waves ($\sin \boldsymbol{k}\cdot\boldsymbol{x}$ or $\cos \boldsymbol{k}\cdot\boldsymbol{x}$ satisfy second-order equations).

The appearance of couples of eigenvalues λ_k, $-\lambda_k$ corresponding to the eigenfunctions φ_k, φ_k^* also permits us to verify that $\rho(q, p, t)$ remains real. Indeed we may write (1.3.18) in the form

$$\rho(p, q, t) = \sum_k [d_k e^{-i\lambda_k t}\varphi_k(p, q) + d_{-k}e^{i\lambda_k t}\varphi_k^*(p, q)] \tag{10}$$

where d_{-k} is the coefficient which corresponds to $-\lambda_k$ in the expansion.

Therefore, if ρ is real at time $t = 0$,

$$d_{-k} = d_k^* \tag{11}$$

and ρ will remain real for all time.

Instead of separating the real and the imaginary parts of φ_k as in (1.4.8), it is also interesting to write φ_k in the form

$$\varphi_k = A_k e^{i\chi_k} \tag{12}$$

where A_k is the modulus of φ_k

$$A_k = (\varphi_k\varphi_k^*)^{1/2} \tag{13}$$

and χ_k a phase angle. The operator L is a first-order differential operator. Therefore, also taking account of (1.4.6), we verify immediately that

$$LA_k = 0 \tag{14}$$

and

$$L\chi_k = (1/i)\lambda_k \tag{15}$$

Relation (1.4.14) is especially interesting. It shows that each eigenfunction φ_k of the Liouville operator furnishes an *invariant of the motion*, A_k [see (1.3.2) and (1.1.2)]. The determination of the eigenfunctions of the Liouville operator is therefore closely related to the problem of finding the invariants of the motion for the given Hamiltonian.

5. Non-Interacting Particles

Let us now consider a few simple situations for which we may explicitly calculate the eigenvalues and the eigenfunctions of the Liouville operator.

First let us consider a single particle enclosed in a large cubic box of length L. The Hamiltonian is

$$H = p^2/2m \tag{1}$$

and the Liouville equation (1.2.5) becomes

$$\partial\rho/\partial t + (\boldsymbol{p}/m) \cdot (\partial\rho/\partial\boldsymbol{x}) = 0 \tag{2}$$

corresponding to the Liouville operator [see (1.3.2)]

$$L = -i(\boldsymbol{p}/m) \cdot (\partial/\partial\boldsymbol{x}) \tag{3}$$

Therefore, the eigenfunctions and the eigenvalues will be given here [see (1.3.7)] by

$$-i(\boldsymbol{p}/m) \cdot (\partial\varphi_k/\partial\boldsymbol{x}) = \lambda_k\varphi_k \tag{4}$$

The eigenfunctions and eigenvalues are clearly of the form

$$e^{i\boldsymbol{k}\cdot\boldsymbol{x}} \tag{5}$$

and

$$\lambda_k = \boldsymbol{k} \cdot (\boldsymbol{p}/m) \tag{6}$$

where $\boldsymbol{k}$ is a real vector, the value of which is determined by the boundary conditions. We see that the eigenfunctions are simply plane waves.

In all subsequent chapters we shall be interested only in large systems such that the length of the cubic box $L \to \infty$. Therefore, the choice of the boundary conditions should become irrelevant in this limit and we shall use the simplest possible ones, which are the *periodic* boundary conditions.

In other words we shall require that

$$\varphi_k(\boldsymbol{x}) = \varphi_k(\boldsymbol{x} + \boldsymbol{L}) \tag{7}$$

or

$$e^{i\boldsymbol{k}\cdot\boldsymbol{x}} = e^{i\boldsymbol{k}\cdot(\boldsymbol{x}+\boldsymbol{L})} \tag{8}$$

We then have

$$\boldsymbol{k} = (2\pi/L)\boldsymbol{n} \tag{9}$$

where $\boldsymbol{n}$ is a vector whose components are integers. For a finite volume we have a discrete spectrum which becomes continuous in the limit $L \to \infty$. It may immediately be seen that the φ_k form an orthogonal set.

Because the Hamiltonian does not depend on the coordinates, the Liouville operator contains the momentum as a parameter.[1] It is then more convenient to define the normalization of the eigenfunctions by

$$\int |\varphi_k|^2 \, dq = 1 \tag{10}$$

instead of by (1.3.12). This gives us the normalized eigenfunctions

$$\varphi_k = (1/L^{3/2})\, e^{i\boldsymbol{k}\cdot\boldsymbol{x}} \tag{11}$$

These eigenfunctions form a complete set in configuration space $\boldsymbol{x}$ (but not in phase space). We can, however, still represent an arbitrary function $f(\mathbf{p}, \boldsymbol{x})$ as a superposition of eigenfunctions $\varphi_k(\boldsymbol{x})$, but with coefficients which are now functions of the momenta [see (1.3.13) and (1.3.14)]:

[1] We could also write the eigenfunctions and eigenvalues of (1.5.4) in the form (see Appendix II)

$$\varphi_k \sim e^{i\boldsymbol{k}\cdot\boldsymbol{x}}\, \delta(\boldsymbol{p} - \boldsymbol{p}_0)$$
$$\lambda_k = \boldsymbol{k}\cdot(\boldsymbol{p}_0/m)$$

but here we shall always drop the Dirac delta function $\delta(\mathbf{p} - \mathbf{p}_0)$ and treat the momentum as a parameter.

$$f(\boldsymbol{p}, \boldsymbol{x}) = \sum_k a_k(\boldsymbol{p}) \varphi_{\boldsymbol{k}}(\boldsymbol{x})$$
$$= 1/L^{3/2} \sum_k a_k(\boldsymbol{p}) e^{i\boldsymbol{k} \cdot \boldsymbol{x}} \tag{12}$$

with

$$a_k(\boldsymbol{p}) = 1/L^{3/2} \int f(\boldsymbol{p}, \boldsymbol{x}) e^{-i\boldsymbol{k} \cdot \boldsymbol{x}} d\boldsymbol{x} \tag{13}$$

These formulae correspond to the usual Fourier expansion of $f(\boldsymbol{p}, \boldsymbol{x})$. Similarly (1.3.18) now becomes [using the letter ρ_k for the Fourier coefficients and (1.5.6)]

$$\rho(\boldsymbol{p}, \boldsymbol{x}, t) = (2\pi/L)^3 \sum_k \rho_k(\boldsymbol{p}) e^{-i\boldsymbol{k} \cdot (\boldsymbol{p}/m)t} e^{i\boldsymbol{k} \cdot \boldsymbol{x}}$$
$$= (2\pi/L)^3 \sum_k \rho_k(\boldsymbol{p}) e^{i\boldsymbol{k} \cdot [\boldsymbol{x} - (\boldsymbol{p}/m)t]} \tag{14}$$

We have introduced

$$\rho_k = (L^{3/2}/8\pi^3)\, a_k$$

for convenience, as will be seen later in this book.[1] Equation (1.5.14) simply expresses the fact that ρ is an arbitrary function of $\boldsymbol{p}$ and of $\boldsymbol{x} - (\boldsymbol{p}/m)t$

$$\rho(\boldsymbol{p}, \boldsymbol{x}, t) = \rho[\boldsymbol{p}, \boldsymbol{x} - (\boldsymbol{p}/m)t] \tag{15}$$

This treatment is easily extended to the case of a number N of molecules which are not interacting. The Hamiltonian is then the sum of the kinetic energies of the molecules

$$H = \sum_{j=1}^{N} p_j^2/2m \tag{16}$$

and the Liouville operator (1.5.3) becomes

$$L = -i \sum_{j=1}^{N} (\boldsymbol{p}_j/m) \cdot (\partial/\partial \boldsymbol{x}_j) \tag{17}$$

The eigenfunctions are the products

$$\varphi_{\{k\}} = [1/L^{3N/2}] e^{i\Sigma \boldsymbol{k}_j \cdot \boldsymbol{x}_j} \tag{18}$$

where $\{\boldsymbol{k}\}$ is an abbreviation for the set $\boldsymbol{k}_1, \cdots, \boldsymbol{k}_N$. Similarly the

[1] This definition allows us to eliminate the factors of 2π in the transition from sums to integrals [see (4.2.4)].

eigenvalues are the sums

$$\lambda_{\{k\}} = \sum_j \boldsymbol{k}_j \cdot (\boldsymbol{p}_j/m) \tag{19}$$

Finally the expansion (1.5.14) becomes the multiple Fourier expansion

$$\begin{aligned} &\rho(\boldsymbol{p}_1, \cdots \boldsymbol{p}_N, \boldsymbol{x}_1 \cdots \boldsymbol{x}_N, t) \\ &= (2\pi/L)^{3N} \sum_{\{k\}} \rho_{\{k\}}(\{\boldsymbol{p}\}) e^{i \sum_j \boldsymbol{k}_j \cdot [\boldsymbol{x}_j - (\boldsymbol{p}_j/m)t]} \end{aligned} \tag{20}$$

The situation we have treated in this paragraph is of course without any real interest. Indeed, if we neglect the interaction between the molecules, we cannot expect an approach to statistical equilibrium. The reason we have studied this situation in some detail is that it will later form the starting point for a treatment of interactions by means of perturbation techniques.

6. Action-Angle Variables

The simplifying feature in the last paragraph was the occurrence of the momenta but not of the coordinates in the Hamiltonian (1.5.1). In many cases it is possible to bring the Hamiltonian into such a form after a change of variables. Let us first consider the case of a one-dimensional harmonic oscillator. The Hamiltonian is

$$H = (p^2/2m) + (kq^2/2) \tag{1}$$

where k is the spring constant related to the frequency ν of the oscillator by $\omega = 2\pi\nu$

$$\nu = 1/2\pi(k/m)^{1/2} \qquad \text{or} \qquad \omega = (k/m)^{1/2} \tag{2}$$

Instead of q, p let us introduce the new variables J, α defined by

$$\begin{aligned} q &= (2J/\pi m\omega)^{1/2} \sin \alpha \\ p &= (2m\omega J/\pi)^{1/2} \cos \alpha \end{aligned} \tag{3}$$

This transformation is quite similar to the transformation from Cartesian to polar coordinates; α is called the angle variable and J, which is the corresponding momentum, the action variable (see Fig. 1.6.1.) With these variables (1.6.1) takes the simple form

$$H = \omega J \tag{4}$$

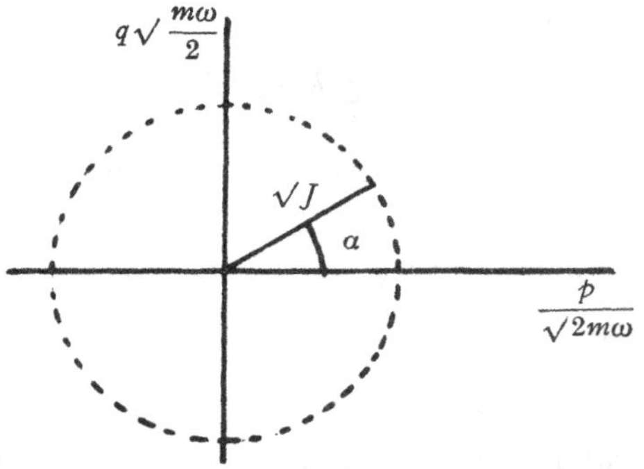

Fig. 1.6.1 Transformation from Cartesian coordinates p, q to action-angle variables J, α for a harmonic oscillator [(see formula (1.6.3)].

The general method for obtaining the action-angle variables is described in detail in most textbooks of analytical mechanics (cf. especially the excellent presentation by Goldstein, **1953**). Let us summarize it very briefly.

Consider a situation in which the Hamiltonian does not depend on the coordinates Q_i but only on the momenta P_i. Such coordinates are called *cyclic*. The equations of motion (1.1.1) are then

$$dQ_i/dt = \partial H/\partial P_i, \qquad dP_i/dt = -\partial H/\partial Q_i = 0 \tag{5}$$

These equations show that the P_i are constants while the coordinates Q_i vary linearly with time,

$$Q_i = \beta_i t + \gamma_i$$

In order to obtain the transformation from the initial set of variables q_i, p_i to the cyclic variables Q_i, P_i, it is useful to introduce the generating function $W(q_1 \cdots q_s, P_1 \cdots P_s)$ which depends both on the old coordinates and on the new momenta. It may be shown (see for example Goldstein, **1953**) that

$$p_i = \partial W/\partial q_i, \qquad Q_i = \partial W/\partial P_i \tag{6}$$

while the Hamiltonian is not affected by this time-independent transformation,

$$\begin{aligned} H(q_1 \cdots p_s) &= H(P_1 \cdots P_s) \\ &= \text{constant} \end{aligned} \tag{7}$$

We shall restrict our discussion to the case in which we have separation of the degrees of freedom $H = \sum_i H_i$ so that

$$W = \sum_i W_i, \qquad W_i = W_i(q_i, P_i) \tag{8}$$

The action-angle variables correspond to a special choice of the cyclic variables P, Q.

The action variables J_i are defined by

$$J_i = (1/2\pi) \oint P_i dq_i \tag{9}$$

where the integration is carried out over a complete period of oscillation or rotation of q_i. Using (1.6.6) this relation may also be written

$$J_i = (1/2\pi) \oint [\partial W_i(q_i, P_i)/\partial q_i] dq_i \tag{10}$$

Expressing the P_i in this way in terms of the J_i, the characteristic function W may be written

$$W = W(q_1 \cdots q_s, J_1 \cdots J_s) \tag{11}$$

while the Hamiltonian is a function of the J_i alone

$$H = H(J_1 \cdots J_s) \tag{12}$$

The generalized coordinates conjugate to J_i are known as the angle variables α_i and are furnished by equations (1.6.6):

$$\alpha_i = \partial W/\partial J_i \tag{13}$$

The equations of motion for the α_i are deduced from (1.6.5)

$$d\alpha_i/dt = \partial H/\partial J_i, \qquad dJ_i/dt = 0 \tag{14}$$

$$\alpha_i = \omega_i t + \delta_i \tag{15}$$

The important point which characterizes this special choice of cyclic variables is that the ω_i are the angular frequencies associated with the periodic motion. That is, if τ_i is the period associated with the motion, then the variation of α_i during a period of oscillation is

$$\Delta\alpha_i = \omega_i \tau_i = 2\pi \tag{16}$$

The frequencies ω_i are directly related to the Hamiltonian by

$$\omega_i = \partial H/\partial J_i \tag{17}$$

or

$$H = \sum_{i=1}^{s} \int_0^{J_i} \omega_i(J_i)\, dJ_i \tag{18}$$

As a consequence of the definition of α_i, q_i is a periodic function of α_i. Therefore a function of q_i may be developed in a Fourier series in α_i,

$$f(q_i) = \sum_m \varphi_m e^{im\alpha_i} \tag{19}$$

To illustrate this procedure let us go back to the case of the harmonic oscillator. Using (1.6.1), (1.6.6), and (1.6.7) we obtain for the generating function W the equation

$$(1/2m)(\partial W/\partial q)^2 + (k/2)q^2 = \beta \tag{20}$$

where β is the constant value of the Hamiltonian. This equation can be integrated to give

$$W = \sqrt{(mk)} \int dq \sqrt{[(2\beta/k) - q^2]} \tag{21}$$

We only need $\partial W/\partial q$ to calculate the action variables (1.6.10)

$$J = (1/2\pi) \oint (\partial W/\partial q)\, dq = 1/2\pi \sqrt{(mk/2\pi)} \oint dq \sqrt{[(2\beta/k) - q^2]} \tag{22}$$

The substitution

$$q = \sqrt{(2\beta/k)} \sin\theta \tag{23}$$

reduces this integral to

$$J = \beta/\pi \sqrt{(m/k)} \int_0^{2\pi} \cos^2\theta\, d\theta = \beta \sqrt{(m/k)} \tag{24}$$

Therefore [see (1.6.2)]

$$\beta = H = \omega J \tag{25}$$

in agreement with (1.6.4). To determine the angle variable we use (1.6.13) as well as (1.6.21) and (1.6.25). We have

$$\begin{aligned} \alpha = \partial W/\partial J &= (\partial W/\partial \beta)(\partial \beta/\partial J) = \sqrt{(k/m)}[\partial W/\partial \beta] \\ &= \int \{1/\sqrt{[(2\beta/k) - q^2]}\} dq = \text{arc sin}\, q\sqrt{(k/2\beta)} \end{aligned} \tag{26}$$

If we solve (1.6.26) in terms of q we recover the first of the relations (1.6.3). Using (1.6.1) as well we obtain the second.

The characteristic feature of the harmonic oscillator is the fact that the frequency ν is independent of J, whereas for all other systems it is still a function of the action variable.

Let us now express Liouville's theorem in action-angle variables.

7. Liouville's Theorem in Action-Angle Variables

We shall first consider the case of a single degree of freedom. The Hamiltonian depends only on the action variable J but is independent of the angle variable. Therefore the Liouville operator is simply [see (1.3.2) and (1.6.17)]

$$L = -i(\partial H/\partial J)(\partial/\partial\alpha) = -i\omega(J)(\partial/\partial\alpha) \tag{1}$$

The situation is therefore very similar to that of non-interacting particles, which we studied in § 5 [see (1.5.3)]. The action is a constant of motion exactly as is the momentum if we neglect the interactions.

The equation for the eigenfunctions is again [see (1.5.4)]

$$-i\omega(\partial\varphi_n/\partial\alpha) = \lambda_n\varphi_n \tag{2}$$

We have to look for *periodic solutions* [see (1.6.19)]

$$\varphi_n(\alpha) = \varphi_n(\alpha + 2\pi) \tag{3}$$

They are

$$\varphi_n = [1/(2\pi)^{1/2}]e^{in\alpha} \tag{4}$$

where n is an integer. The corresponding eigenvalue is

$$\lambda_n = n\omega \tag{5}$$

The factor $(2\pi)^{-1/2}$ has been introduced to ensure the normalization [see (1.5.10)]

$$\int_0^{2\pi} |\varphi_n|^2 \, d\alpha = 1 \tag{6}$$

The essential difference from § 5 is that here the geometrical volume L^3 does not enter into the boundary conditions [compare (1.5.8) with (1.7.3)]. The normalization of the eigenfunctions here, therefore, is also independent of L^3.

This discussion extends easily to the case of many degrees of freedom if the Hamiltonian is a sum of independent contributions [see (1.6.18)]:

$$H = \sum_{j=1}^{s} H(J_j) \tag{7}$$

Each action J_j is then separately a constant of the motion. The eigenfunctions are the products

$$\varphi_{\{n\}} = [1/(2\pi)^{s/2}] \exp\,(i \sum_{j=1}^{s} n_j \alpha_j) \tag{8}$$

with the corresponding eigenvalue

$$\lambda_{\{n\}} = n_1 \omega_1 + \cdots + n_s \omega_s \tag{9}$$

Relation (1.5.20) becomes

$$\begin{aligned} &\rho(J_1 \cdots J_s, \alpha_1 \cdots \alpha_s, t) \\ &= 1/(2\pi)^{s/2} \sum_{\{n\}} \rho_{\{n\}}(J_1 \cdots J_s) \exp\,[i \sum_j n_j(\alpha_j - \omega_j t)] \end{aligned} \tag{10}$$

with

$$\begin{aligned} &\rho_{\{n\}}(J_1 \cdots J_s) \\ &= 1/(2\pi)^{s/2} \int_0^{2\pi} \cdots \int_0^{2\pi} \rho(J_1 \cdots J_s, \alpha_1 \cdots \alpha_s, t=0)\, e^{-i \sum n_j \alpha_j} d\alpha_1 \cdots d\alpha_s \end{aligned}$$

We may note that in action-angle variables the Liouville equation becomes

$$(\partial \rho / \partial t) + \sum_j \omega_j (\partial \rho / \partial \alpha_j) = 0 \tag{11}$$

whose general solution is

$$\rho = \rho(J_1 \cdots J_s,\ \alpha_1 - \omega_1 t, \ldots, \alpha_s - \omega_s t) \tag{12}$$

Clearly (1.7.10) is simply the Fourier expansion of (1.7.12).

8. Liouville's Equation in Interaction Representation

In §5 and §7 we studied systems in which the Hamiltonian is composed of a sum of independent terms. We shall now consider the real physical situation in which the Hamiltonian contains, in addition, a potential energy which couples the different degrees of freedom and permits the system to reach a state of statistical equilibrium. We shall use the language of action-angle variables. Nothing would be changed if we should use $\boldsymbol{x}$, $\boldsymbol{p}$ instead of the action-angle variables, as in §5.

We now consider a Hamiltonian of the form

$$H = H_0(J_1 \cdots J_s) + \lambda V(J_1 \cdots J_s, \alpha_1 \cdots \alpha_s) \tag{1}$$

where H_0 is again, as in (1.7.7), the sum of contributions due to the different degrees of freedom, while V depends both on the action and the angle variables. The parameter λ is a dimensionless quantity characterizing the strength of the potential energy, which we shall find useful in classifying terms in our later developments.

To the decomposition (1.8.1) of the Hamiltonian clearly corresponds a similar decomposition of the Liouville operator (1.3.2). Formula (1.3.1) then becomes

$$i(\partial\rho/\partial t) = (L_0 + \lambda\delta L)\rho \tag{2}$$

where the operators L_0 and δL are given by

$$L_0 = -i \sum_j \omega_j (\partial/\partial\alpha_j) \tag{3}$$

$$\delta L = i \sum_j [(\partial V/\partial\alpha_j)(\partial/\partial J_j) - (\partial V/\partial J_j)(\partial/\partial\alpha_j)] \tag{4}$$

Equation (1.8.2) is of course much more difficult to study than the "unperturbed" equation corresponding to $\lambda = 0$.

We have seen that in this case the solution is given by (1.7.10) where the $\rho_{\{n\}}(J_1 \cdots J_s)$ are arbitrary functions of the action variables but independent of time. Let us now look for a solution of the form

$$\rho(J_1 \cdots J_s, \alpha_1 \cdots \alpha_s, t) = [1/(2\pi)^{s/2}] \sum_{\{n\}} \rho_{\{n\}}(J_1 \cdots J_s, t) \exp [i \sum_j n_j(\alpha_j - \omega_j t)] \tag{5}$$

The Fourier coefficients $\rho_{\{n\}}(J_1 \cdots J_s, t)$ are now functions of time.

The advantage of the form (1.8.5) is that the effects of H_0 and V are separated. The time appears twice — in the exponentials because of H_0 and in the $\rho_{\{n\}}$ because of V.

Substituting (1.8.5) into (1.8.2) we have

$$i(\partial/\partial t) \sum_{\{n\}} \rho_{\{n\}} \exp [i \sum_j n_j(\alpha_j - \omega_j t)] = (L_0 + \lambda\delta L) \sum_{\{n\}} \rho_{\{n\}} \exp [i \sum n_j(\alpha_j - \omega_j t)] \tag{6}$$

We now make use of (1.8.3). This permits us to eliminate L_0 in (1.8.6) and to write it in the simpler form

$$i\sum_{\{n\}}(\partial\rho_{\{n\}}/\partial t)\exp[i\sum_j n_j(\alpha_j-\omega_j t)] = \lambda\delta L\sum_{\{n\}}\rho_{\{n\}}\exp[i\sum n_j(\alpha_j-\omega_j t)$$

We then multiply on the left by $\exp(-i\sum n_l\alpha_l)$, and integrate over all α. Taking account of the orthogonality properties of the eigenfunctions (1.7.8) we obtain

$$i(\partial/\partial t)\rho_{\{n\}}(J_i\cdots J_s, t) = \lambda\sum_{\{n'\}} e^{i\sum n_j\omega_j t}\langle\{n\}|\delta L|\{n'\}\rangle e^{-i\sum n'_j\omega_j t}\rho_{\{n'\}}(J_1\cdots J_s, t) \quad (7)$$

where we use the matrix notation

$$\langle\{n\}|\delta L|\{n'\}\rangle = [1/(2\pi)^s]\int_0^{2\pi}\cdots\int_0^{2\pi} d\alpha_1\cdots d\alpha_s\, e^{-i\sum n_j\alpha_j}\,\delta L\, e^{i\sum n'_j\alpha_j} \quad (8)$$

The equations (1.8.7) are the analogue of the familiar quantum mechanical equations [see (1.3.19)]:

$$i(dc_n/dt) = \sum_{n'} e^{iE_n t/\hbar}\langle n|V|n'\rangle e^{-iE_{n'}t/\hbar}c_{n'} \quad (9)$$

where $\langle n|V|n'\rangle$ is the matrix element of the perturbation. However, in the quantum mechanical case the amplitudes c_n are only functions of time, while the $\rho_{\{n\}}$ are also functions of the actions $J_1\cdots J_s$. It should also be noticed that $\langle\{n\}|\delta L|\{n'\}\rangle$ still contains the operator $\partial/\partial J$ [see (1.8.4) and (1.8.8)]. These differences are clearly due to the fact that the Schrödinger equation is an equation in configuration space alone, while the Liouville equation has as independent variables both the coordinates and the momenta. The set of equations (1.8.7) is of course equivalent to (1.8.2) but permits, as we shall see, a much more detailed description of the time evolution of the system.

As the unperturbed Hamiltonian H_0 or the unperturbed Liouville operator L_0 has been partly eliminated from (1.8.7) (it appears only in the oscillating exponentials), we shall call (1.8.7) the Liouville equation in the *interaction representation* by analogy with the terminology used in quantum mechanics. Equation (1.8.7) will play a fundamental role throughout this book.

9. Example

As a simple exercise let us write equation (1.8.7) for a harmonic oscillator of charge ε in a uniform electrical field $E(t)$, which may be time-dependent. The Hamiltonian is clearly [see (1.6.1) and (1.6.3)]

$$\begin{aligned} H &= (p^2/2m)+(k/2)q^2+\lambda\varepsilon q E(t) \\ &= \omega J+\lambda(1/2i)(e^{i\alpha}-e^{-i\alpha})J^{1/2}f(t) \end{aligned} \tag{1}$$

We have put all quantities independent of J and α into $f(t)$. The perturbed Liouville operator is therefore [see (1.8.4)]

$$\begin{aligned} \delta L &= (f(t)/2)\,[i(e^{i\alpha}+e^{-i\alpha})J^{1/2}(\partial/\partial J)-(1/2J^{1/2})\,(e^{i\alpha}-e^{-i\alpha})\,(\partial/\partial\alpha)] \\ &= (if(t)/2)J^{1/2}\{e^{i\alpha}[(\partial/\partial J)-(1/2iJ)(\partial/\partial\alpha)] \\ &\quad + e^{-i\alpha}[(\partial/\partial J)+(1/2iJ)\,(\partial/\partial\alpha)]\} \end{aligned} \tag{2}$$

The only non-vanishing matrix elements (1.8.8) will therefore be those in which n and n' differ by one:

$$\begin{aligned} \langle n|\delta L|n-1\rangle &= (i/2)J^{1/2}f(t)\,\{(\partial/\partial J)-[(n-1)/2J]\} \\ \langle n|\delta L|n+1\rangle &= (i/2)J^{1/2}f(t)\{(\partial/\partial J)+[(n+1)/2J]\} \\ \langle n|\delta L|n'\rangle &= 0 \quad \text{for} \quad n' \neq \begin{matrix} n+1 \\ n-1 \end{matrix} \end{aligned} \tag{3}$$

Therefore the equations (1.8.7) take the relatively simple form

$$\begin{aligned} \partial\rho_n/\partial t &= \lambda/2e^{-i\omega t}J^{1/2}f(t)\,\{(\partial/\partial J)+[(n+1)/2J]\}\rho_{n+1} \\ &\quad + \lambda/2e^{i\omega t}J^{1/2}f(t)\,\{(\partial/\partial J)-[(n-1)/2J]\}\,\rho_{n-1} \end{aligned} \tag{4}$$

We shall not study these equations in detail because the problem is sufficiently simple to be solved directly by using the canonical equations corresponding to (1.9.1).

The important point to notice is that the Fourier expansion of δL in α directly determines which matrix elements $\langle n|\delta L|n'\rangle$ will appear; to a term proportional to exp $(\pm im\alpha)$ in V will correspond a non-vanishing matrix element $\langle n|\delta L|n\mp m\rangle$.

10. Discussion of the Interaction Representation

Let us go back to the Liouville equation (1.8.2) and write

$$\rho = e^{-itL_0}\tilde{\rho} \tag{1}$$

We obtain then for $\tilde{\rho}$, the equation

$$i(\partial\tilde{\rho}/\partial t) = \lambda e^{itL_0}\delta L e^{-itL_0}\tilde{\rho} \tag{2}$$

This is the general form of the Liouville equation in interaction representation. The special form (1.8.7) is obtained by expanding $\tilde{\rho}$ in eigenfunctions of the unperturbed Liouville operator according to (1.8.5).

The operator exp (itL_0) is a displacement operator in the angle variables. Indeed, using (1.8.3) we see that

$$e^{itL_0}f(\alpha) = e^{t\omega(\partial/\partial\alpha)}f(\alpha) = f(\alpha+\omega t) \tag{3}$$

We see therefore that e^{itL_0} "displaces" the value of α by an amount ωt. The great advantage of the expansion in eigenfunctions of L_0 itself is that the operator exp (itL_0) is then, of course *diagonal*:

$$\langle n|e^{itL_0}|n'\rangle = e^{in\omega t}\delta_{nn'} \tag{4}$$

On the contrary, the operator δL is then non-diagonal (see the example treated in §9). In the language used in quantum mechanics one could say that δL introduces transitions from some values of $\{n\}$ to others $\{n'\}$ whenever $\langle\{n\}|\delta L|\{n'\}\rangle$ is different from zero.

This last feature will prove to be decisive for our whole method. We shall indeed see that these transitions may be considered to describe changes in the "state" of the system when this state is defined in terms of all correlations which exist between the degrees of freedom. The development of this picture will lead us to the formulation of a dynamics of correlations which may be considered to be the main object of this monograph.

11. Time-Dependent and Time-Independent Perturbation Theory

The method used in §8 corresponds to the *time-dependent* perturbation theory of quantum mechanics. We may also use the well-known time-independent perturbation method (the so-called Rayleigh-Schrödinger method, see for example Bohm, 1951). We then have to determine the eigenvalues and the eigenfunctions of the equation

$$(L_0+\lambda\delta L)\varphi_k = \lambda_k\varphi_k \tag{1}$$

The standard method is to expand φ_k and λ_k in powers of λ,

$$\begin{aligned}\varphi_k &= \varphi_k^{(0)} + \lambda\varphi_k^{(1)} + \cdots \\ \lambda_k &= \lambda_k^{(0)} + \lambda\lambda_k^{(1)} + \cdots\end{aligned} \tag{2}$$

to substitute these expansions in (1.11.1), and to obtain successive equations by equating coefficients of the same powers of λ on both sides of the equation.

We shall give an example of the application of the time-independent theory in Appendix II. We shall in general use the time-dependent method.

It may be interesting to note that the very possibility of using time-independent perturbation techniques means that there exist, as shown by (1.4.14), invariants of the motion which may be expanded in powers of the coupling constant λ. Indeed to the expansion (1.11.2) of φ_k corresponds the expansion of the invariant A_k in (1.4.14).

$$A_k = A_k^{(0)} + \lambda A_k^{(1)} + \cdots \tag{3}$$

The existence of such invariants is at first very surprising because an important theorem due to Poincaré (1892; see also Fermi, 1923) shows that for a large class of mechanical systems with interacting degrees of freedom there exist no analytical invariants in the coupling constant besides the trivial ones (energy, total momentum, and total angular momentum).

In Chapter 14 we shall present a more detailed discussion of this important point. We shall show there that invariants such as (1.11.3) really exist in the limit of a large system as long as the influence of the "boundaries" of the volume in which the system is enclosed may be neglected.

CHAPTER 2

Anharmonic Solids

1. Hamiltonian

As long as one neglects the anharmonic forces, a solid is a collection of independent harmonic oscillators.

Such an approximation is sufficient for many equilibrium problems but fails for all problems dealing with the approach of the solid to equilibrium. Here the anharmonic forces which couple the different oscillators play the central role. The anharmonic solid is the first example we shall study using the method of Chapter 1.

The interest in this example is not confined to the field of solid state physics; in many other problems we also have to deal with collections of interacting harmonic oscillators (e.g., the noise problem). Similar methods can be applied to all these problems.

We first need the explicit expression for the Hamiltonian. This problem is treated in all books dealing with lattice vibrations (see for example the excellent presentation by Peierls, 1955, or Kittel, 1953). For convenience we shall give a brief summary of it.

We shall consider the simple case of a linear chain in which N atoms, each of mass m, are situated on a line. The extension of the results to real three-dimensional lattices will be indicated at the end of this section.

We assume that the potential energy is minimum when the atoms are regularly spaced, the distance between successive neighbors being a. We note by u_n the displacement of the nth atom from its equilibrium position. The potential energy U may then be expanded in a Taylor series:

$$U-U_0 = \tfrac{1}{2}\sum_{nn'} A_{nn'}u_n u_{n'} + \tfrac{1}{6}\sum_{nn'n''} B_{nn'n''}\, u_n u_{n'} u_{n''} + \cdots \tag{1}$$

where U_0 is the potential energy in equilibrium. By definition

$$\begin{aligned} A_{nn'} &= A_{n'n} \\ B_{nn'n''} &= B_{n'nn''} = \cdots \end{aligned} \tag{2}$$

Moreover since the forces depend only on the distances between the molecules, the coefficients $A_{nn'}$, $B_{nn'n}$, . . . can only depend on the differences $|n-n'|$, $|n-n''|$, and we may also use the notation $A_{n-n'}$ or $A(n-n')$.

If we neglect the anharmonic terms, the equations of motion are

$$m(d^2u_n/dt^2) = -\sum_{n'} A_{nn'}u_{n'} \tag{3}$$

As the forces do not change if we displace all atoms by the same amount in the same direction, we have

$$\sum_{n'} A_{n-n'} = 0 \tag{4}$$

Let us introduce *normal coordinates* q_k related to u_n by

$$u_n = \sum_k q_k e^{ikna} \tag{5}$$

The reality condition for u_n imposes the condition

$$q_k^* = q_{-k} \tag{6}$$

We shall again use periodic boundary conditions to fix the permissible values of k [see (1.5.7)]; we imagine the atoms arranged on a large circumference so that the last atom $n = N$ is again at a distance a from the first. This corresponds to the so-called *cyclic condition.*

$$u_{n+N} = u_n \tag{7}$$

Relation (2.1.5) shows that this imposes the condition

$$e^{ikNa} = 1 \tag{8}$$

which in turn requires that

$$k = m(2\pi/Na) \tag{9}$$

where m is an integer.

On the other hand k occurs only in the combination exp $(ikna)$;

therefore u_n does not change if we add to k a multiple of $2\pi/a$. We may therefore restrict k to the interval

$$-\pi/a < k \leqq \pi/a \tag{10}$$

in which there are N values of k satisfying (2.1.9). Any other values of k can be reduced to this interval by addition or subtraction of a multiple of $2\pi/a$. The interval (2.1.10) is called a *Brillouin zone.*

We may note that the difference between two successive permitted values of k is, according to (2.1.9), of order $1/N$. When $N \to \infty$, the permitted values of k form a continuous set, or, as one generally puts it, k has a *continuous spectrum.*

Let us now express the energy of the system in terms of the normal coordinates q_k. We have, neglecting the anharmonic terms,

$$\begin{aligned} E &= (m/2) \sum_n \dot{u}_n^2 + \tfrac{1}{2} \sum_{n,n'} A(n-n') u_n u_{n'} \\ &= (m/2) \sum_n \sum_{kk'} \dot{q}_k \dot{q}_{k'}^* \, e^{i(k-k')na} \\ &\quad + \tfrac{1}{2} \sum_{nn'} \sum_{kk'} A(n-n') q_k q_{k'}^*, \, e^{ikna-ik'n'a} \end{aligned} \tag{11}$$

Consider the sum

$$S = \sum_n e^{i(k-k')na} \tag{12}$$

where the summation is extended over the whole lattice. Now suppose we displace each lattice point by the lattice distance a. This changes each lattice point into the next without altering the sum S. On the other hand, each term of S is multiplied by $\exp[i(k-k')a]$. Therefore,

$$S = S' \, e^{i(k-k')a} \tag{13}$$

This means that either $S = 0$ or $(k-k')a$ is a multiple of 2π:

$$k-k' = m(2\pi/a) \tag{14}$$

where m is an integer. Vectors K (for one-dimensional lattices, K is a scalar), which satisfy (2.1.14) or

$$e^{iKna} = 1 \tag{15}$$

again form a lattice [a sum of quantities satisfying (2.1.14) or (2.1.15) will satisfy the same equation]. This lattice is known as the "reciprocal lattice." Two vectors $\boldsymbol{k}$ are "equivalent" if they differ by a lattice vector of the reciprocal lattice. Thus $S = 0$ whenever k or k' are equivalent.

We now return to the evaluation of (2.1.11). Using the results just derived, we see that

$$(m/2) \sum_{n} \sum_{kk'} \dot{q}_k \dot{q}^*_{k'} e^{i(k-k')na} = (mN/2) \sum_k \dot{q}_k \dot{q}^*_k \tag{16}$$

The second part of (2.1.11) is transformed in the same way. Introducing the notation

$$G(k) = \sum_{n-n'} A(n-n') e^{ik(n-n')a} \tag{17}$$

we immediately obtain

$$\begin{aligned} E &= (mN/2) \sum_k [|\dot{q}_k|^2 + (G(k)/m)\, |q_k|^2] \\ &= (mN/2) \sum_{k'} (|\dot{q}_k|^2 + \omega_k^2 |q_k|^2) \end{aligned} \tag{18}$$

where $\omega_k = [G(k)/m]^{1/2}$ is the frequency associated with the normal mode k. We see that the energy is the sum of independent contributions due to the normal modes of the lattice. We now transform (2.1.18) from normal coordinates to action-angle variables. As we have seen [see (1.6.3)] this simply corresponds to a transition from Cartesian to polar coordinates. Because we are using complex coordinates q_k here, the transition to action-angle variables takes the form

$$\begin{aligned} q_k &= [1/(2Nm)^{1/2}]\, \{(J_k/\omega_k)^{1/2} e^{i\alpha_k} + (J_{-k}/\omega_{-k})^{1/2} e^{-i\alpha_{-k}}\} \\ \dot{q}_k &= [i/(2Nm)^{1/2}]\, \{(J_k\omega_k)^{1/2} e^{i\alpha_k} + (J_{-k}\omega_{-k})^{1/2} e^{-i\alpha_{-k}}\} \end{aligned} \tag{19}$$

with

$$\omega_k = \omega_{-k} \tag{20}$$

Substituting in (2.1.18) we obtain the Hamiltonian

$$H = \sum_k \omega_k J_k \tag{21}$$

which is the expected form for a collection of harmonic oscillators [see (1.6.4)].

Let us now consider the anharmonic contribution to the potential energy in (2.1.1). We shall retain only the cubic term

$$V = \frac{1}{6} \sum_{nn'n''} B_{nn'n''} u_n u_{n'} u_{n''} \tag{22}$$

There also exist of course terms of fourth and higher order, but their treatment would be completely similar.

Introducing normal coordinates by formula (2.1.5) we first have

$$V = \frac{1}{6} \sum_{kk'k''} b(k, k', k'') q_k q_{k'} q_{k''} \tag{23}$$

with

$$b(k, k', k'') = \sum_{nn'n''} B_{nn'n''} e^{i(kn+k'n'+k''n'')a} \tag{24}$$

The reality condition for V gives

$$b(k, k', k'') = b^*(-k, -k', -k'') \tag{24'}$$

Exactly as in the case of the sum (2.1.12) one shows that

$$b(k, k', k'') = 0 \tag{25}$$

except if $k+k'+k''$ is either zero or a vector of the reciprocal lattice.

The formal similarity of condition (2.1.25) to the conservation of momentum in quantum mechanics is obvious. Using (2.1.19) we put (2.1.23) in the form

$$\begin{aligned} V = \sum_{kk'k''} V_{kk'k''} \{ & (J_k J_{k'} J_{k''}/\omega_k \omega_{k'} \omega_{k''})^{1/2} e^{i(\alpha_k+\alpha_{k'}+\alpha_{k''})} \\ & + (J_k J_{k'} J_{-k''}/\omega_k \omega_{k'} \omega_{-k''})^{1/2} e^{i(\alpha_k+\alpha_{k'}-\alpha_{-k''})} \\ & + (J_k J_{-k'} J_{k''}/\omega_k \omega_{-k'} \omega_{k''})^{1/2} e^{i(\alpha_k-\alpha_{-k'}+\alpha_{k''})} \\ & + (J_{-k} J_{k'} J_{k''}/\omega_{-k} \omega_{k'} \omega_{k''})^{1/2} e^{i(-\alpha_{-k}+\alpha_{k'}+\alpha_{k''})} \} \\ & + \text{four other terms obtained by } k \to -k,\ k' \to -k',\ k'' \to -k'' \} \end{aligned} \tag{26}$$

with

$$V_{kk'k''} = \frac{1}{6} (2Nm)^{-3/2} b(k, k', k'') \tag{27}$$

We have as usual the reality condition

$$V_{kk'k''} = V^*_{-k\,-k'\,-k''} \tag{28}$$

We may also write (2.1.26) in the more compact form

$$V = \sum_{kk'k''} (J_k J_{k'} J_{k''}/\omega_k \omega_{k'} \omega_{k''})^{1/2} \{V_{kk'k''}\, e^{i(\alpha_k+\alpha_{k'}+\alpha_{k''})} + 3V_{kk'-k''}\, e^{i(\alpha_k+\alpha_{k'}-\alpha_{k''})} + \text{c.c.}^{1}\} \tag{29}$$

In the case of three-dimensional lattices, all these expressions remain valid but the single vector $\boldsymbol{k}$ now means both the wave vector $\boldsymbol{k}$ and the polarization s. (There are three directions of polarization for a given value of $\boldsymbol{k}$.) We shall not go into greater detail (see for example Peierls, **1955**; Kittel, **1953**), and we shall very often use the terminology corresponding to one-dimensional lattices.

Summarizing, we see that the Hamiltonian may be written in the form

$$H = H_0 + \lambda V \tag{30}$$

where H_0 is given by (2.1.21) and V by (2.1.29). We are now ready for the application of the perturbation technique we described in Chapter 1, § 8.

2. The Liouville Operator

Using (1.8.3) and (1.8.4) we easily obtain from (2.1.30) the Liouville operator of an anharmonic solid

$$L_0 = -i \sum \omega_k (\partial/\partial\alpha_k) \tag{1}$$

$$\begin{aligned} \delta L = i \sum_{kk'k''} & (J_k J_{k'} J_{k''}/\omega_k \omega_{k'} \omega_{k''})^{1/2} \{V_{kk'k''}\, e^{i(\alpha_k+\alpha_{k'}+\alpha_{k''})} \\ & \times \{i[(\partial/\partial J_k) + (\partial/\partial J_{k'}) + (\partial/\partial J_{k''})] \\ & - \tfrac{1}{2}[(1/J_k)(\partial/\partial\alpha_k) + (1/J_{k'})(\partial/\partial\alpha_{k'}) + (1/J_{k''})(\partial/\partial\alpha_{k''})]\} \\ & + 3V_{kk'-k''}\, e^{i(\alpha_k+\alpha_{k'}-\alpha_{k''})} \{i[(\partial/\partial J_k) + (\partial/\partial J_{k'}) - (\partial/\partial J_{k''})] \\ & - \tfrac{1}{2}[(1/J_k)(\partial/\partial\alpha_k) + (1/J_{k'})(\partial/\partial\alpha_{k'}) + (1/J_{k''})(\partial/\partial\alpha_{k''})]\} + \text{c.c.}\} \end{aligned} \tag{2}$$

We now calculate the matrix element $\langle\{n'\}|\delta L|\{n\}\rangle$. Because δL contains only contributions of the form

$$e^{i(\pm\alpha_k \pm \alpha_{k'} \pm \alpha_{k''})} \tag{3}$$

we may immediately expect [see Chapter 1, § 9 and (2.1.29)] that

[1] The letters c.c. always mean complex conjugate.

the only non-vanishing matrix elements will be

$$\langle n_k\, n_{k'}\, n_{k''}\,\{n\}'|\delta L|n_k \pm 1\;\, n_{k'} \pm 1\;\, n_{k''} \pm 1\{n\}'\rangle \tag{4}$$

with $\pm k \pm k' \pm k'' = 0$ (or equal to a vector of the reciprocal lattice) in which only three Fourier indices k, k', k'' are changed.

Let us write the non-vanishing matrix element (2.2.4) in detail. We obtain from (2.2.2)[1]

$$\begin{aligned}&\langle n_k n_{k'} n_{k''}|\delta L|\,n_k - 1\, n_{k'} - 1\, n_{k''} - 1\rangle\\ &= V_{kk'k''}(J_k J_{k'} J_{k''}/\omega_k\omega_{k'}\omega_{k''})^{1/2}\,[(n_k - 1/2J_k) + (n_{k'} - 1/2J_{k'})\\ &\quad + (n_{k''} - 1/2J_{k''}) - (\partial/\partial J_k) - (\partial/\partial J_{k'}) - (\partial/\partial J_{k''})]\end{aligned} \tag{5}$$

This term comes from the contribution of exp $[i(\alpha_k + \alpha_{k'} + \alpha_{k''})]$ in (2.2.2).

We shall find it useful to use expressions in which the factor $(J_k J_{k'} J_{k''}/\omega_k\omega_{k'}\omega_{k''})^{1/2}$ of (2.2.5) is commuted to the right of the operator $[(\partial/\partial J_k) + (\partial/\partial J_{k'}) + (\partial/\partial J_{k''})]$. We then obtain

$$\begin{aligned}&\langle n_k n_{k'} n_{k''}|\delta L|n_k - 1\, n_{k'} - 1\, n_{k''} - 1\rangle\\ &= V_{kk'k''}[(n_k/2J_k) + (n_{k'}/2J_{k'}) + (n_{k''}/2J_{k''}) - (\partial/\partial J_k) - (\partial/\partial J_{k'}) - (\partial/\partial J_{k''})]\\ &\quad \times (J_k J_{k'} J_{k''}/\omega_k\omega_{k'}\omega_{k''})^{1/2}\end{aligned} \tag{6}$$

where the operators $\partial/\partial J_k \cdots$ act, as usual, on all functions of the actions situated at their right.

Let us note the following rule which is also true for the other matrix elements: If the factor $(J_k J_{k'} J_{k''}/\omega_k\omega_{k'}\omega_{k''})^{1/2}$ *is written at the left, the Fourier indices* $\{n\}$ *which appear in the matrix element are the initial occupation numbers*; and, similarly, *if this factor is written at the right, the Fourier indices are the final ones.*[2] We find more generally

$$\begin{aligned}&\langle n_k n_{k'} n_{k''}|\delta L|\,n_k \pm 1\, n_{k'} \pm 1\, n_{k''} \pm 1\rangle\\ &= V_{\mp k \mp k' \mp k''}[(n_k/2J_k) + (n_{k'}/2J_{k'}) + (n_{k''}/2J_{k''})\\ &\quad \pm (\partial/\partial J_k) \pm (\partial/\partial J_{k'}) \pm (\partial/\partial J_{k''})](J_k J_{k'} J_{k''}/\omega_k\omega_{k'}\omega_{k''})^{1/2}\end{aligned} \tag{7}$$

[1] The unchanged indices $\{n\}'$ are dropped in (2.2.5) and in the subsequent formulae.

[2] In $\langle \{n\}|\delta L|\{n\}'\rangle$ the $\{n\}'$ are the Fourier indices *before* the transition due to δL, and the $\{n\}$, *after* the transition.

Let us now study the change of the distribution function with time.

3. Formal Solution for the Energy Distribution Function

In this chapter we shall be interested only in the energy distribution function. It is easily seen from (1.8.5) that this is precisely the Fourier coefficient $\rho_{\{0\}}$. Indeed

$$\rho_{\{0\}}(J_1 \cdots J_s, t) = 1/(2\pi)^{s/2} \int_0^{2\pi} \cdots \int_0^{2\pi} d\alpha_1 \cdots d\alpha_s \, \rho(J_1 \cdots J_s, \alpha_1 \cdots \alpha_s, t) \qquad (1)$$

The Fourier coefficient $\rho_{\{0\}}$ therefore has an especially simple and important physical meaning.

Let us solve (1.8.7) by iteration. To illustrate this procedure, consider the following simple example:

$$[\partial y(x, t)/\partial t] = \lambda K(x, t)\, y(x, t)$$

in which K is a time-dependent operator acting on y (or simply a function of x) and λ is some parameter. The initial value of y is

$$y(x, 0) = y_0(x)$$

Replacing $y(x, t)$ on the right-hand side by y_0 we obtain

$$y(x, t) = y_0 + \lambda \int_0^t dt_1 K(x, t_1)\, y_0(x)$$

This is used as a basis for a new iteration

$$y(x, t) = y_0 + \lambda \int_0^t dt_1 K(x, t_1) y_0 + \lambda^2 \int_0^t dt_1 \int_0^{t_1} dt_2 K(x, t_1) \times K(x, t_2)\, y_0 + \cdots$$

The correctness of this solution is easily verified by differentiation. In this way we obtain $y(x, t)$ as a power expansion in λ.

Let us now apply this iteration technique to (1.8.7). We have

$$\begin{aligned}\rho_{\{0\}}(t) = {} & \rho_{\{0\}}(0) + (\lambda/i) \sum_{\{n'\}} \int_0^t dt_1 \langle \{0\} | \delta L | \{n'\} \rangle e^{-i \Sigma n'_k \omega_k t_1} \rho_{\{n'\}}(0) \\ & + (\lambda/i)^2 \sum_{\{n'\}\{n''\}} \int_0^t dt_1 \int_0^{t_1} dt_2 \langle \{0\} | \delta L | \{n'\} \rangle e^{-i \Sigma n'_k \omega_k t_1} \\ & \times e^{i \sum_k n'_k \omega_k t_2} \langle \{n'\} | \delta L | \{n''\} \rangle e^{-i \sum_k n''_k \omega_k t_2} \rho_{\{n''\}}(0) \\ & + \lambda^3 \cdots \end{aligned} \qquad (2)$$

We may verify that by differentiating (2.3.2) with respect to time we indeed recover (1.8.7).

It is out of the question to try to add the series (2.3.2) exactly; this would be equivalent to an exact solution of the N-body problem involved. Such a solution would be both inconceivably difficult to obtain and useless; even if it were obtainable it would take a geological time to write it down. What we really want to show is that under well-defined conditions $\rho_{\{0\}}(t)$ satisfies a much simpler equation than the initial Liouville equation. From this simpler equation, which we shall call the "master equation," we shall be able to obtain all the information we are interested in.

To proceed we need two kinds of assumptions: (1) an assumption about the initial state of the system; (2) a statement about the kind of terms we want to retain. Here, we shall consider the simple case corresponding to

$$\rho_{\{n\}}(t=0)=0 \qquad \{n\}\neq\{0\} \tag{3}$$

This means that the initial distribution is angle independent; it is called random phase assumption. Note that it is only used at the initial time $t = 0$. We may also say that we consider a "*homogeneous*" system because the distribution function is angle independent. We shall see later (cf. Chapters 8 and 11) that the two kinds of assumptions we have mentioned are *not independent*.

Moreover, (2.3.3) is unnecessarily restrictive, but we use it here because of its simplicity. We shall see in Chapter 8 and 11 that the equations we shall derive are in fact valid for much wider classes of initial conditions (cf. also Appendix III).

Assumption (2.3.3) permits us to write (2.3.2) in a simple form. Indeed, because of the form of the Liouville operator (see § 2) there exists no possibility of going from $\rho_{\{0\}}(t=0)$ to $\rho_{\{0\}}(t)$ in an *odd* number of steps.

For example, taking account of (2.2.4), we see that

$$\langle\{0\}|\delta L|\{n\}\rangle\langle\{n\}|\delta L|\{n'\}\rangle\langle\{n'\}|\delta L|\{0\}\rangle=0 \tag{4}$$

because the changes in the $\{n\}$ introduced by the first two transitions cannot all be compensated by the third. Therefore (2.3.2) reduces to

$$\begin{aligned}
\rho_{\{0\}}(t) &= \rho_{\{0\}}(0) + (\lambda/i)^2 \sum_{\{n'\}} \int_0^t dt_1 \int_0^{t_1} dt_2 \langle\{0\}|\delta L|\{n'\}\rangle \\
&\times e^{-i\sum n'_k \omega_k (t_1-t_2)} \langle\{n'\}|\delta L|\{0\}\rangle \rho_{\{0\}}(0), \\
&+ (\lambda/i)^4 \sum_{\{n'\}\{n''\}\{n'''\}} \int_0^t dt_1 \int_0^{t_1} dt_2 \int_0^{t_2} dt_3 \int_0^{t_3} dt_4 \langle\{0\}|\delta L|\{n'\}\rangle \\
&\times e^{-i\sum n'_k \omega_k (t_1-t_2)} \langle\{n'\}|\delta L|\{n''\}\rangle e^{-i\sum n''_k \omega_k (t_2-t_3)} \\
&\times \langle\{n''\}|\delta L|\{n'''\}\rangle e^{-i\sum n'''_k \omega_k (t_3-t_4)} \langle\{n'''\}|\delta L|\{0\}\rangle \rho_{\{0\}}(0) + \lambda^6 \cdots
\end{aligned} \tag{5}$$

We must now decide what kind of terms we want to retain in (2.3.5). As we shall see in §5 there exists a characteristic time t_{int} which may be considered as the "duration" of the interaction between normal modes. For times long with respect to this characteristic time

$$t \gg t_{\text{int}} \tag{6}$$

the series (2.3.5) has a very simple time dependence. As we shall show it is formed by contributions of the form

$$\begin{aligned}
&(\lambda^{2n} t^n) \qquad n = 1, 2, \ldots \\
&\lambda(\lambda^{2n} t^n) \qquad \lambda^2(\lambda^{2n} t^n)
\end{aligned} \tag{7}$$

Their meaning is the following: We may expect that the energy distribution will approach its equilibrium value according some relaxation law

$$e^{-t/t_r} \tag{8}$$

where t_r is a characteristic relaxation time.

Now in this part of the book (until Chapter 5) we shall consider only the case of *weakly coupled systems*. This means that the strength of the interaction λ is very small or that the relaxation time t_r is very long. It may be expected that for small λ [1]

$$t_r \sim 1/\lambda^2 \tag{9}$$

We shall later verify that this is so. By comparison with (2.3.8)

[1] Recall the Born approximation in quantum mechanics. The dimensions in (2.3.9) are obviously wrong. What we want to stress is the λ dependence. The complete expression for the relaxation time is for example given in (3.3.36).

we see that we have to expect terms of the form $(\lambda^2 t)^n$. On the other hand contributions of the form $\lambda^m(\lambda^2 t)^n$, $m \geqq 1$ are λ dependent corrections to the relaxation time and may be neglected for weakly coupled systems.

What we shall do now is "extract" the $(\lambda^2 t)^n$ contributions using a diagram technique.

4. Diagram Technique

If the phase distribution function $\rho(J_1 \cdots J_s, \alpha_1 \cdots \alpha_s, t)$ is independent of the angle variable α_l corresponding to the normal mode l, the only Fourier coefficients $\rho_{\{n\}}$ that will be different from zero are those corresponding to $n_l = 0$. Similarly, a Fourier coefficient $\rho_{n_l n_m}$ will only be different from zero for distributions corresponding to the existence of phase correlations

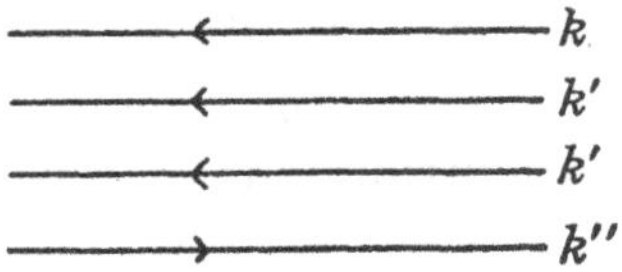

Fig. 2.4.1 Representation of the state $n_k = 1$, $n_{k'} = 2$, $n_{k''} = -1$, $\{n'\} = 0$.

for the normal modes l and m. To represent such a state, corresponding to the eigenfunction $e^{i(n_l\alpha_l+n_k\alpha_k)}$ let us draw n_l directed lines l, and n_m directed lines m; lines directed from right to left correspond to positive n_l, while negative values of n_l are represented by lines oriented from left to right.

For example, the eigenfunction $e^{i(\alpha_k+2\alpha_{k'}-\alpha_l)}$ will be represented by Fig. 2.4.1.

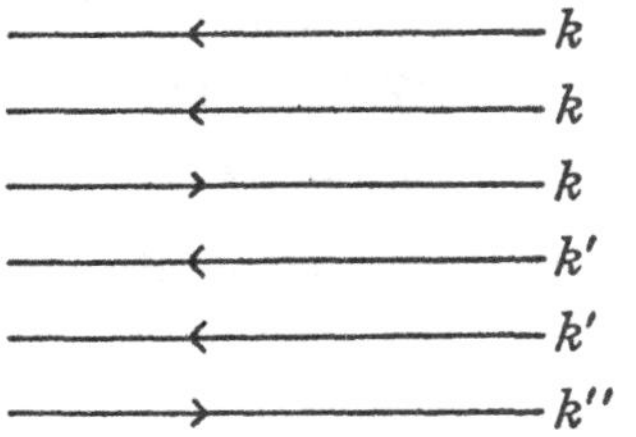

Fig. 2.4.2 Equivalent representation of the state $n_k = 1$, $n_{k'} = 2$, $n_{k''} = -1$, $\{n'\} = 0$.

This representation stresses the formal analogy between the Fourier indices n_l and occupation numbers in quantum mechanics; the k's correspond to momenta and the n_k's to number of particles ($n_k > 0$) or holes ($n_k < 0$).

We adopt the convention that lines corresponding to the same normal mode k may be added algebraically. Therefore the state represented by Fig. 2.4.1 might also be represented by Fig. 2.4.2.

We shall, of course, always use a minimum number of lines.

The state $\{n\} = 0$ corresponding to an eigenfunction $\varphi_{\{0\}}$, which is independent of the angle variables corresponds to a homogeneous situation for all normal modes. Here it plays somewhat the role of the vacuum in quantum field theory, while the states $\{n\} \neq 0$ then correspond to "excited states."

The interactions between the normal modes as expressed by δL therefore have a double role: (1) they modify the phase relations: the value of $\{n\}$ and therefore the number of corresponding lines is modified; (2) they modify the energy distribution (each matrix element of δL is still an operator with respect to the actions J).

As shown by (2.2.7) the effect of the operator δL is to modify the initial state by the addition or destruction of three lines.

The simplest situation occurs when either the initial or the

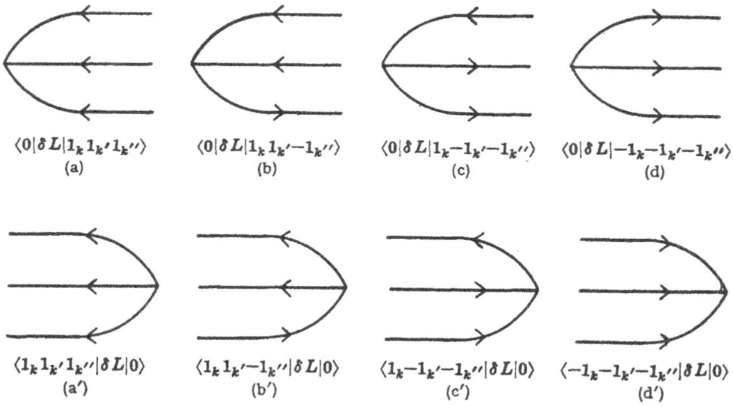

Fig. 2.4.3 Basic diagrams containing a single vertex. (1_k means $n_k = 1$, and so on.)

final state corresponds to $\{n\} = 0$. We then obtain the diagrams of Fig. 2.4.3 (we always read diagrams from right to left).

The diagrams of Fig. 2.4.3 are of two types: diagrams (a), (b), (c), and (d) correspond to a *destruction of a correlation* between the normal modes k, k', k'', while diagrams (a′), (b′), (c′), and (d′) correspond to the *creation of such a correlation.*

Any non-vanishing matrix element, $\langle\{n\}|\delta L|\{n'\}\rangle$ can be obtained from the diagrams of Fig. 2.4.3 by adding a set of lines. For example, the diagram of Fig. 2.4.4 corresponds to the matrix element $\langle 1_k 1_{k'} 1_{k''} 2_l|\delta L|2_l\rangle$. It expresses the creation of a correlation between the normal modes k, k', and k''.

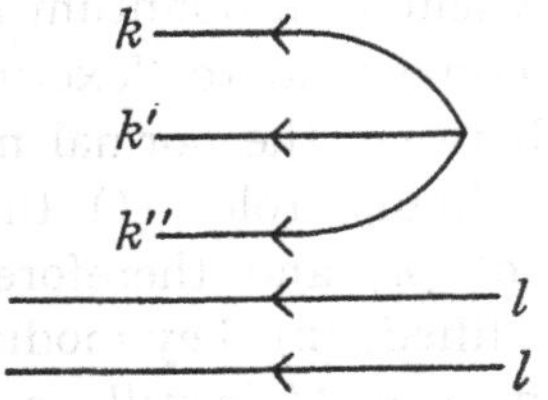

Fig. 2.4.4 Diagram corresponding to $\langle 1_k 1_{k'} 1_{k''} 2_l|\delta L|2_l\rangle$.

The matrix element $\langle 3_k 2_{k'} - 4_{k''}|\delta L|2_k 3_{k'} - 3_{k''}\rangle$ may be represented in the two ways shown in Fig. 2.4.5. In this case the normal modes k, k', k'', are correlated *both* before and after the interaction. Such a transition corresponds therefore to what may be called the *propagation* of the correlation.

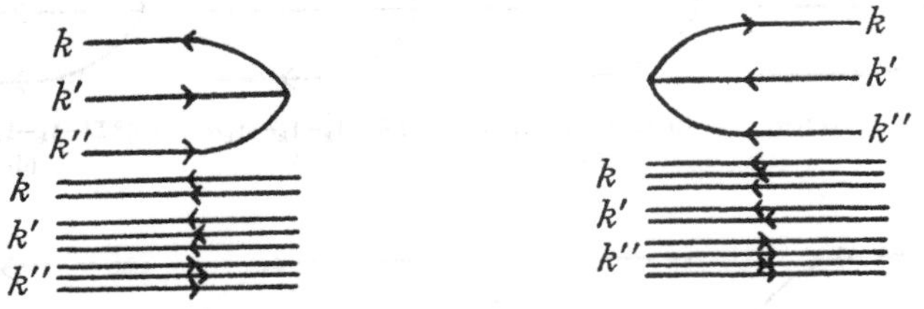

Fig. 2.4.5 Possible representations of $\langle 3_k 2_{k'} - 4_{k''}|\delta L|2_k 3_{k'} - 3_{k''}\rangle$.

The fact that two possible diagram representations correspond to a given matrix element (as in Fig. 2.4.5) is of no importance. The only thing that matters are the successive values of the set $\{n\}$ of the Fourier indices.

To a single matrix element corresponds a diagram with a single intersection or *vertex*. One could say that such a diagram corresponds to an *"elementary process" of the dynamics of correlations in the solid.*

In the series (2.3.5) a term of order λ^m will be represented by a diagram with m vertices. An especially important role will be played by sequences of two matrix elements starting and finishing with $\{n\} = 0$. Such sequences correspond to the existence of correlations between three normal modes over a limited period of time. These sequences of two inverse transitions will be, according to the rules we have given, represented by the diagrams of Fig. 2.4.6. The four diagrams of Fig. 2.4.6. differ by the nature of the

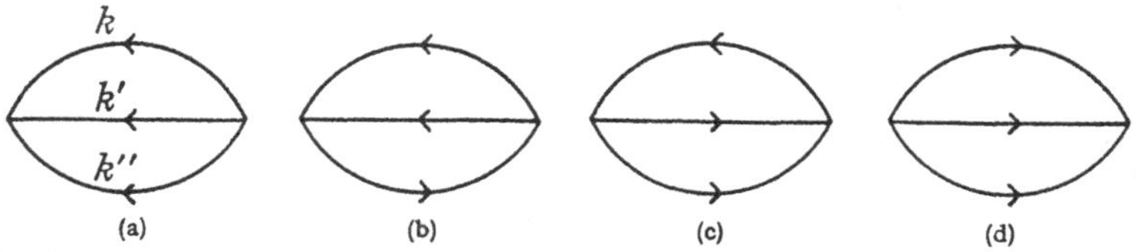

Fig. 2.4.6 Diagrams corresponding to sequences of two transitions starting and finishing with $\{n\} = 0$.

intermediate state: for (*b*) $n_k = 1$, $n_{k'} = 1$, $n_{k''} = -1$, and for (*c*), $n_k = 1$, $n_{k'} = -1$, $n_{k''} = -1$. We shall often use an abbreviated notation for the sum of diagrams which differ only by the directions of their lines. For example, the diagram of Fig. 2.4.7 will represent the sum of the four diagrams of Fig. 2.4.6. Diagrams of the form represented in Fig. 2.4.6 or Fig. 2.4.7 will be called *"cycles."*

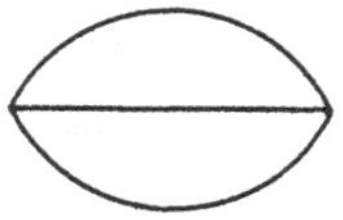

Fig. 2.4.7 Sum of the diagrams of Fig. 2.4.6.

Such cycles represent the simplest collision processes that may occur between normal modes. Indeed in the standard collision process a homogeneous beam falls on the scattering center and one is interested in the velocity distribution function for times much longer than the duration of the collision (for a detailed study of the scattering theory see Chapter 6). The analogy with one of the processes represented in Fig. 2.4.6 is complete. Here also we start with a homogeneous situation and calculate the energy distribution ρ_0 after the process.

The explicit evolution of the cycle is equivalent to the calculation of the cross section due to the interaction of three normal modes.

5. Second-Order Contributions

Let us now consider, in a systematic way, the different contributions that appear in (2.3.5). The second-order contribution is given by the diagrams of Fig. 2.4.6. To each diagram of this type corresponds a time integral

$$\int_0^t dt_1 \int_0^{t_1} dt_2\, e^{i\alpha(t_1-t_2)} \tag{1}$$

with

$$\alpha = \pm\omega_k \pm\omega_{k'} \pm\omega_{k''} \tag{2}$$

We shall be especially interested in the contribution of such integrals for large t [cf. (2.3.6)].

We shall consider two methods. The first permits a better understanding of the physical meaning of asymptotic time integration. The second is more compact and is of special importance for the more complicated cases we shall have to consider later.

Let us introduce the variables

$$\tau = t_1 - t_2, \qquad T = t_1 + t_2 \tag{3}$$

The Jacobian of the transformation is 1/2, while for a fixed value of τ, the limits for T are τ and $2t-\tau$; therefore the integral (2.5.1) becomes

$$\tfrac{1}{2}\int_0^t d\tau\, e^{i\alpha\tau} \int_\tau^{2t-\tau} dT = t\int_0^t d\tau\, e^{i\alpha\tau} - \int_0^t d\tau\tau\, e^{i\alpha\tau} \tag{4}$$

We have to introduce (2.5.4) into (2.3.5). Since all n'_k are equal to 1 or 0, the summation in (2.3.5) is in fact a summation over k, k', k''. In the limit of a large lattice, such a summation over the wave vectors becomes an *integration* (see § 1 of this chapter).

Moreover we can use as a variable $\alpha = \pm\omega_k \pm \omega_{k'} \pm \omega_{k''}$, taking into account the dispersion relation which expresses the frequency as a function of the wave vector. Therefore, rather than (2.5.4),

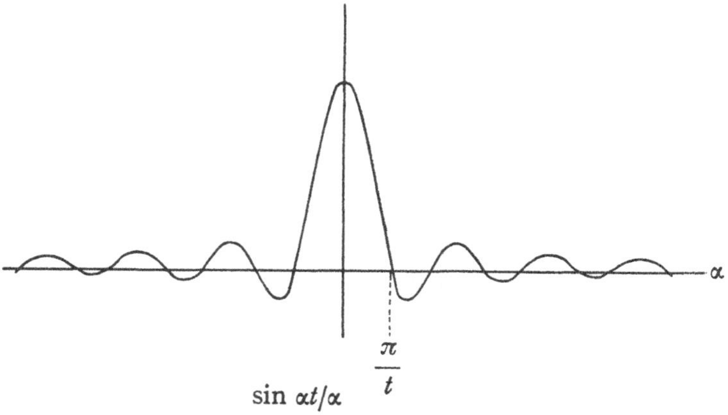

Fig. 2.5.1.

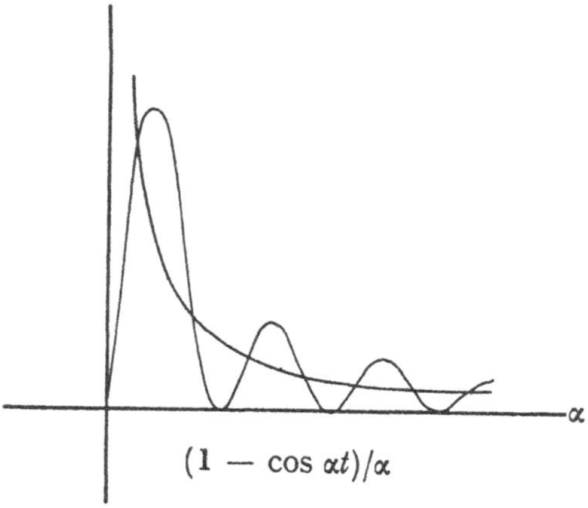

Fig. 2.5.2.

we have to consider an expression of the form

$$t \int d\alpha f(\alpha) \int_0^t d\tau\, e^{i\alpha\tau} - \int d\alpha f(\alpha) \int_0^t d\tau\tau\, e^{i\alpha\tau} \tag{5}$$

Let us consider the first term. We have

$$\int d\alpha f(\alpha) \int_0^t d\tau\, e^{i\alpha\tau} = \int d\alpha f(\alpha) (e^{i\alpha t}-1)/(i\alpha)$$
$$= \int d\alpha f(\alpha) (\sin \alpha t/\alpha) + i \int d\alpha f(\alpha) (1-\cos \alpha t)/\alpha \quad (6)$$

and we must find the asymptotic limit of this expression for large t. The functions $\sin \alpha t/\alpha$ and $(1-\cos \alpha t)/\alpha$ behave as indicated in Figs. 2.5.1 and 2.5.2.

For large t, the function $\sin \alpha t/\alpha$ will be appreciably different from 0 in a very small region around the origin: $-(\pi/t) \leqq \alpha \leqq \pi/t$. If the function $f(\alpha)$ is slowly varying over an interval $\delta\alpha$ such that $(\delta\alpha)^{-1} \ll t$, we may replace $f(\alpha)$ by $f(0)$ in the interval $[-(\pi/t), \pi/t]$. We then obtain

$$\int d\alpha f(\alpha) \sin \alpha t/\alpha \approx f(0) \int_{-\infty}^{+\infty} d\alpha \sin \alpha t/\alpha$$
$$\approx f(0)\, 2 \int_0^{\infty} du \sin u/u = \pi f(0) \quad (7)$$

where we have used the well-known formula (see for example Dwight, 1947):

$$\int_0^{\infty} \sin u/u \, du = \pi/2$$

On the other hand, by the definition of the Dirac delta function, we also have:

$$\int f(\alpha)\, \delta(\alpha)\, d\alpha = f(0) \quad (8)$$

In other words, for large t, the function $\sin \alpha t/\alpha$ acts as a Dirac delta function.

As for the second term on the right-hand side of (2.5.6), we may notice that the function $(1-\cos \alpha t)/\alpha$ vanishes for $\alpha \to 0$. When α is different from 0, it behaves like the function $1/\alpha$ on which are superposed oscillations described by the function $(\cos \alpha t/\alpha)$. The amplitudes of these oscillations decrease very rapidly and they can again be neglected provided that the function $f(\alpha)$ is slowly varying over intervals $(\delta\alpha)^{-1} \ll t$. Thus we have for large t:

$$\begin{aligned}(1-\cos \alpha t)/\alpha &\to 1/\alpha \qquad \text{for } \alpha \neq 0 \\ &\to 0 \qquad \text{for } \alpha \to 0\end{aligned}$$

which is by definition the *principal part* of $1/\alpha$, denoted by $\mathscr{P}(1/\alpha)$.

We have, therefore,

$$t \int d\alpha f(\alpha) \int_0^t d\tau\, e^{i\alpha\tau} = t \int d\alpha f(\alpha)[\pi\delta(\alpha)+i\mathscr{P}(1/\alpha)]$$
$$= t \int d\alpha f(\alpha)\, \pi\delta_+(\alpha) \tag{9}$$

where we have introduced the so-called δ_+ function, defined by

$$\pi\delta_+(\alpha) = \pi\delta(\alpha)+i\mathscr{P}(1/\alpha) \tag{10}$$

From our discussion we see that (2.5.9) is valid only for times t such that

$$t \gg (\delta\alpha)^{-1} \tag{11}$$

$\delta\alpha$ being the interval over which the function $f(\alpha)$ is slowly varying. The time

$$t_{\text{int}} \approx (\delta\alpha)^{-1} \tag{12}$$

appears as the duration of the three phonons' interactions or the collision time. It is only for times long with respect to t_c that the asymptotic form of the time integration may be used.

The situation is characteristic for scattering or radiation theory (see for example Bohm, 1951). In all these cases, asymptotic expressions are only valid for times *long with respect to some characteristic time which measures the duration of the interactions.* We may immediately guess the order of magnitude of the characteristic time t_{int}. Indeed, the range of the interactions is of the order of the lattice distance a and the waves travel with a velocity of the order of the velocity of sound, c. Therefore we may expect the time t_{int} to be

$$t_{\text{int}} \approx a/c \approx 1/\omega_D \tag{13}$$

where ω_D is the Debye frequency of the lattice. This estimation may be verified as follows. As we shall see later in §7 the integral $\int d\alpha f(\alpha)$ in (2.5.5) is in fact a sum of the form

$$\sum_{kk'k''} (|V_{kk'k''}|^2/\omega_k\omega_{k'}\omega_{k''})\, J_k J_{k'} J_{k''}$$
$$= \sum_{kk'k''} (|V_{kk'k''}|^2/\omega_k^2\omega_{k'}^2\omega_{k''}^2)\, E_k E_{k'} E_{k''} \quad (14)$$

which in the limit of a large crystal becomes an integral. Now the expression $|V_{kk'k''}|^2/(\omega_k^2\omega_{k'}^2\omega_{k''}^2)$ may be considered in the first approximation as a constant (see Peierls, 1955). Similarly in a system not too far from equilibrium E_k may also be considered as a constant independent of k. Therefore the variation of $f(\alpha)$ in (2.5.5) with α will be determined by the transition from the sum to the integral, that is, by the form of the group velocity $d\omega/dk$. This expression is equal to the velocity of sound for acoustical wave lengths and begins to vary appreciably for wave lengths of the order of lattice distances. A characteristic frequency corresponding to such a wave length is precisely the Debye frequency ω_D.

Let us now consider the second integral in (2.5.5). We have, using (2.5.9),

$$\int_0^t d\tau\,\tau e^{i\alpha\tau} = (1/i)(d/d\alpha)\int_0^t d\tau\, e^{i\alpha\tau}$$
$$= (1/i)\,\pi\delta'_+(\alpha) \quad (15)$$

with

$$\delta'_+(\alpha) = (d/d\alpha)\,\delta_+(\alpha) \quad (16)$$

The complete expression (2.5.5) may therefore be writen

$$\lambda^2 t \int d\alpha f(\alpha)\,\pi\delta_+(\alpha) - (\lambda^2/i)\int d\alpha f(\alpha)\,\pi\delta'_+(\alpha) \quad (17)$$

The second term corresponds to a correction to the scattering cross section. As we are interested in weakly coupled systems, only terms of the order $\lambda^2 t \approx t/t_r$ have to be kept here. However in the general theory the second term has to be retained (see Chapter 11 and Chapter 12).

The method we have used till now is not very practical for more complicated time integrations. Let us now briefly indicate an alternative approach which is of wider application; let us go back to (2.5.1). By straightforward integration we have

$$\int_0^t dt_1 \int_0^{t_1} dt_2\, e^{i\alpha(t_1-t_2)} = -(t/i\alpha) + [(e^{i\alpha t}-1)/(i\alpha)^2] \quad (18)$$

We therefore have to evaluate the integral [see (2.5.5)]:

$$\int d\alpha f(\alpha)\{-(t/i\alpha)+[(e^{i\alpha t}-1)/(i\alpha)^2]\} \tag{19}$$

The integration extends over the real axis.

It is important to note that the term in brackets is regular for $\alpha \to 0$,

$$-(t/i\alpha)+[(e^{i\alpha t}-1)/(i\alpha)^2] \to \tfrac{1}{2}t^2 \qquad \text{for } \alpha \to 0$$

Assuming also that $f(\alpha)$ has no singularities on the real axis, we may shift the integration path by adding to α a small imaginary quantity ε.[1] This quantity has to be choosen positive to ensure that $e^{i(\alpha+i\varepsilon)t} = e^{(i\alpha-\varepsilon)t}$ goes to zero for large t.

The bracket now becomes

$$-[t/i(\alpha+i\varepsilon)]+\{(e^{i(\alpha+i\varepsilon)t}-1)/[i(\alpha+i\varepsilon)]^2\} \tag{20}$$

We first let t become very large and then let $\varepsilon \to 0$. Because of the factor $e^{-\varepsilon t}$ the exponential vanishes and we are left with

$$-[t/i(\alpha+i\varepsilon)]-\{1/[i(\alpha+i\varepsilon)]^2\} \tag{21}$$

Again the second term represents a correction to the scattering cross section and has to be neglected.

The first term gives

$$i\int d\alpha f(\alpha)/(\alpha+i\varepsilon)=\lim_{\varepsilon\to 0} i\int d\alpha f(\alpha)[\alpha/(\alpha^2+\varepsilon^2)]+\lim_{\varepsilon\to 0}\int d\alpha f(\alpha)[\varepsilon/(\alpha^2+\varepsilon^2)] \tag{22}$$

where the integration is over the real axis. Taking into account the fact that

$$\begin{aligned}\lim_{\varepsilon\to 0}[\alpha/(\alpha^2+\varepsilon^2)] &= 1/\alpha \qquad \text{if } \alpha \neq 0\\ &= 0 \qquad \text{if } \alpha = 0\end{aligned}$$

we have

$$\lim_{\varepsilon\to 0} i\int d\alpha f(\alpha)[\alpha/(\alpha^2+\varepsilon^2)] = i\int d\alpha f(\alpha)\mathscr{P}(1/\alpha) \tag{23}$$

[1] We make use here of Cauchy's fundamental theorem of integration in the complex plane; see for example, Margenau and Murphy, 1943.

On the other hand if we call $(-b, a)$ the domain of integration, we have

$$\lim_{\varepsilon\to 0}\int_{-b}^{+a} d\alpha f(\alpha)[\varepsilon/(\alpha^2+\varepsilon^2)] = \lim_{\varepsilon\to 0} f(0)\,[\arctan(a/\varepsilon) - \arctan(-b/\varepsilon)]$$
$$+ \lim_{\varepsilon\to 0} \varepsilon \int_{-b}^{+a} d\alpha\{f'(0)+(\alpha/2)f''(0)+\cdots\}[\alpha/(\alpha^2+\varepsilon^2)] \qquad (24)$$

Now the first term gives $\pi f(0)$ (remember $\varepsilon > 0$), while the second vanishes. We see, therefore, that $\lim_{\varepsilon\to 0} -i/(\alpha+i\varepsilon)\ (1/\pi)$ is a representation of the function $\delta_+(\alpha)$ as well as

$$\lim_{T\to\infty} [(\sin \alpha T/\alpha)+i(1-\cos \alpha T/\alpha)](1/\pi)$$

We shall also use the δ_- function defined by [see (2.5.10)

$$\pi\delta_-(\alpha) = \pi\delta(\alpha) - i\mathscr{P}(1/\alpha) = \pi\delta_+^*(\alpha) \qquad (25)$$

We have

$$\pi\delta_-(\alpha) = 1/i(\alpha - i\varepsilon) = \int_0^\infty e^{-i(\alpha-i\varepsilon)\tau}\,d\tau \qquad (26)$$

It should be noted that $\delta(\alpha)$ is an even function of α while $\mathscr{P}(1/\alpha)$ is an odd function.

The singular functions δ_- or δ_+ play an important role in all subsequent calculations. We shall discuss their physical meaning in more detail in Chapter 6. Adding (2.5.10) and (2.5.25) and taking account of (2.5.26) we obtain the representation of the δ function:

$$\delta(\alpha) = 1/2\pi \int_{-\infty}^{+\infty} e^{i\alpha\tau}\,d\tau \qquad (27)$$

Let us now consider more briefly the higher order contributions.

6. Fourth-Order Contributions

In this case we have two possibilities: either (1) the four vertices form two cycles separated in time (see Fig. 2.6.1), or (2) we have overlapping cycles (see Fig. 2.6.2).

Fig. 2.6.1 Two successive cycles. (Diagrams in which wave vectors are not explicitly written represent sums over all possible processes.)

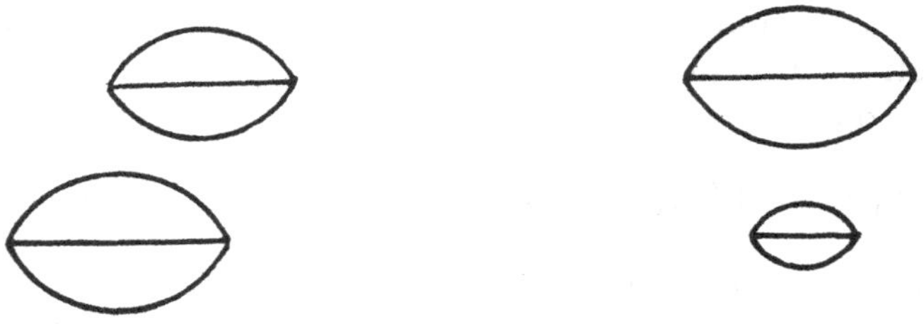

(a) (b)

Fig. 2.6.2 Overlapping cycles.

Let us introduce the following definition which will be useful in the general theory. We shall call a diagram a "*diagonal fragment*" when none of the intermediate states (in $\{n\}$) is identical with the initial state. In Fig. 2.6.1 we have a succession of two diagonal fragments corresponding to two independent collisions, each involving three normal modes. In the case of Fig. 2.6.2 each of the fourth-order diagrams forms a single diagonal fragment. It is now very easy to show that the diagrams of Fig. 2.6.1 give a contribution of order $(\lambda^2 t)^2$ while each diagonal fragment of Fig. 2.6.2. gives a contribution almost of order $\lambda^4 t$ and has to be neglected for weakly coupled systems. These statements are special cases of general theorems we shall discuss in Chapter 8. Indeed, a time integral [see (2.5.1)] of the form

$$\int_0^t dt_1 \int_0^{t_1} dt_2 \int_0^{t_2} dt_3 \int_0^{t_3} dt_4 \, e^{i\alpha(t_1-t_2)} \, e^{i\beta(t_3-t_4)} \tag{1}$$

corresponds to Fig. 2.6.1. Using the method of § 5 (2.5.18) through (2.5.21), we obtain by straightforward integration

$$(t^2/2)\delta_+(\alpha)\delta_+(\beta)\pi^2 \tag{2}$$

On the other hand, the same method applied to Fig. 2.6.2 gives at most a single factor t. Because each vertex gives a factor of λ, these terms are then of order $\lambda^4 t$ and have to be neglected here.

These results may be understood in the following intuitive way. As we have seen, we are interested in times much longer than the duration of a collision. We therefore obtain the correct time behavior by replacing (2.6.1) by

$$\int_0^t dt_1 \int_0^{t_1} dt_2 \int_0^{t_2} dt_3 \int_0^{t_3} dt_4 \, \delta(t_1-t_2)\,\delta(t_3-t_4) \sim t^2 \tag{3}$$

where we have simply replaced the duration of each collision by a δ function. The same procedure, when applied to Fig. 2.6.2, gives, for diagram (a),

$$\int_0^t dt_1 \int_0^{t_1} dt_2 \int_0^{t_2} dt_3 \int_0^{t_3} dt_4 \, \delta(t_1-t_2)\,\delta(t_2-t_3)\,\delta(t_3-t_4) \sim t \qquad (4)$$

Indeed each of the three time intervals t_4-t_3, t_3-t_2, t_2-t_1 has to be of the order of the duration of a collision and may be replaced by a δ function. We have a similar expression for diagram (b). Fig. 2.6.3 explicitly shows the time relationship for the

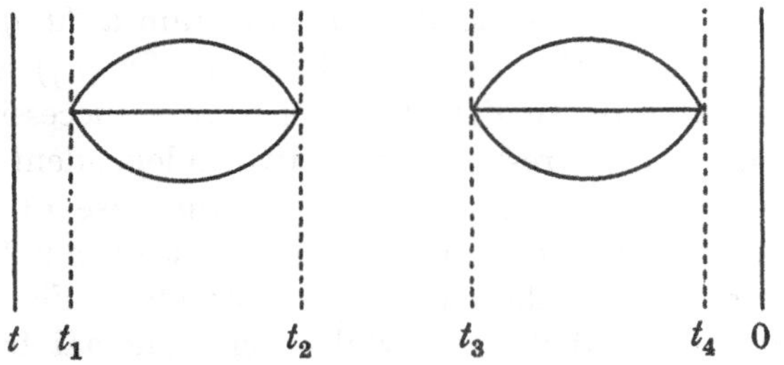

Fig. 2.6.3 Time relationships for the diagram (2.6.1).

diagram of Fig. 2.6.1; the time intervals t_1-t_2 t_3-t_4 are very short. The time integration extends over the three time intervals $0-t_4$ t_3-t_2 t_1-t, related by the condition that their sum has the given value t. We are therefore left with two independent

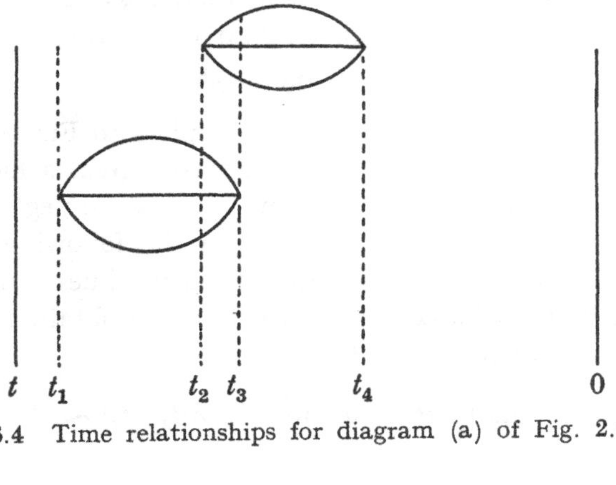

Fig. 2.6.4 Time relationships for diagram (a) of Fig. 2.6.2.

time intervals of order t and this gives us the result $\sim t^2$ of (2.6.2). On the other hand, the time relationships for diagram (a) of Fig. 2.6.2 are shown in Fig. 2.6.4. We are left with a single independent time interval of order t, and the result after integration is $\sim t$. We shall prove in Chapter 8, §5 that the asymptotic time contribution of n diagonal fragments whatever their structure is $\sim t^n$. The results we have discussed here are special cases of this general theorem.

7. Master Equation

The discussion we have presented in detail for the second- and fourth-order terms in (2.3.5) may be easily extended to the general term λ^{2n}. All the contributions proportional to λ^{2n} consist of n cycles arranged in all possible ways. However, the only contributions of order $\lambda^{2n}t^n$ come from a succession of n disconnected cycles separated by macroscopic time intervals. Each of these cycles corresponds to a "collision" between normal modes. The relevant term in λ^{2n} is therefore

$$\lambda^{2n}\sum_{\{n\}_1}\cdots\sum_{\{n\}_n}\int_0^t dt_1\int_0^{t_1} dt_2\cdots\int_0^{t_{2n-1}} dt_{2n}\,\langle 0|\delta L|\{n\}_1\rangle\langle\{n\}_1|\delta L|0\rangle$$
$$\times\langle 0|\delta L|\{n\}_2\rangle\langle\{n\}_2|\delta L|0\rangle\cdots\langle 0|\delta L|\{n\}_n\rangle\langle\{n\}_n|\delta L|0\rangle$$
$$\times\, e^{i\alpha_1(t_1-t_2)}e^{i\alpha_2(t_3-t_4)}\ldots e^{i\alpha_n(t_{2n-1}-t_{2n})} \tag{1}$$

The time integral can be performed in the way described above. We consider as integration variables $\tau_i = t_i - t_{i+1}$. The integrals over τ_i lead to $\pi\delta_+(\alpha_i)$. The remaining integrals give $t^n/n!$. Equation (2.3.5) therefore becomes

$$\rho_{\{0\}}(t) = \rho_{\{0\}}(0) - \lambda^2 t\sum_{\{n'\}}\langle\{0\}|\delta L|\{n'\}\rangle\pi\delta_-(\textstyle\sum n'_k\omega_k)$$
$$\times\langle\{n'\}|\delta L|\{0\}\rangle\,\rho_{\{0\}}(0)$$
$$+(\lambda^4t^2/2)\sum_{\{n'\}\{n''\}}\langle\{0\}|\delta L|\{n'\}\rangle\pi\delta_-(\textstyle\sum n'_k\omega_k)\langle\{n'\}|\delta L|\{0\}\rangle$$
$$\times\langle\{0\}|\delta L|\{n''\}\rangle\pi\delta_-(\textstyle\sum n''_k\omega_k)\langle\{n''\}|\delta L|\{0\}\rangle\,\rho_{\{0\}}(0)$$
$$-\lambda^6t^3/3!\sum\cdots \tag{2}$$

Let us introduce the operator

$$O_0 = -\sum_{\{n'\}} \langle\{0\}|\delta L|\{n'\}\rangle \pi\delta_-(\sum n'_k\omega_k) \langle\{n'\}|\delta L|\{0\}\rangle \tag{3}$$

The index 0 of O_0 recalls that the initial and final state are both $\{0\}$.

The series (2.7.2) may then be written

$$\begin{aligned}\rho_{\{0\}}(t) &= \rho_{\{0\}}(0)+\lambda^2 t O_0 \rho_{\{0\}}(0)+(\lambda^4 t^2/2)(O_0)^2\rho_{\{0\}}(0)\\ &\quad +\cdots+[(\lambda^2 t)^n/n!](O_0)^n\rho_{\{0\}}(0)+\cdots \\ &= e^{\lambda^2 t O_0}\rho_{\{0\}}(0)\end{aligned} \tag{4}$$

Differentiating with respect to time we obtain the so-called *master equation* valid for weakly coupled systems (Brout and Prigogine, 1956).

$$\partial\rho_{\{0\}}/\partial t = \lambda^2(O_0)_2\rho_{\{0\}} \tag{5}$$

This equation is valid for all weakly coupled systems. In each case one simply has to use the appropriate values of the Liouville matrix elements.

Let us write (2.7.3) for anharmonic solids. Using the results of § 5 we have

$$\begin{aligned}O_0 = -\pi\sum_{kk'k''} &\{\delta_+(\omega_k+\omega_{k'}+\omega_{k''})\langle 0|\delta L|-1_k-1_{k'}-1_{k''}\rangle\\ &\times\langle -1_k-1_{k'}-1_{k''}|\delta L|0\rangle\\ &+3\delta_+(\omega_k+\omega_{k'}-\omega_{k''})\langle 0|\delta L|-1_k-1_{k'}1_{k''}\rangle\langle -1_k-1_{k'}1_{k''}|\delta L|0\rangle\\ &+3\delta_+(-\omega_k-\omega_{k'}+\omega_{k''})\langle 0|\delta L|1_k1_{k'}-1_{k''}\rangle\langle 1_k1_{k'}-1_{k''}|\delta L|0\rangle\\ &+\delta_+(-\omega_k-\omega_{k'}-\omega_{k''})\langle 0|\delta L|1_k1_{k'}1_{k''}\rangle\langle 1_k1_{k'}1_{k''}|\delta L|0\rangle\}\end{aligned} \tag{6}$$

These four contributions correspond to the four diagrams of Fig. (2.4.6). The summation over k, k', k'' has to be understood as an integration in order to give a meaning to the δ functions.

This expression may be considerably simplified using the explicit expressions for the matrix elements given in § 2. For example [see (2.2.6) and (2.2.7)],

$$\begin{aligned}
&\langle 0|\delta L|-1_k-1_{k'}-1_{k''}\rangle\langle -1_k-1_{k'}-1_{k''}|\delta L|0)\rangle \\
&= V_{kk'k''}[-(\partial/\partial J_k)-(\partial/\partial J_{k'})-(\partial/\partial J_{k''})](J_kJ_{k'}J_{k''}/\omega_k\omega_{k'}\omega_{k''})^{1/2}(V^*_{kk'k''}) \\
&\quad\times(J_kJ_{k'}J_{k''}/\omega_k\omega_{k'}\omega_{k''})^{1/2}[(\partial/\partial J_k)+(\partial/\partial J_{k'})+(\partial/\partial J_{k''})] \\
&= -(|V_{kk'k''}|^2/\omega_k\omega_{k'}\omega_{k''})[(\partial/\partial J_k)+(\partial/\partial J_{k'})+(\partial/\partial J_{k''})]J_kJ_{k'}J_{k''} \\
&\quad\times[(\partial/\partial J_k)+(\partial/\partial J_{k'})+(\partial/\partial J_{k''})] \qquad (7)
\end{aligned}$$

We proceed in the same way with the other terms of (2.7.6). We have to take the following expression into account

$$\delta(\omega_k+\omega_{k'}+\omega_{k''}) = \delta(-\omega_k-\omega_{k'}-\omega_{k''}) = 0 \qquad (8)$$

because all frequencies are positive. We also have to use the fact that the principal part that appears in δ_+ [see (2.5.10)] is an odd function of its argument. After a few calculations one finds in this way

$$\begin{aligned}
O_0 = 6\pi \sum_{kk'k''} &\delta(\omega_k+\omega_{k'}-\omega_{k''})(|V_{kk'-k''}|^2/\omega_k\omega_{k'}\omega_{k''}) \\
&\times[(\partial/\partial J_k)+(\partial/\partial J_{k'})-(\partial/\partial J_{k''})]J_kJ_{k'}J_{k''} \\
&\times[(\partial/\partial J_k)+(\partial/\partial J_{k'})-(\partial/\partial J_{k''})] \qquad (9)
\end{aligned}$$

The operators act on everything on their right. It is easy to verify that O_0 is a self-adjoint operator. Indeed by partial integration

$$\int dJ\,\varphi_1(J)[O_0\varphi_2(J)] = \int dJ\,\varphi_2(J)[O_0\varphi_1(J)] \qquad (10)$$

In his fundamental work on the thermal conductivity of solids Peierls (1929) has postulated equations (2.7.5) and (2.7.9) on the basis of arguments similar to the classical arguments of Boltzmann for gases.

Let us now study the properties of equation (2.7.5).

8. H-Theorem

Using equations (2.7.5) and (2.7.9) we may easily obtain an H-theorem[1] for the distribution function $\rho_{\{0\}}$

$$(d/dt)\int dJ\,\rho_{\{0\}}\ln\rho_{\{0\}} \leqq 0 \qquad (1)$$

[1] We call (2.8.1) an H-theorem because of its obvious analogy with the H-theorem established by Boltzmann for the velocity distribution function [see for example Chapman and Cowling (1939)].

We have, indeed,

$$(d/dt)\int dJ\,\rho_{\{0\}} \ln \rho_{\{0\}} = \lambda^2 \int dJ (1+\ln \rho_{\{0\}}) O_0\, \rho_{\{0\}}$$

$$= -6\pi\lambda^2 \sum_{kk'k''} \delta(\omega_k+\omega_{k'}-\omega_{k''}) (|V_{kk'-k''}|^2/\omega_k \omega_{k'} \omega_{k''})$$

$$\times \int dJ\, J_k J_{k'} J_{k''} (1/\rho_{\{0\}})[(\partial\rho_{\{0\}}/\partial J_k)+(\partial\rho_{\{0\}}/\partial J_{k'})-(\partial\rho_{\{0\}}/\partial J_{k''})]^2 \quad (2)$$

This expression will be equal to zero only when statistical equilibrium is reached, that is when

$$\rho_{\{0\}} = f(H_0) = f(\sum \omega_k J_k) \quad (3)$$

Because of the assumption of weak coupling this statistical equilibrium corresponds to the *unperturbed* Hamiltonian H_0; $f(H_0)$ is an arbitrary function which is fixed by the initial conditions. Indeed to the initial distribution $\rho_{\{0\}}(J_1 \cdots J_N, t = 0)$ corresponds a well-defined distribution of energy

$$f(H_0) = \int_{\sum_k \omega_k J_k = H_0} dJ\, \rho_{\{0\}}(J_1 \cdots J_N, t = 0)$$

$$= \int dJ\, \delta(\sum \omega_k J_k - H_0) \rho_{\{0\}} \quad (4)$$

The integration has to be performed over values of the action variables such that $\sum_k \omega_k J_k = H_0$. This distribution $f(H_0)$ is then constant in time (see also Chapter 5, § 2). Indeed, using (2.7.10) we have

$$df(H_0)/dt = \int dJ\, \rho_{\{0\}} O_0 \delta(\sum \omega_k J_k - H_0) \quad (5)$$

Now (2.7.9) shows that this expression contains the factor

$$\delta(\omega_k+\omega_{k'}-\omega_{k''})[(\partial/\partial J_k)+(\partial/\partial J_{k'})-(\partial/\partial J_{k''})]\delta(\sum \omega_k J_k - H_0)$$

$$\sim x\delta(x) \qquad \text{with} \qquad x = \omega_k+\omega_{k'}-\omega_{k''}$$

which clearly vanishes. It is interesting to notice that the conservation of energy is therefore ensured by the function $\delta(\omega_k+\omega_{k'}-\omega_{k''})$ involving the frequencies of the normal modes. It is therefore closely related to the quantum mechanical form of energy conservation (see Chapter 13). Let us briefly consider the relation of equation (2.7.5) to random processes.

9. Fokker-Planck Equation and Master Equation

We shall first define what we call a random process $y(t)$ (see especially M. C. Wang and G. E. Uhlenbeck, 1945).[1] In such a process the variable y does not depend in a deterministic or causal way on the time t, but one observes different successive values of y if the process is repeated. One often calls a random process a *purely* random process when the successive values of y are not correlated at all. For example, the successive numbers which appear in a game of dice correspond to a purely random process. In more complicated situations there exists only a correlation between two successive events. The probability of a succession of three events i, j, k may then be expressed in terms of the probability of i, and of the transitions $i \rightarrow j$, $j \rightarrow k$. Each of the transition probabilities depends only on the two states involved and not on the previous history of the system. Such processes are called *Markoff processes* or *Markoff chains*.

It may be shown that the probability distribution $p(y, t)$ corresponding to the random variable y satisfies a continuity equation which reduces in simple cases to a partial differential equation of the diffusion type. This partial differential equation is also called the *Fokker-Planck* equation and is of the form

$$\partial p(y, t)/\partial t = (\partial/\partial y)\{-[\overline{(\Delta y)}/\Delta\tau] + \tfrac{1}{2}\,\partial/\partial y(\overline{\Delta y^2}/\Delta\tau\}p(y, t) \qquad (1)$$

where $\overline{(\Delta y)}$ is the average change of y during the time $\Delta\tau$; the ratio $\overline{(\Delta y)}/\Delta\tau$ has to be taken in the sense

$$\lim_{\Delta\tau\to 0} \overline{(\Delta y)}/\Delta\tau \qquad (2)$$

$\overline{(\Delta y^2)}$ is defined in a similar manner. In the simple case when

$$\overline{(\Delta y)}/\Delta\tau = 0, \qquad \tfrac{1}{2}\overline{(\Delta y^2)}/\Delta\tau = D \qquad (3)$$

the Fokker-Planck equation reduces to the ordinary diffusion equation

$$\partial p/\partial t = D(\partial^2 p/\partial y^2) \qquad (4)$$

[1] The paper by Wang and Uhlenbeck as well as other important papers on this subject are reprinted in Wax, (1954).

The quantity $(\overline{\Delta y^n})$ is called the nth *transition moment*. For several independent variables $y_1, \cdots, y_n$ the Fokker-Planck equation is

$$\partial p/\partial t = \sum_j (\partial/\partial y_j)\{-[(\overline{\Delta y_j})/\Delta\tau] + \tfrac{1}{2}\sum_{j'} (\partial/\partial y_{j'})[\overline{\Delta y_j \Delta y_{j'}})/\Delta\tau]\}p \tag{5}$$

The characteristic feature of the Fokker-Planck equation is that only first and second transition moments appear.

It is immediately clear that the master equation (2.7.5)

$$\begin{aligned}\partial \rho_{\{0\}}/\partial t = \lambda^2\, 6\pi \sum_{kk'k''} \delta(\omega_k+\omega_{k'}-\omega_{k''})[|V_{kk'-k''}|^2/\omega_k\omega_{k'}\omega_{k''}] \\ \times[(\partial/\partial J_k)+(\partial/\partial J_{k'})-(\partial/\partial J_{k''})]J_k J_{k'} J_{k''} \\ \times[(\partial/\partial J_k)+(\partial/\partial J_{k'})-(\partial/\partial J_{k''})]\rho_{\{0\}}\end{aligned} \tag{6}$$

corresponds to a Markowian process. Indeed the change of $\rho_{\{0\}}$ at any given time depends only on the value of $\rho_{\{0\}}$ *at that time* and is independent of its previous history.

We shall now put (2.9.6) into the form (2.9.5) and prove in this way that the master equation is the Fokker-Planck equation for the distribution of the energy.

Let us use as independent variables the energies $E_k = \omega_k J_k$. We then have

$$\begin{aligned}\partial \rho_{\{0\}}/\partial t = \lambda^2 \sum_{kk'k''} C_{kk'k''}\delta(\omega_k+\omega_{k'}-\omega_{k''}) \\ \times[\omega_k(\partial/\partial E_k)+\omega_{k'}(\partial/\partial E_{k'})-\omega_{k''}(\partial/\partial E_{k''})] \\ \times E_k E_{k'} E_{k''}[\omega_k(\partial/\partial E_k)+\omega_{k'}(\partial/\partial E_{k'})-\omega_{k''}(\partial/\partial E_{k''})]\rho_{\{0\}}\end{aligned} \tag{7}$$

with

$$C_{kk'k''} = 6\pi(|V_{kk'-k''}|^2/\omega_k^2\omega_{k'}^2\omega_{k''}^2) \tag{8}$$

In equation (2.9.7) $\omega_{k''}$ is associated with $-k''$ [see (2.9.8)]. To obtain more symmetrical expressions let us use the notation $\omega_{-k} = -\omega_k$. In this way (2.9.7) takes the symmetrical form

$$\begin{aligned}\partial \rho_{\{0\}}/\partial t = \lambda^2 \sum_{kk'k''} \delta(\omega_k+\omega_{k'}+\omega_{k'})C_{kk'k''} \\ \times[\omega_k(\partial/\partial E_k)+\omega_{k'}(\partial/\partial E_{k'})+\omega_{k''}(\partial/\partial E_{k''})]E_k E_{k'} E_{k''} \\ \times[\omega_k(\partial/\partial E_k)+\omega_{k'}(\partial/\partial E_{k'})+\omega_{k''}(\partial/\partial E_{k''})]\rho_{\{0\}}\end{aligned} \tag{9}$$

We want to put this equation into the following form [see (2.9.5)]:

$$\partial\rho_{\{0\}}/\partial t = \sum_k (\partial/\partial E_k)\{-(\overline{\Delta E_k})/\Delta\tau + \tfrac{1}{2}\sum_{k'} (\partial/\partial E_{k'})[(\overline{\Delta E_k \Delta E_{k'}})/\Delta\tau]\}\rho_{\{0\}} \tag{10}$$

Let us calculate $(\overline{\Delta E_k})/\Delta\tau$. We have, by definition,

$$\begin{aligned}(\overline{\Delta E_k})/\Delta\tau &= \{[\overline{E_k(t+\Delta\tau)}] - [\overline{E_k(t)}]\}/\Delta\tau \\ &= (1/\Delta\tau) \int dE\, E_k[\rho_{\{0\}}(t+\Delta\tau) - \rho_{\{0\}}(t)] \\ &= \lambda^2 \int dE\, E_k O_0 \rho_{\{0\}} + O(\Delta\tau)\end{aligned} \tag{11}$$

We now replace the operator O_0 by its explicit value (2.9.9) and calculate the average (2.9.11).

We may suppose that at the time t all energies had well-defined values $E_1 \cdots E_s$, i.e., that $\rho_{\{0\}}(t)$ is a product of δ functions. In this way we immediately obtain

$$\begin{aligned}(\overline{\Delta E_k})/\Delta\tau = (3\lambda^2) \sum_{k'k''} C_{kk'k''}\,\delta(\omega_k+\omega_{k'}+\omega_{k''})\,\omega_k \\ \times [\omega_{k'} E_{k'} E_{k''} + \omega_{k''} E_k E_{k'} + \omega_k E_{k'} E_{k''}]\end{aligned} \tag{12}$$

A similar calculation gives

$$\begin{aligned}(\overline{\Delta E_k \Delta E_{k'}})/\Delta\tau &= \{[\overline{E_k(t+\Delta\tau) - E_k(t)][E_{k'}(t+\Delta\tau) - E_{k'}(t)]}\}/\Delta\tau \\ &= 12\lambda^2 \sum_{k''} C_{kk'k''}\,\delta(\omega_k+\omega_{k'}+\omega_{k''})\,\omega_k\omega_{k'} E_k E_{k'} E_{k''}\end{aligned} \tag{13}$$

$$(\overline{\Delta E_k})^2/\Delta\tau = 6\lambda^2 \sum_{k'k''} C_{kk'k''}\,\omega_k^2\,\delta(\omega_k+\omega_{k'}+\omega_{k''}) E_k E_{k'} E_{k''} \tag{14}$$

It is easy to verify that if we substitute (2.9.12) through (2.9.14) into (2.9.10), we recover the master equation (2.9.9).

We see, therefore, that starting with a purely mechanical description of the system we have obtained an equation corresponding to a Markowian random process. We shall discuss in greater detail the significance of this result later. Nevertheless, we want to make the following remarks here.

Our description is only valid in the limit of a large system. It is only then that the time integration in conjunction with a continuous spectrum gives rise to the contribution $(\lambda^2 t)^n$ which we have retained.

For small systems we would not have the right to replace the summation over the wave vectors by an integration, and we would always keep only oscillating functions of time. No equation like the master equation, expressing a systematic approach to equilibrium, is to be expected.

It is the energy distribution function $\rho_{\{0\}}$ and not the complete phase distribution function ρ which satisfies the irreversible master equation. This will appear to be the general situation; we shall always derive transport equations only for some Fourier components of the phase distribution function ρ and never for ρ itself.

An interesting application of our formalism lies in the theory of Brownian motion. Starting with the master equation (2.9.7) or (2.9.9) we could study the evolution of a single normal mode when all others have reached their equilibrium. However, we shall postpone this discussion to Chapter 3, where we shall study such situations without assuming homogeneity [see (2.3.3)].

CHAPTER 3

Brownian Motion

1. Basic Equations

In this chapter we shall study the approach to equilibrium of a single normal mode in a situation in which all other degrees of freedom are in equilibrium or have at least achieved a homogeneous, angle independent distribution. We consider in some detail the case of an anharmonic solid characterized by the Hamiltonian which we introduced in Chapter 2, § 1. However, the method we shall use applies as well to practically all problems involving the interaction of an anharmonic oscillator with a "thermostat" of any nature to which it is weakly coupled. Such a thermostat might be, for example, a gas or black body radiation (Prigogine and Leaf, 1959; Leaf, 1959).

We now have to introduce initial conditions which generalize the conditions (2.3.3).

Let us call l the normal mode which is out of equilibrium. We shall assume that the only Fourier coefficients which are different from zero correspond to $\rho_{\{0\}}$ or to $\rho_{n_l, 0 \cdots 0}$ where n_l refers to the normal mode l.

In other words, we consider distribution functions of the form

$$\rho(J_l, \alpha_l, J_1 \cdots J_{l-1}, J_{l+1} \cdots J_s) \tag{1}$$

which depend on the single angle variable α_l and for which, therefore,

$$\rho_{n_l\{n'\}} = 0 \qquad \text{unless } \{n'\} = 0 \tag{2}$$

We may say that we retain the exact phase relations for the single normal mode l while we use the random phase approximation for all other normal modes.

The exact dependence of ρ on the action variables $J_1, J_2, \ldots$ is irrelevant for the moment. In §3 we shall introduce a supplementary condition which expresses the fact that (3.1.1) corresponds to an equilibrium distribution for all normal modes except l.

We shall assume that (3.1.1) or (3.1.2) is valid for *all times* $t \geqq 0$. Indeed in the approximation of weak coupling we may expect that the angle dependence of one single degree of freedom will introduce negligible phase correlations for the others. We shall prove this statement later. (See Chapter 8 and Appendix III.)

Let us consider the time change of ρ_{n_l}. Using (1.8.7) as well as the diagrams of Chapter 2, § 4, we see that ρ_{n_l} can be connected in a single transition only to $\rho_{n_l \pm 1_k \pm 1_{k'} \pm 1_{k''}}$ (where l may be equal to k, k', or k''). In diagram form the corresponding equation may be written

$$\frac{\partial \rho_{n_l}}{\partial t} =$$

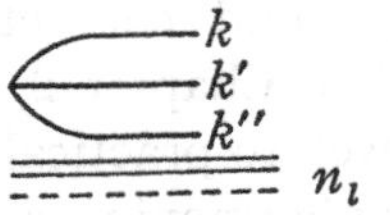

$$\rho_{n_l \pm 1_k \pm 1_{k'} \pm 1_{k''}} \qquad (3)$$

Next let us consider the time change of $\rho_{n_l \pm 1_k \pm 1_{k'} \pm 1_{k''}}$. Here we must distinguish between two possibilities: either we may connect $\rho_{n_l \pm 1_k \pm 1_{k'} \pm 1_{k''}}$ back to ρ_{n_l}, or we may connect it to some other Fourier coefficient

$$\rho_{n_l \pm 1_k \pm 1_{k'} \pm 1_{k''} \pm 1_m \pm 1_{m'} \pm 1_{m''}}$$

This is expressed in diagram form in (3.1.4).

$$\partial \rho_{n_l \pm 1_k \pm 1_{k'} \pm 1_{k''}} / \partial t$$

$=$ 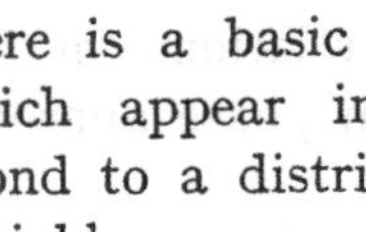

$\rho_{n_l} +$ 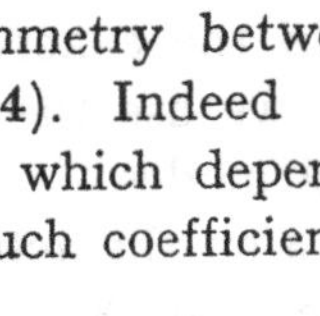

$$\rho_{n_l \pm 1_k \cdots \pm 1_{m''}} \qquad (4)$$

Because of the physical situation that we are investigating, there is a basic dissymmetry between the two contributions which appear in (3.1.4). Indeed $\rho_{n_l \pm 1_k \cdots \pm 1_{m''}}$ would correspond to a distribution which depends explicitly on the angle variables $\alpha_k, \cdots, \alpha_{m''}$. Such coefficients are assumed to be zero according to (3.1.2).

The remarkable feature here is that when we neglect these

contributions in (3.1.4), the two equations (3.1.3) and (3.1.4) reduce to a closed system for the unknown

$$\rho_{n_l}$$

Let us write the equations corresponding to this closed system more explicitly. We have

$$\partial \rho_{n_l}/\partial t = (\lambda/i) \sum_{kk'k''} e^{-i(\pm\omega_k\pm\omega_{k'}\pm\omega_{k''})t} \langle n_l|\delta L|n_l\pm 1_k\pm 1_{k'}\pm 1_{k''}\rangle \rho_{n_l\pm 1_k\pm 1_{k'}\pm 1_{k''}} \tag{5}$$

$$\partial \rho_{n_l\pm 1_k\pm 1_{k'}\pm 1_{k''}}/\partial t = (\lambda/i)\, e^{i(\pm\omega_k\pm\omega_{k'}\pm\omega_{k''})t} \times \langle n_l \pm 1_k \pm 1_{k'} \pm 1_{k''}|\delta L|n_l\rangle \rho_{n_l} \tag{6}$$

We now eliminate $\rho_{n_l\pm 1_k\pm 1_{k'}\pm 1_{k''}}$ between (3.1.5) and (3.1.6) to obtain a closed equation for ρ_{n_l} alone:

$$\partial \rho_{n_l}/\partial t = -\lambda^2 \sum_{kk'k''} \langle n_l|\delta L|n_l\pm 1_k\pm 1_{k'}\pm 1_{k''}\rangle e^{-i(\pm\omega_k\pm\omega_{k'}\pm\omega_{k''})t} \times \int_0^t dt_1\, e^{i(\pm\omega_k\pm\omega_{k'}\pm\omega_{k''})t_1} \langle n_l\pm 1_k\pm 1_{k'}\pm 1_{k''}|\delta L|n_l\rangle \rho_{n_l}(t_1) \tag{7}$$

In this method $\rho_{n_l\pm 1_k\pm 1_{k'}\pm 1_{k''}}$ appears as a kind of intermediate state corresponding to the appearance of a correlation among the normal modes k, k', k'' during the collision.

As long as we are interested in the weakly coupled case, we may consider ρ_{n_l} as a constant in the integration over a time of the order of the duration of the interaction between the normal modes. Using (2.5.25) as well, we obtain the basic equation

$$\partial \rho_{n_l}/\partial t = \lambda^2 O_{n_l} \rho_{n_l} \tag{8}$$

with [see (2.7.6)]

$$O_{n_l} = -\pi \sum_{kk'k''} \langle n_l|\delta L|n_l-1_k-1_{k'}-1_{k''}\rangle \delta_+(\omega_k+\omega_{k'}+\omega_{k''}) \langle n_l-1_k-1_{k'}-1_{k''}|\delta L|n_l\rangle + \cdots + \cdots + \cdots \tag{9}$$

where the three other terms are similar and correspond to the intermediate states $1_k - 1_{k'} - 1_{k''}$; $1_k 1_{k'} - 1_{k''}$; $1_k 1_{k'} 1_{k''}$. The basic diagram which corresponds to (3.1.8) is given by Fig. 3.1.1.

It differs from the diagram involved in the master equation (2.7.5) by the presence of the n_l lines.

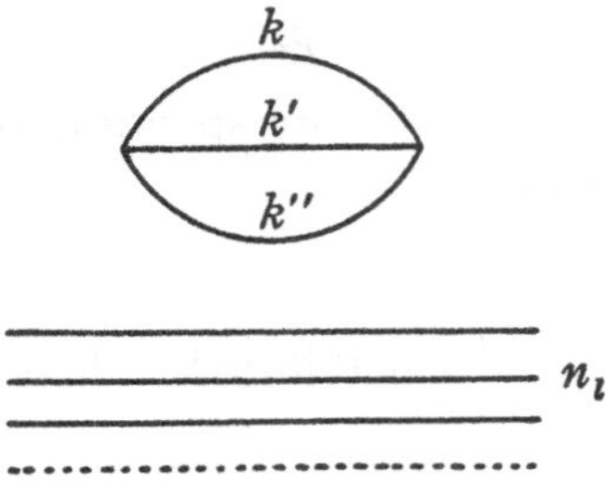

Fig. 3.1.1 Diagram corresponding to (3.1.8).

It is easy to obtain the explicit expression for this operator, exactly as we did for O_0 in Chapter 2, § 7. For example [see (2.7.7), (2.7.9), and (2.2.7)],

$$\begin{aligned}
&\langle n_l|\delta L|n_l-1_k-1_{k'}-1_{k''}\rangle\langle n_l-1_k-1_{k'}-1_{k''}|\delta L|n_l\rangle\\
&\qquad = -|V_{kk'k''}|^2[-(n_k\delta_{kl}/2J_k)-(n_{k'}\delta_{k'l}/2J_{k'})-(n_{k''}\delta_{k''l}/2J_{k''})\\
&\qquad +(\partial/\partial J_k)+(\partial/\partial J_{k'})+(\partial/\partial J_{k''})](J_kJ_{k'}J_{k''}/\omega_k\omega_{k'}\omega_{k''})\\
&\qquad \times[(n_k\delta_{kl}/2J_k)+(n_{k'}\delta_{k'l}/2J_{k'})+(n_{k''}\delta_{k''l}/2J_{k''})\\
&\qquad\qquad +(\partial/\partial J_k)+(\partial/\partial J_{k'})+(\partial/\partial J_{k''})]
\end{aligned}\tag{10}$$

The Kronecker δ takes account of the fact that in both the initial and the final states $n_k = n_l$ for $k = l$ and $n_k = 0$ for $k \neq l$.

Adding all contributions we obtain [see (2.7.9)]

$$\begin{aligned}
O_{n_l} &= O_0 - n_l^2\, 6\pi \sum_{kk'k''}(|V_{kk'-k''}|^2/\omega_k\omega_{k'}\omega_{k''})\,\delta(\omega_k+\omega_{k'}-\omega_{k''})\\
&\qquad\qquad \times(J_kJ_{k'}J_{k''}/4J_l^2)(\delta_{kl}+\delta_{k'l}+\delta_{k''l})^2\\
&+2in_l\{\sum_{kk'k''}(|V_{kk'k''}|^2/\omega_k\omega_{k'}\omega_{k''})\,\mathscr{P}(1/\omega_k+\omega_{k'}+\omega_{k''})(\delta_{kl}+\delta_{k'l}+\delta_{k''l})\\
&\qquad\qquad \times\{[(\partial/\partial J_k)+(\partial/\partial J_{k'})+(\partial/\partial J_{k''})][J_kJ_{k'}J_{k''}/2J_l]\\
&+3\sum_{kk'k''}(|V_{kk'-k''}|^2/\omega_k\omega_{k'}\omega_{k''})\,\mathscr{P}(1/\omega_k+\omega_{k'}-\omega_{k''})(\delta_{kl}+\delta_{k'l}+\delta_{k''l})\\
&\qquad\qquad \times\{[(\partial/\partial J_k)+(\partial/\partial J_{k'})-(\partial/\partial J_{k''})][J_kJ_{k'}J_{k''}/2J_l]\}\}
\end{aligned}\tag{11}$$

This expression is of the form

$$O_{n_l} = O_0 - n_l^2M_l^2 + in_lF_l \tag{12}$$

Only O_0 is an operator; F_l and M_l^2 are ordinary functions of the J_l. (The derivatives $(\partial/\partial J_k)$ in F_l do *not* act on anything outside the brackets). M_l^2 is positive; for this reason we have used the "square" notation.

The form (3.1.12) used in conjunction with the H-theorem of Chapter 2 will permit us to describe in detail the approach to equilibrium.

2. Approach to Equilibrium

For a weakly coupled system the equilibrium properties will be determined by the unperturbed Hamiltonian alone. Therefore, at equilibrium [see (2.8.3)]

$$\rho = f(H_0) = f(\sum \omega_k J_k) \tag{1}$$

This implies that at equilibrium the following two properties hold

$$\rho_0 = f(H_0) \tag{2}$$

$$\rho_{n_l} = 0 \tag{3}$$

The first relation has already been derived in Chapter 2, § 8 by means of the usual H-theorem. We shall now prove the second property (3.2.3) which expresses the disappearance of the phase relations that initially existed for the normal mode l.

Let us multiply (3.1.8) by $\rho_{n_l}^*$ and the complex conjugate equation by ρ_{n_l} and add the two equations. Taking account of (3.1.12), we obtain

$$\partial|\rho_{n_l}|^2/\partial t = \lambda^2[\rho_{n_l}^* O_0\, \rho_{n_l} + \rho_{n_l} O_0\, \rho_{n_l}^* - 2n_l^2 M_l^2 |\rho_{n_l}|^2] \tag{4}$$

We have in this way eliminated the imaginary part $in_l F_l$ of (3.1.12). Let us call φ_m and μ_m the eigenfunctions and the eigenvalues of O_0.

$$O_0 \varphi_m = \mu_m \varphi_m \tag{5}$$

We have seen [see (2.7.10)] that O_0 is a self-adjoint operator. Its eigenvalues are therefore real. Moreover the validity of the H-theorem (cf. Chapter 2, § 8) indicates that the eigenvalues are all negative or zero. Indeed,

$$\mu_m = \int dJ\, \varphi_m^* O_0\, \varphi_m$$

using (2.7.9) and integrating by parts we obtain

$$\mu_m = -6\pi \sum_{kk'k''} (|V_{kk'-k''}|^2/\omega_k\omega_{k'}\omega_{k''})\,\delta(\omega_k+\omega_{k'}-\omega_{k''})$$
$$\times \int dJ J_k J_{k'} J_{k''} |[(\partial\varphi_m/\partial J_k)+(\partial\varphi_m/\partial J_{k'})-(\partial\varphi_m/\partial J_{k''})]|^2 \leqq 0 \quad (6)$$

Let us expand ρ_{n_l} in terms of the eigenfunctions φ_m

$$\rho_{n_l} = \sum_m a_m^{(n_l)} \varphi_m \quad (7)$$

We then substitute this expression into (3.2.4) and integrate over the action variables:

$$(\partial/\partial t)\int dJ|\rho_{n_l}|^2 = 2\lambda^2 \sum_m |a_m^{(n_l)}|^2 \mu_m - 2\lambda^2 n_l^2 \int dJ M_l^2 |\rho_{n_l}|^2 \leqq 0 \quad (8)$$

Each of these two parts is *separately negative*. The equality will hold only if

$$|\rho_{n_l}|^2 = 0 \quad (9)$$

This implies in turn[1]

$$\rho_{n_l} = 0 \quad (10)$$

The difference in behavior between ρ_0 and ρ_{n_l} [see (3.2.2) and (3.2.3)] is closely related to the conservation of normalization (see Chapter 1, § 4)

$$(d/dt)\int dJ d\alpha \rho = (d/dt)\int dJ \rho_0 = 0 \quad (12)$$

On the other hand the integral

$$\int dJ \rho_{n_l}$$

[1] In the sense that if (3.2.9) is satisfied, the average value of any function $f(\alpha)$ of α is equal to that calculated from (3.2.10). Indeed, taking account of the reality condition, we have

$$f(\alpha) = \sum_k f_k e^{ik\alpha}$$
$$\overline{[f(\alpha)]} = \sum_k f_k \varrho_k^* = \sum_{k\geqq 0} |f_k||\rho_k| \cos(\varphi_k - \Phi_k) \quad (11)$$

where we have written

$$f_k = |f_k| \exp i\varphi_k, \qquad \varrho_k = |\rho_k| \exp i\Phi_k.$$

depends on time and vanishes at equilibrium. Here we meet two aspects of the approach to equilibrium: the evolution of the energy distribution towards the equilibrium distribution, and the disappearance of the initial phase relations.

Let us now write the equations in a more explicit form and in this way follow in detail the approach to equilibrium.

3. Statistical Theory of Brownian Motion

We shall now introduce the assumption of statistical equilibrium for all degrees of freedom except l. We shall call "Brownian motion" the evolution of a few degrees of freedom of a large system towards statistical equilibrium in situations in which all other degrees of freedom remain in equilibrium. This is the problem we want to treat now.

A convenient description of a degree of freedom in statistical equilibrium is provided by the canonical distribution involving the equilibrium temperature T (see for example Tolman, 1938). We then have

$$\begin{aligned}\rho &= f(J_l, \alpha_l, t) \prod_{k \neq l} g_k(J_k) \\ &= [1/\sqrt{(2\pi)}] \sum_{n_l} f_{n_l}(J_l, t)\, e^{in_l(\alpha_l - \omega_l t)} \prod_k g_k(J_k)\end{aligned} \tag{1}$$

where g_k corresponds to the canonical equilibrium distribution for all degrees of freedom except l,

$$g_k(J_k) = (\omega_k/\boldsymbol{k}T) \exp(-\omega_k J_k/\boldsymbol{k}T) \tag{2}$$

and where $\boldsymbol{k}$ is the universal Boltzmann constant. We have

$$\int_0^\infty dJ_k g_k(J_k) = 1 \tag{3}$$

Clearly in equilibrium

$$\bar{J}_k = \boldsymbol{k}T/\omega_k \tag{4}$$

This corresponds to the equipartition of energy among normal modes

$$\bar{E}_k = \omega_k \bar{J}_k = \boldsymbol{k}T \tag{5}$$

We also have for the equilibrium value of the energy fluctuations

$$\overline{[E_k - \bar{E}_k]^2} = (kT)^2 \tag{6}$$

According to (3.3.1) we replace ρ_{n_l} in equation (3.1.8) by $f_{n_l} \prod_{k \neq l} g_k(J_k)$ and integrate over all action variables except J_l.

Using (2.7.9) and (3.1.11), we find after a few manipulations:

$$\begin{aligned} \partial f_{n_l}/\partial t = \lambda^2 \omega_l^2 kTC_l[kT(\partial/\partial E_l)E_l(\partial/\partial E_l)f_{n_l} + (\partial/\partial E_l)E_l f_{n_l} \\ -(n_l^2/4E_l)kTf_{n_l}] - in_l\lambda^2 \Delta\omega_l \end{aligned} \tag{7}$$

with

$$\begin{aligned} \Delta\omega_l = -kT\sum_{k'k''}\{\mathscr{P}[1/(\omega_{k'}+\omega_{k''}+\omega_l)\cdot[(\omega_{k'}+\omega_{k''})/(\omega_l\omega_{k'}^2\omega_{k''}^2] \\ \times[|V_{lk'k''}|^2+|V_{k'lk''}|^2+|V_{k'k''l}|^2] \\ +\mathscr{P}[1/(\omega_l+\omega_{k'}-\omega_{k''})]\cdot[(\omega_{k'}-\omega_{k''})/\omega_l\omega_{k'}^2\omega_{k''}^2] \\ \times[|V_{lk'-k''}|^2+|V_{k'l-k''}|^2] \\ -\mathscr{P}[1/(\omega_{k'}+\omega_{k''}-\omega_l)]\cdot[(\omega_{k'}+\omega_{k''})/\omega_l\omega_{k'}^2\omega_{k''}^2]|V_{k'k''-l}|^2\} \end{aligned} \tag{8}$$

and

$$\begin{aligned} C_l = 6\pi\sum_{k'k''}(1/\omega_l^2\omega_{k'}^2\omega_{k''}^2)\{|V_{lk'-k''}|^2\delta(\omega_l+\omega_{k'}-\omega_{k''}) \\ +|V_{k'l-k''}|^2\delta(\omega_{k'}+\omega_l-\omega_{k''})+|V_{k'k''-l}|^2\delta(\omega_{k'}+\omega_{k''}-\omega_l)\} \end{aligned} \tag{9}$$

We use the notation $\Delta\omega_l$ for the lengthy expression (3.3.8), because, as we shall now see, it corresponds to a shift in the frequencies due to the interaction. Indeed, let us deduce from (3.3.7) the equations for the distribution function $f(J, \alpha, t)$. We have [see (3.3.1)] (henceforth we shall drop the index l)

$$\begin{aligned} \partial f/\partial t &= -[1/\sqrt{(2\pi)}]\sum_n in\omega f_n\, e^{in(\alpha-\omega t)} + 1/\sqrt{(2\pi)}\sum_n \partial f_n/\partial t\, e^{in(\alpha-\omega t)} \\ &= -\omega(\partial f/\partial\alpha) + 1/\sqrt{(2\pi)}\sum_n \partial f_n/\partial t\, e^{in(\alpha-\omega t)} \end{aligned} \tag{10}$$

Therefore, using (3.3.7), we obtain

$$\begin{aligned})\partial f/\partial t) + (\omega+\lambda^2\Delta\omega)(\partial f/\partial\alpha) \\ = \lambda^2\omega^2 kTC[kT(\partial/\partial E)E(\partial/\partial E)f + (\partial/\partial E)Ef + (kT/4E)(\partial^2 f/\partial\alpha^2)] \end{aligned} \tag{11}$$

We see indeed that $\Delta\omega$ corresponds to a small shift in the frequencies due to the principal part in the δ_+ function [see (2.5.10)]. As it will play no role in our subsequent considerations, we shall drop it from now on. (One may also use, if desired, a "renormalized frequency," which includes $\Delta\omega$).

For angle independent distributions this equation reduces to the form

$$\partial f/\partial t = \eta(\partial/\partial E)(Ef + \boldsymbol{k}TE\, \partial f/\partial E) \tag{11'}$$

where η may be called the coefficient of friction. Such an equation has been derived by Kramers (1940, equation 14) on the basis of a phenomenological theory of Brownian motion in the limit of a small coefficient of friction. We shall come back to the comparison between the phenomenological theory of Brownian motion and our mechanical approach in § 6 as well as in Chapter 5, § 4. Let us first study, in more detail, the approach to equilibrium as described by the equation

$$\begin{aligned}(\partial f/\partial t)+\omega(\partial f/\partial \alpha) = \lambda^2\omega^2 C\boldsymbol{k}T[\boldsymbol{k}T(\partial^2/\partial E^2)\, Ef \\ +(\partial/\partial E)(E-\boldsymbol{k}T)f+(\boldsymbol{k}T/4E)(\partial^2 f/\partial\alpha^2)]\end{aligned} \tag{12}$$

or by the corresponding equation for the Fourier components [e.g., (3.3.7), neglecting the term in $\Delta\omega$]. We introduce into (3.3.7) the reduced variables

$$x = E/\boldsymbol{k}T \qquad \tau = (\lambda^2\omega^2 C\boldsymbol{k}T)t = t/t_r \tag{13}$$

We then obtain

$$\partial f_n/\partial\tau = (\partial^2/\partial x^2)xf_n + (\partial/\partial x)(x-1)f_n - (n^2/4x)f_n \tag{14}$$

This equation may be transformed into a standard equation for Laguerre polynomials. For convenience we have summarized, in Appendix I, the properties of Laguerre polynomials that we shall use here. Let us introduce χ_n instead of f_n through the relation

$$f_n = e^{-x}x^{n/2}e^{n\tau/2}\chi_n \tag{15}$$

With this substitution, equation (3.3.14) becomes

$$\partial\chi_n/\partial\tau = \{x(\partial^2/\partial x^2)+(n+1-x)(\partial/\partial x)-n\}\chi_n \tag{16}$$

Let us first look for solutions of the form

$$e^{-m\tau}P(x) \tag{17}$$

We then obtain for $P(x)$ the equation

$$x(d^2P/dx^2)+(n+1-x)(dP/dx)-(n-m)P = 0 \tag{18}$$

This is precisely the equation for the associated Laguerre polynomial $L_{(m)}^{(n)}(x)$ (see Appendix I). Therefore for a given value of n, the functions $e^{-m\tau} L_m^{(n)}(x)$, where m is a positive integer, $m \geqq n$, are special solutions of (3.3.16) and the general solution is given by

$$\chi_n = \sum_{m=n}^{\infty} e^{-m\tau} \alpha_m^{(n)} L_m^{(n)}(x) \tag{19}$$

where $\alpha_m^{(n)}$ are constants whose values are determined by the initial conditions. As we see by (3.3.15) this solution satisfies the boundary condition

$$f_n \to 0 \qquad \text{for} \quad x \to \infty \tag{20}$$

It is indeed a polynomial in x multiplied by the exponential e^{-x}.

Using (3.3.15) again we have for the time change of the distribution function $f(J, \alpha, t)$

$$f=(1/\sqrt{(2\pi)}) \sum_{n=-\infty}^{\infty} f_n e^{in(\alpha-\omega t)} = (1/\sqrt{(2\pi)})f_0+(1/\sqrt{(2\pi)}) \sum_{n=1}^{\infty} (f_n e^{in(\alpha-\omega t)}+\text{c.c.})$$

$$=(1/\sqrt{(2\pi)}) \sum_{m=0}^{\infty} e^{-x} e^{-m\tau} \alpha_m^{(0)} L_m^{(0)}(x)$$

$$+(1/\sqrt{(2\pi)}) \sum_{n=1}^{\infty} \sum_{m=n}^{\infty} e^{-x} x^{n/2} e^{-[m-n/2]\tau} L_m^{(n)}(x) \, [\alpha_m^{(n)} e^{in(\alpha-\omega t)}+\text{c.c.}] \tag{21}$$

We immediately verify that for $t \to \infty$

$$f \to e^{-x} \tag{22}$$

which is indeed the equilibrium distribution.

We are especially interested in the so-called *fundamental solution* (or Green's function) $f(x, \alpha, t \,|x_0, \alpha_0, 0)$ which, for $t \to 0$, reduces to a δ function

$$\underset{t\to 0}{f(x, \alpha, t \,|x_0, \alpha_0, 0)} = \delta(x-x_0)\,\delta(\alpha-\alpha_0) \tag{23}$$

Once the fundamental solution is known, the distribution function corresponding to an arbitrary initial condition $\Phi(x, \alpha)$ for $t = 0$ is expressed by

$$f(x, \alpha, t) = \iint dx_0 \, d\alpha_0 \, f(x, \alpha, t \,|\, x_0, \alpha_0, 0) \, \Phi(x_0, \alpha_0) \tag{24}$$

Indeed, because of (3.3.23), we have

$$\begin{aligned} \underset{t\to 0}{f}(x, \alpha, t) &= \iint dx_0\, d\alpha_0\, \delta(\alpha-\alpha_0)\, \delta(x-x_0)\, \Phi(x_0, \alpha_0) \\ &= \Phi(x, \alpha) \end{aligned} \tag{25}$$

We shall now fix the constants $\alpha_m^{(n)}$ to satisfy the initial condition (3.3.23).

Using the expression

$$\delta(\alpha-\alpha_0) = 1/2\pi \sum_{n=-\infty}^{+\infty} e^{in(\alpha-\alpha_0)}$$

for the δ function, condition (3.3.23) becomes

$$[1/\sqrt{(2\pi)}] \sum_{n=-\infty}^{+\infty} f_n e^{in\alpha} = \delta(x-x_0)(1/2\pi) \sum_{n=-\infty}^{+\infty} e^{in(\alpha-\alpha_0)} \tag{26}$$

This implies the initial condition

$$f_n = \delta(x-x_0)[1/\sqrt{2\pi})]\, e^{-in\alpha_0} \tag{27}$$

which gives, using (3.3.21),

$$e^{-x} x^{n/2} \sum_{m=n}^{\infty} \alpha_m^{(n)} L_m^{(n)}(x) = \delta(x-x_0)[1/\sqrt{(2\pi)}]\, e^{-in\alpha_0} \tag{28}$$

Now (see Appendix I) the functions

$$[(m-n)!/(m!)^3]^{1/2}\, e^{-x/2} x^{n/2} L_m^{(n)}(x) \tag{29}$$

form a normalized orthogonal set. We therefore obtain from (3.2.28), in the usual way, multiplying both sides by $x^{n/2} L_{(m)}^{(n)}(x)$ and integrating over x

$$\alpha_m^{(n)} = (m-n)!/(m!)^3\, x_0^{n/2} L_m^{(n)}(x_0)[1/\sqrt{(2\pi)}]\, e^{-in\alpha_0} \tag{30}$$

The fundamental solution is, therefore (see 3.3.21)

$$\begin{aligned} f(x, \alpha, t \mid x_0, \alpha_0, 0) &= (e^{-x}/2\pi) \sum_{m=0}^{\infty} e^{-m\tau}\, 1/(m!)^2 L_m^{(0)}(x_0) L_m^{(0)}(x) \\ &+ e^{-x}/\pi \sum_{n=1}^{\infty} (x x_0 e^{\tau})^{n/2} \cos n(\alpha-\alpha_0-\omega t) \sum_{m\geq n}^{\infty} e^{-m\tau}(m-n)!/(m!)^3 L_m^{(n)}(x_0) L_m^{(n)}(x) \end{aligned} \tag{31}$$

This formula represents a complete solution to the problem of

Brownian motion. It is possible to write the solution in a much more compact form, using a formula due to Myller-Lebedoff (see Appendix I) which states

$$e^{-(x+x_0)/2}(xx_0)^{n/2}a^{-n/2}\sum_{m\geqq n}^{\infty}(m-n)!/(m!)^3 L_m^{(n)}(x)L_m^{(n)}(x_0)a^m$$
$$= [1/(1-a)]\exp\{-[(x+x_0)/2][(1+a)/(1-a)]\}I_n[2\sqrt{(xx_0a)}/(1-a)] \tag{32}$$

where I_n is the Bessel function of order n and of purely imaginary argument. We may in this way perform the summation over m.

Using (3.3.31) and (3.3.32) we see that the time evolution of the Fourier components of the distribution function $f(x, \alpha, t)\,|x_0, \alpha_0, 0)$ is given (apart from the phase factor $e^{-in\alpha_0}$) by the closed formula

$$f_n(x, t\,|\,x_0, 0)$$
$$= [1/\sqrt{(2\pi)}(1-e^{-\tau})]\exp[-(x+x_0e^{-\tau})/(1-e^{-\tau})]I_n[2\sqrt{(xx_0e^{-\tau})}/1-e^{-\tau}] \tag{33}$$

The final step is to perform the summation over n to obtain f. Here we have to use the identity (see Appendix I)

$$\tfrac{1}{2}e^{z\cos u} = \sum_{n=1}^{\infty} I_n(z)\cos nu + \tfrac{1}{2}I_0(z) \tag{34}$$

We obtain in this way the final formula for the fundamental solution

$$f(x, \alpha, t\,|\,x_0, \alpha_0, 0) = [1/2\pi(1-e^{-\tau})]$$
$$\times \exp -[(x+e^{-\tau}x_0-2\sqrt{(xx_0e^{-\tau})}\cos(\alpha-\alpha_0-\omega t))/(1-e^{-\tau})] \tag{35}$$

The relations (3.3.33) or (3.3.35) show very clearly how statistical equilibrium is reached.

In particular one sees that after times of order $\tau = 1$ or [see (3.3.13)]

$$t_r \sim 1/\lambda^2\omega^2 CkT \tag{36}$$

the initial conditions x_0, α_0 are "forgotten." This is precisely the relaxation time of the normal mode that we have considered. Let us now study the evolution of the Fourier components f_n in more detail.

4. Evolution of the Fourier Components

The Bessel functions $I_n(z)$ of imaginary argument and integral order are always positive (see Appendix I). They increase with z and satisfy the relations

$$I_n(0) = 0 \qquad n \neq 0 \tag{1}$$

$$I_0(0) = 1 \tag{2}$$

$$I_n(z) < I_m(z) \qquad n > m \tag{3}$$

The relation (3.4.1) shows that f_n, as given by (3.3.33), will, for $n \neq 0$, vanish for sufficiently long times. This is in agreement with the results derived in §2 of this chapter. On the other hand, f_0 will approach its equilibrium value e^{-x}.

The inequality (3.4.3) shows that the rapidity of the decay of the Fourier components increases with their order. For τ sufficiently large we may use the formula (see Appendix I)

$$I_n(z) = (\tfrac{1}{2}z)^n/n! \tag{4}$$

When applied to (3.3.33) this gives

$$f_n \sim e^{-n\tau/2} \tag{5}$$

Higher Fourier components will thus decay more rapidly than lower ones. Such behavior is familiar from the diffusion equation or the Fourier heat equation.

With the help of (3.3.33) we may calculate the evolution of any transition moment:

$$\overline{x^r}^{x_0} = \int_0^\infty dx\, x^r f_0(x, t \,|\, x_0, 0) \tag{6}$$

The corresponding integrals are given in Appendix I, § 11. The index x_0 in $\overline{x^r}^{x_0}$ indicates that for $t = 0$, $x = x_0$. The most important moments are

$$\overline{x}^{x_0} = 1 - e^{-\tau} + x_0 e^{-\tau} \tag{7}$$

$$\overline{\left(x - \overline{x}^{x_0}\right)^2}^{x_0} = (1-e^{-\tau})(1+2x_0 e^{-\tau} - e^{-\tau}) \tag{8}$$

These relations show how the equipartition of energy and the equilibrium value for energy fluctuations are reached [see also (3.3.5) and (3.3.6)].

Let us finally consider the relaxation time which will be, as we have seen in (3.3.36), of order $(\lambda^2\omega^2CkT)^{-1}$. It is indeed natural that the relaxation time decreases with increasing temperature. This will increase the average energy of each normal mode and therefore the energy transmitted in each interaction.

It is generally assumed that as a first approximation, the coefficient C in (3.3.36) may be considered to be independent of the wavelength (see for example Peierls, 1955). To this approximation $t_r \sim \omega^{-2}$. This means that a high-frequency normal mode will interact more strongly with the other degrees of freedom and adjust itself more rapidly than a low-frequency acoustical mode.

The parameter C also determines the contribution of anharmonic forces to the specific heat (see Peierls, 1955, formula 2.5.1). Therefore, we also have

$$1/t_r \sim \omega^2 kT(C_v - C_v^h) \tag{9}$$

where $C_v - C_v^h$ is the difference between the actual specific heat and its value in the absence of anharmonic forces. (For more details see Prigogine and Philippot, 1957, formula A.6). Such a simple formula can, however, be expected only in the range of *classical* statistical mechanics.

5. Brownian Motion in Displacement and Velocity

Let us now turn to the comparison between our theory and the usual phenomenological treatment of Brownian motion (Mazur, 1959). For this purpose it is useful to transform our equation from action-angle variables back to normal coordinates. Let us introduce the change of variables [see (1.6.3)]

$$\begin{aligned} \omega q &= \sqrt{(2E/m)} \sin \alpha \\ v &= \sqrt{(2E/m)} \cos \alpha \end{aligned} \tag{1}$$

and study the Brownian motion in the variables q, v. Equation (3.3.12) is easily transformed into these variables

$$\begin{aligned}\partial f/\partial t &= -v(\partial f/\partial q)+\omega^2 q(\partial f/\partial v)+(\beta/2)\{[\partial/\partial(\omega q)](\omega q f)+(\partial/\partial v)(fv)\\ &\quad +(\beta \boldsymbol{k}T/2m)[(\partial^2 f/\partial(\omega q)^2)+(\partial^2 f/\partial v^2)]\}\\ &= (\partial/\partial \omega q)[f(\beta\omega q/2-\omega v)]+(\partial/\partial v)[f(\beta v/2+\omega^2 q)]\\ &\quad +(\beta \boldsymbol{k}T/2m)[(\partial^2 f/\partial(\omega q)^2)+(\partial^2 f/\partial v^2)]\end{aligned} \tag{2}$$

We have used (ωq) as the independent variable to emphasize the symmetry between ωq and v. Also

$$\beta = \lambda^2 C\omega^2 \boldsymbol{k}T \tag{3}$$

We may easily write down the fundamental solution of (3.5.2). Using the fact that

$$dq dv = (1/m)\, dE d\alpha = (\boldsymbol{k}T/m)\, dx d\alpha \tag{4}$$

as well as (3.3.24) we obtain

$$f(q, v, t|q_0, v_0, 0) = (m/\boldsymbol{k}T)f(x, \alpha, t|x_0, \alpha_0, 0) \tag{5}$$

We now substitute the expressions for x, x_0, α, α_0 in terms of q, q_0, v, v_0 into (3.3.35) and obtain, after an elementary calculation,

$$\begin{aligned}f(q, 0, t|q_0, v_0, 0) &= m/[2\pi \boldsymbol{k}T(1-e^{-\tau})]\\ &\quad \times \exp\{-(m/2\boldsymbol{k}T)[(v-\bar{v})^2+\omega^2(q-\bar{q})^2]/[1-e^{-\tau}]\}\end{aligned} \tag{6}$$

with

$$\bar{v} = e^{-\tau/2}(v_0 \cos \omega t - \omega q_0 \sin \omega t) \tag{7}$$

$$\omega\bar{q} = e^{-\tau/2}(v_0 \sin \omega t + \omega q_0 \cos \omega t) \tag{8}$$

An alternative derivation of (3.5.6) is due to Mazur (1959) and Higgins (1960). Instead of solving (3.3.12) and then expressing the solution in terms of the variables q, v, they have solved (3.5.2) directly. This can be done in a straightforward way but we refer to the original publications for more details.

The interesting point is that the process appears to be Gaussian in the variables q, v. Moreover, comparison between (3.5.2) and (2.9.5) shows that here too we have a Markoff process whose Fokker-Planck equation is precisely (3.5.2) with the transition moments

$$\overline{\Delta\omega q}/\Delta\tau = \omega v - (\beta\omega q/2), \qquad \overline{\Delta v}/\Delta\tau = -\omega^2 q - \beta v/2 \tag{9}$$

$$\overline{\Delta(\omega q)^2}/\Delta\tau = \boldsymbol{k}T\beta/m, \qquad \overline{\Delta v^2}/\Delta\tau = \boldsymbol{k}T\beta/m \tag{10}$$

In the action-angle variables we also have a Markowian process [with the Fokker-Planck equation (3.3.12)] which is not Gaussian.[1]

Gaussian Markowian processes have attracted a great deal of attention from the general point of view of random processes. Indeed many important problems, such as that of noise in electrical networks, belong to this category. In the last section therefore we shall briefly compare our treatment to the usual method (see for example Wang and Uhlenbeck, 1945).

6. Comparison with the Phenomenological Theory of Brownian Motion

The basic equations of the usual phenomenological theory for the Brownian motion of a harmonic oscillator are

$$dq/dt = v \tag{1}$$

$$dv/dt = -\beta v - \omega^2 q + A(t) \tag{2}$$

The solution of these two equations is found by attributing well-defined statistical properties to the random force $A(t)$. It is assumed (see Wang and Uhlenbeck, 1945) that

$$\overline{A(t)} = 0 \tag{3}$$

$$\overline{A(t_1)A(t_2)} \sim \delta(t_1 - t_2) \tag{4}$$

and that the distribution of the values of $A(t)$ follows a Gaussian form.

It is precisely condition (3.6.4) which cannot be satisfied in action-angle variables (Mazur, private communication) because then the "force" becomes the work $\boldsymbol{v} \cdot \boldsymbol{A}$ and is therefore proportional to the velocity of the particle. As the velocity will change only gradually the values of the force at two nearby times will no longer be uncorrelated. This is the reason why the random process will not be Gaussian in action-angle variables.

We shall not give any details here because the development

[1] Because the basic distribution (3.3.35) does not have a Gaussian form in these variables.

of the usual theory of random processes is outside the scope of this book. (There exist many excellent presentations; see for example Wang and Uhlenbeck, 1945; Mazur, 1954).

Starting with equations (3.6.1) through (3.6.4) it may be shown that the fundamental solution is given by the Gaussian distribution

$$f(q, v, t \mid q_0, v_0, t = 0) = (g_{vv}g_{qq} - g_{vq}^2)^{1/2} \times 1/2\pi$$
$$\times \exp\{-\tfrac{1}{2}g_{vv}(v-\bar{v})^2 + g_{vq}(v-\bar{v})(q-\bar{q}) - \tfrac{1}{2}g_{qq}(q-\bar{q})^2\} \quad (5)$$

with mean values and variances

$$\bar{v}^{v_0 q_0} = v_0 e^{-\frac{1}{2}\beta t}\cos(\omega_1 t+\delta)\cos^{-1}\delta - \omega q_0 e^{-\frac{1}{2}\beta t}\sin\omega_1 t\cos^{-1}\delta \quad (6)$$

$$\overline{\omega q}^{v_0 q_0} = v_0 e^{-\frac{1}{2}\beta t}\sin\omega_1 t\cos^{-1}\delta + \omega q_0 e^{-\frac{1}{2}\beta t}\cos(\omega_1 t-\delta)\cos^{-1}\delta \quad (7)$$

$$\overline{(v-\bar{v})^2}^{v_0 q_0} = g_{qq}/(g_{vv}g_{qq} - g_{qv}^2)$$
$$= (kT/m)\{1 - e^{-\beta t}\cos^{-2}\delta\,[\cos^2(\omega_1 t+\delta) + \sin^2\omega_1 t]\} \quad (8)$$

$$\overline{\omega^2(q-\bar{q})^2}^{v_0 q_0} = \omega^2 g_{vv}/(g_{vv}g_{qq} - g_{vq}^2)$$
$$= (kT/m)\{1 - e^{-\beta t}\cos^{-2}\delta\,[\cos^2(\omega_1 t-\delta) + \sin^2\omega_1 t]\} \quad (9)$$

$$\overline{\omega(v-\bar{v})(q-\bar{q})}^{v_0 q_0} = -\omega g_{vq}/(g_{vv}g_{qq} - g_{vq}^2)$$
$$= (2kT/m)e^{-\beta t}\sin^2\omega_1 t\sin\delta\cos^{-2}\delta \quad (10)$$

Here

$$\omega_1^2 = \omega^2 - \beta^2/4 \quad (11)$$

and

$$\delta = \arctan(\beta/2\omega_1) \quad (12)$$

The formulae are written for the "underdamped" case $(\omega^2 > (\beta^2/4))$. The fundamental solution (3.6.5) satisfies the diffusion equation first derived by Kramers (Kramers, 1940; see also Chandrasekhar, 1945, and Wax, 1954)

$$\partial f/\partial t = -v(\partial f/\partial q) + \omega^2 q(\partial f/\partial v)$$
$$+\beta[(\partial/\partial v)(fv) + (kT/m)(\partial^2 f/\partial v^2)] \quad (13)$$

This equation has to be compared to our equation (3.5.2). The basic difference lies in the fact that our equation is symmetric in the displacement and velocity, whereas (3.6.13) is not. The displacement q does not enter at all in the last term of the right-hand side of (3.6.13); it appears only in the "streaming" term.

We have, however, not yet introduced into the phenomenological theory an assumption corresponding to our "*weak coupling.*" This will mean here [see (3.6.11) and (3.6.12)] that $\beta \ll \omega$ or $\delta \ll 1$, corresponding to a strongly underdamped oscillator. For the comparison with our theory, we shall therefore neglect terms of order δ, $\delta^2, \ldots$ in the mean values and the variances. We then obtain, instead of (3.6.5) through (3.6.10),

$$\overline{v}^{v_0 q_0} = v_0 e^{-\frac{1}{2}\beta t} \cos \omega_1 t - \omega q_0 e^{-\frac{1}{2}\beta t} \sin \omega_1 t \tag{14}$$

$$\omega \overline{q}^{v_0 q_0} = v_0 e^{-\frac{1}{2}\beta t} \sin \omega_1 t + \omega q_0 e^{-\frac{1}{2}\beta t} \cos \omega_1 t \tag{15}$$

$$\overline{(v-\bar{v})^2}^{v_0 q_0} = (kT/m)(1-e^{-\beta t}) \tag{16}$$

$$\overline{\omega^2 (q-\bar{q})^2}^{v_0 q_0} = kT/m(1-e^{-\beta t}) \tag{17}$$

$$\overline{(v-\bar{v})(q-\bar{q})}^{v_0 q_0} = 0 \tag{18}$$

If we now substitute these results into (3.6.5), the fundamental solution becomes identical with (3.5.6). *In this sense we may say that for the weakly coupled case the two theories coincide.* This is in agreement with the remark we made in § 3. We have seen that our equation (3.3.11′) coincides with the equation derived by Kramers for the evolution of the energy distribution in the limit of a small friction coefficient.

However, the agreement between the two approaches is only realized in a limiting sense for sufficiently small values of λ. What do we have to expect in the general case?

For larger values of λ the simplicity of the Brownian motion problem is lost. Indeed the "Brownian particle" then perturbs the equilibrium state of the other degrees of freedom. We then have to use the methods developed in the later chapters of this book.

We have considered here only the case of a harmonic oscillator. However all our calculations may be extended to other situations corresponding to an arbitrary field of force using the appropriate angle-action variables. We shall not go into details about these calculations (Prigogine, 1959, see also Thor Bak and K. Anderson, 1960) but one may show that here the results are not the same *even in the limit of weak coupling* as obtained by Kramers' theory.

The origin of this difference is briefly the following: We derive the Fokker-Planck equation by an application of a perturbation technique to the unperturbed motion. The nature of the unperturbed motion and therefore the nature of the force field in which the particle moves is therefore crucial.

On the contrary in the phenomenological theory of Brownian motion it is assumed (see Kramers, 1940; Wang and Uhlenbeck, 1945) that the equations of motion are

$$\begin{aligned} dq/dt &= v \\ dv/dt &= -\beta v + F(q) + A(t) \end{aligned} \tag{19}$$

These equations generalize equations (3.6.1) and (3.6.2) to an arbitrary force field. The main point is that it is assumed that the statistical properties of the random force $A(t)$ [see (3.6.3), (3.6.4)] are independent of F.

Equation (5.5.1) therefore expresses a kind of linear superposition of the effect of the forces F and A. This superposition is not valid in general.

CHAPTER 4

Weakly Coupled Gases

1. Liouville Operator

We now turn to the case of gases. The Hamiltonian will be assumed to be of the usual form

$$H = \sum_j (p_j^2/2m) + \lambda \sum_{j<n} V_{jn}(|\mathbf{x}_j - \mathbf{x}_n|) \tag{1}$$

where λV_{jn} is the interaction potential between particles j and n. We consider central forces depending only on the distance between j and n.

The interaction potential V_{jn} will be expanded in the Fourier series

$$V_{jn} = (2\pi/L)^3 \sum_l V_l \exp i\mathbf{l}\cdot(\mathbf{x}_j - \mathbf{x}_n) \tag{2}$$

Because of our assumption of central forces, V_l depends only on the absolute value of $\mathbf{l}$.

The perturbed Liouville operator is here [see (1.8.4)]

$$\begin{aligned} \delta L &= i \sum_{jn} \partial V_{jn}/\partial \mathbf{x}_j \cdot [(\partial/\partial \mathbf{p}_j) - (\partial/\partial \mathbf{p}_n)] \\ &= i(2\pi/L)^3 \sum_{j,n} \sum_l e^{i\mathbf{l}\cdot(\mathbf{x}_j - \mathbf{x}_n)} V_l \, i\mathbf{l} \cdot [(\partial/\partial \mathbf{p}_j) - (\partial/\partial \mathbf{p}_n)] \end{aligned} \tag{3}$$

In the Hamiltonian (4.1.1) the potential energy depends on the coordinates alone and not on the velocities. From this point of view the situation is therefore simpler than for anharmonic solids (see Chapter 2, § 1).

The matrix element $\langle\{k\}|\delta L|\{k'\}\rangle$ [see (1.8.8)], using the eigenfunction (1.5.18), is here

$$\begin{aligned} \langle\{k\}|\delta L|\{k'\}\rangle &= i(1/L^3)^N \int d\mathbf{x}_1 \cdots d\mathbf{x}_N \, e^{-i\sum \mathbf{k}_r \cdot \mathbf{x}_r} \, \delta L \, e^{i\sum \mathbf{k}'_r \cdot \mathbf{x}_r} \\ &= i(2\pi/L)^3 (1/L^3)^N \sum_{jn} \sum_l \int d\mathbf{x}_1 \cdots d\mathbf{x}_N \, e^{-i\sum \mathbf{k}_r \cdot \mathbf{x}_r} e^{i\mathbf{l}\cdot(\mathbf{x}_j - \mathbf{x}_n)} \\ &\qquad \times V_l \, i\mathbf{l} \cdot [(\partial/\partial \mathbf{p}_j) - (\partial/\partial \mathbf{p}_n)] \, e^{i\sum \mathbf{k}'_r \cdot \mathbf{x}_r} \end{aligned} \tag{4}$$

We now make use of the identity

$$\int dx\, e^{i\alpha \cdot x} = L^3 \delta_\alpha^{\mathrm{Kr}} \tag{5}$$

where $\delta_\alpha^{\mathrm{Kr}}$ is the Kronecker δ-function[1] ($\delta_\alpha^{\mathrm{Kr}} = 0$ if $\alpha \neq 0$, $\delta_\alpha^{\mathrm{Kr}} = 1$ if $\alpha = 0$). We then perform the integrations over $x_1 \ldots x_N$ in (4.14):

$$\langle\{k\}|\delta L|\{k'\}\rangle = i(2\pi/L)^3 \sum_{jn}\sum_{l} V_l\, il[(\partial/\partial p_j)-(\partial/\partial p_n)]$$

$$\times \delta^{\mathrm{Kr}}(k'_j - k_j + l)\, \delta^{\mathrm{Kr}}(k'_n - k_n - l) \prod_{r\neq j,n} \delta^{\mathrm{Kr}}(k'_r - k_r) \tag{6}$$

The only non-vanishing matrix elements are therefore of the form

$$\langle\{k\}|\delta L|k_j - l, k_n + l, \{k'\}\rangle = i(2\pi/L)^3 V_l\, il\cdot[(\partial/\partial p_j)-(\partial/\partial p_n)] \tag{7}$$

All indices k have kept their values except the two indices k_j, k_n. This is a consequence of the binary interaction which we use in (4.1.1). Moreover we have the *law of conservation of wave vectors*

$$k'_n + k'_j = k_n + k_j \tag{8}$$

which is due to the fact that the potential energy V_{nj} depends only on the distance between n and j and is therefore *invariant with respect to translation.*[2]

The conservation law (4.1.8) plays an essential role in our theory.

The Liouville equation in interaction representation (1.8.7) becomes, using (4.1.7), [see also (1.5.20)]

$$\partial\rho_{\{k\}}/\partial t = \lambda(2\pi/L)^3 \sum_{jn}\sum_{l} e^{i(k_j\cdot v_j + k_n\cdot v_n)t}\, V_l$$

$$\times il\cdot[(\partial/\partial p_j)-(\partial/\partial p_n)]\, e^{-i[(k_j-l)\cdot v_j + (k_n+l)\cdot v_n]t}\, \rho_{k_j-l,\, k_n+l,\, \{k'\}} \tag{9}$$

The invariance with respect to translation gives rise to momentum conservation. Indeed, let us introduce as variables the momentum of the center of gravity of the couple n, j, and their relative

[1] We also use the notation $\delta^{\mathrm{Kr}}(\alpha)$.

[2] We have $V(|x_j - x_n|) = V(|(x_j + a) - (x_n + a)|)$.

momentum

$$\boldsymbol{P} = \tfrac{1}{2}(\boldsymbol{p}_j+\boldsymbol{p}_n) \tag{10}$$

$$\boldsymbol{p} = \boldsymbol{p}_j-\boldsymbol{p}_n \tag{11}$$

We have

$$(\partial/\partial \boldsymbol{p}_j)-(\partial/\partial \boldsymbol{p}_n) = 2(\partial/\partial \boldsymbol{p}) \tag{12}$$

Therefore, the operator in (4.1.9) acts only on the relative momentum and $\boldsymbol{P}$ disappears by cancellation of the two exponentials in (4.1.9). The change in $\rho_{\{k\}}$ due to the interaction between j and n is expressed in terms of the other Fourier components taken with the *same* momentum $\boldsymbol{P}$. We could also say that the equations (4.1.9) are "diagonal" in $\boldsymbol{P}$. Both the conservation of momentum and the conservation of wave vectors are consequences of the invariance of the potential energy with respect to translation. We shall see in Chapter 13, § 3 that the same situation is true for quantum mechanics.

2. Master Equation for Weakly Coupled Gases — Diagrams[1]

We may now proceed, exactly as in Chapter 2, to derive the master equation for the velocity distribution $\rho_{\{0\}}(\boldsymbol{v}_1, \ldots, \boldsymbol{v}_N, t)$ of a weakly coupled gas from (4.1.9). We shall once more use the initial condition

$$\rho_{\{k\}}\,(t = 0) = 0 \qquad \{k\} \neq 0 \tag{1}$$

and retain only terms of the form $(\lambda^2 t)^n$.

In this way we again obtain the general master equation (2.7.5) for weakly coupled systems. We need only replace the Liouville matrix elements in expression (2.7.3) by their values as derived in § 1 of this chapter. Here we get (Brout and Prigogine, 1956)

[1] In most expressions used in this book, we shall make no distinction between a velocity $\boldsymbol{v}$ and a momentum $\boldsymbol{p}$. In other words, we shall generally take

$$m = 1.$$

However, in formulae in which dimensions are important, this convention will not be used.

$$\partial\rho_{\{0\}}/\partial t = \lambda^2\pi/(2\pi/L)^6 \sum_{jn}\sum_{l} V_l i\boldsymbol{l}\cdot[(\partial/\partial \boldsymbol{v}_j)-(\partial/\partial \boldsymbol{v}_n)]\,\delta[\boldsymbol{l}\cdot(\boldsymbol{v}_j-\boldsymbol{v}_n)]$$
$$\times V_{-l}i(-\boldsymbol{l})\cdot[(\partial/\partial \boldsymbol{v}_j)-(\partial/\partial \boldsymbol{v}_n)]\rho_{\{0\}} \tag{2}$$
$$= (\lambda^2 8\pi^4/L^3)\sum_{jn}\int d\boldsymbol{l}\,|V_l|^2\boldsymbol{l}\cdot[(\partial/\partial \boldsymbol{v}_j)-(\partial/\partial \boldsymbol{v}_n)]\,\delta[\boldsymbol{l}\cdot(\boldsymbol{v}_j-\boldsymbol{v}_n)]$$
$$\times\ \boldsymbol{l}\cdot[(\partial/\partial \boldsymbol{v}_j)-(\partial/\partial \boldsymbol{v}_n)]\,\rho_{\{0\}} \tag{3}$$

In the transition from (4.2.2) to (4.2.3) we have replaced the summation over $\boldsymbol{l}$ by an integration valid in the limit $L \to \infty$. Using periodic boundary conditions we see [cf. (1.5.9)] that the difference between two permitted values of l_x, l_y or l_z is $2\pi/L$. Therefore, the density of states is $(L/2\pi)^3$ and we have, for a large system,[1]

$$(8\pi^3/L^3)\sum_{l} \to \int d\boldsymbol{l} \tag{4}$$

Again, as in the case of anharmonic solids, it is very useful to associate with each state and with each transition a well-defined diagram. With the transitions corresponding to the matrix element (4.1.7) we shall associate the diagram of Fig. 4.2.1.

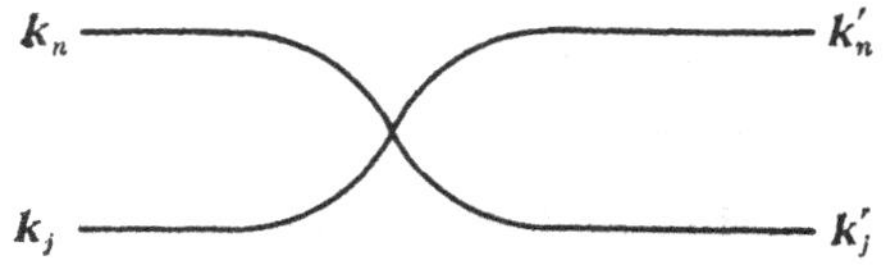

Fig. 4.2.1 Diagram associated with (4.1.7).

The diagram of Fig. 4.2.2 is therefore the basic diagram which corresponds to the master equation (4.2.3).

This is, again, as in the case of solids, a "*diagonal*" diagram

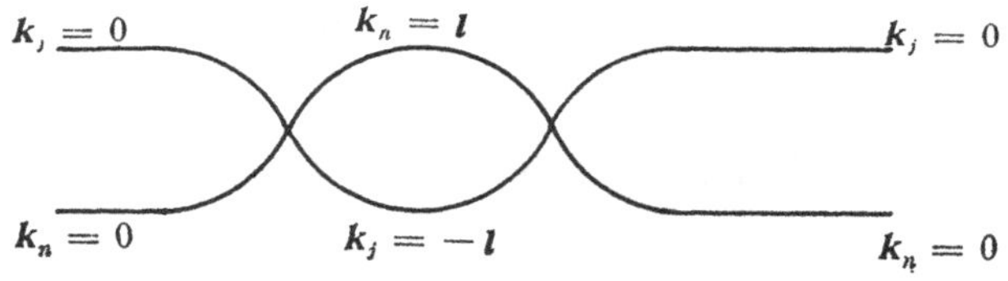

Fig. 4.2.2 Diagram associated with the master equation (4.2.3).

[1] This question is dealt with in all books on solid state physics, or quantum mechanics; see for example Kittel, 1953; Peierls, 1955; Bohm, 1951.

(see Fig. 2.4.7) in which the initial value of the wave vectors $\boldsymbol{k}_j = \boldsymbol{k}_n = 0$ is restored after two transitions. If the distribution function $\rho(\boldsymbol{x}_1 \cdots \boldsymbol{x}_N, \boldsymbol{v}_1 \cdots \boldsymbol{v}_N, t)$ does not depend on the coordinates of, for example, particle j, then the only non-vanishing Fourier coefficients of ρ will be those corresponding to $\boldsymbol{k}_j = 0$. Conversely, a Fourier component with $\boldsymbol{k}_j \neq 0$ corresponds to an inhomogeneous distribution function with respect to particle l. To emphasize this distinction between $\boldsymbol{k}_j \neq 0$ and $\boldsymbol{k}_j = 0$ we shall draw lines only for $\boldsymbol{k}_j \neq 0$. This is also the convention we used in Chapter 2, § 4 for harmonic oscillators. Fig. 4.2.2 then reduces to the diagram of Fig. 4.2.3. With this convention we

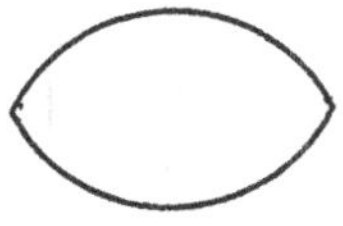

Fig. 4.2.3 Diagram associated with the master equation (4.2.3).

may say that a line represents an "*inhomogeneity*" or a *correlation* which propagates in the system. We shall elaborate the diagram technique and its physical interpretation in Chapter 7, but first we shall proceed with the study of a few especially simple situations (weakly coupled gases, scattering, dilute gases).

3. Boltzmann Equation and Molecular Chaos

The master equation (4.2.3) refers to the velocity distribution of the whole set of N molecules. On the other hand the really meaningful problems arise in connection with the distribution of velocities of only a single molecule, or at most of a few molecules.

Let us call φ_s the velocity distribution function for molecules $1, \ldots, s$ (see Chapter 7, § 1). We first integrate (4.2.3) over all velocities except $\boldsymbol{v}_r$, obtaining

$$\partial\varphi_1(\boldsymbol{v}_r)/\partial t = (8\pi^4\lambda^2/m^2L^3) \sum_j \int d\boldsymbol{v}_j \int d\boldsymbol{l} |V_l|^2 \boldsymbol{l}\cdot(\partial/\partial\boldsymbol{v}_r)\delta[\boldsymbol{l}\cdot(\boldsymbol{v}_r - \boldsymbol{v}_j)] \times \boldsymbol{l}\cdot[(\partial/\partial\boldsymbol{v}_r) - (\partial/\partial\boldsymbol{v}_j)]\varphi_2(\boldsymbol{v}_r, \boldsymbol{v}_j)$$

or

$$\partial\varphi_1(\boldsymbol{v})/\partial t = (8\pi^4\lambda^2/m^2)\, C \int d\boldsymbol{v}' \int d\boldsymbol{l} |V_l|^2 \boldsymbol{l}\cdot(\partial/\partial\boldsymbol{v})\delta(\boldsymbol{l}\cdot\boldsymbol{g}) \times \boldsymbol{l}\cdot[(\partial/\partial\boldsymbol{v}) - (\partial/\partial\boldsymbol{v}')]\varphi_2(\boldsymbol{v}, \boldsymbol{v}') \quad (1)$$

where C is the concentration

$$C = (N/L^3) \tag{2}$$

and $\boldsymbol{g}$ the relative velocity $\boldsymbol{v}-\boldsymbol{v}'$. The usual way to go from equation (4.3.1), which connects φ_1 to φ_2, to an equation for φ_1 alone, is through the introduction of the so-called *molecular chaos assumption*:

$$\varphi_2(\boldsymbol{v}, \boldsymbol{v}') = \varphi_1(\boldsymbol{v})\varphi_1(\boldsymbol{v}') \tag{3}$$

We then obtain the "Boltzmann" equation for φ_1,

$$\partial\varphi_1/\partial t = (8\pi^4\lambda^2/m^2)\, C \int d\boldsymbol{v}' \int d\boldsymbol{l}\, |V_l|^2\, \boldsymbol{l}\cdot\partial/\partial\boldsymbol{v}\, \delta(\boldsymbol{l}\cdot\boldsymbol{g}) \times \boldsymbol{l}\cdot[(\partial/\partial\boldsymbol{v})-(\partial/\partial\boldsymbol{v}')]\varphi_1(\boldsymbol{v})\varphi_1(\boldsymbol{v}') \tag{4}$$

Before we discuss this equation it is interesting to present a few comments about the molecular chaos assumption (4.3.3). This assumption introduces two problems of very different nature: (a) What is the justification for introducing this assumption as an initial condition for $t = 0$; and (b) If molecular chaos is assumed for $t = 0$, will it persist for $t > 0$.

In the discussion of the first problem it is essential to remember that we are dealing here with homogeneous systems. Therefore any deviation from (4.3.3) would mean a velocity correlation between molecules extending over the whole system. Indeed such a correlation would then exist whatever the distance between the particles. This is physically very unlikely because we may expect that any correlation between molecules goes to zero when their distance goes to infinity.

We meet here one of the two conditions which we shall always require for the distribution function at time $t = 0$: all correlations (space or velocity correlations or both) are of finite range.

We shall discuss in Chapter 11 the fundamental importance of this condition for the whole theory. The "molecular chaos" condition (4.3.3) is a special case of this general requirement.

The second condition which we shall impose on the distribution function at the initial time $t = 0$ will be formulated and discussed in Chapter 7.

We may show that if molecular chaos is assumed for $t = 0$

it will indeed persist for $t > 0$. In order to avoid duplication we shall postpone the proof to Chapter 8, §11. An essential step in this proof is the passage to the limit $L \to \infty$, $N \to \infty$ ($c = N/L$, being constant).

The fact that the persistence of molecular chaos can only be valid as an asymptotic property for large systems was first recognized by Kac (1954).

4. Explicit Form of the Boltzmann Equation

Let us write (4.3.4) in a form due to Auer and Tamor (1957), which brings out in an especially clear way its physical significance. (We use the notation φ instead of φ_1, we also take $m = 1$)

$$\partial\varphi/\partial t = \lambda^2 N \int d\boldsymbol{v}'(\partial/\partial \boldsymbol{v})\cdot\langle(\partial V/\partial \boldsymbol{x})\cdot(\partial\tilde{V}/\partial \boldsymbol{x})\rangle\cdot[(\partial/\partial \boldsymbol{v})-(\partial/\partial \boldsymbol{v}')]\varphi\varphi' \tag{1}$$

where $V(|(\boldsymbol{x}-\boldsymbol{x}')|)$ is the potential energy between two particles and $\tilde{V}$ its time average over an unperturbed path

$$\tilde{V} = \int_0^\infty dt\, V(|\boldsymbol{x}-\boldsymbol{x}'-\boldsymbol{g}t|) \tag{2}$$

Here again $\boldsymbol{g}$ is the relative velocity. The bracket $\langle\ \rangle$ means an average over space

$$\langle\partial V/\partial \boldsymbol{x}\cdot\partial\tilde{V}/\partial \boldsymbol{x}\rangle = 1/(L^3)^2\int d\boldsymbol{x}\, d\boldsymbol{x}'(\partial V/\partial \boldsymbol{x})\cdot(\partial\tilde{V}/\partial \boldsymbol{x}) \tag{3}$$

The expression (4.4.3) is a tensor. In explicit form (4.4.1) would read

$$\partial\varphi/\partial t = \lambda^2 N \sum_{s,r=1,2,3}\int d\boldsymbol{v}'(\partial/\partial \boldsymbol{v}_s)\langle(\partial V/\partial \boldsymbol{x}_s)\cdot(\partial\tilde{V}/\partial \boldsymbol{x}_r)\rangle$$
$$[(\partial/\partial \boldsymbol{v}_r)-(\partial/\partial \boldsymbol{v}_r')]\varphi\varphi' \tag{4}$$

The equivalence (4.4.1) and (4.3.4) is easy to verify. Indeed, using (4.1.2) and (4.4.3) we have

$$\langle(\partial V/\partial \boldsymbol{x})\cdot(\partial\tilde{V}/\partial \boldsymbol{x})\rangle = (2\pi)^6/(L^3)^4\int d\boldsymbol{x}\, d\boldsymbol{x}' \sum_{ll'} V_l\, e^{i\boldsymbol{l}\cdot(\boldsymbol{x}-\boldsymbol{x}')}\, i\boldsymbol{l}$$
$$\times\int_0^\infty dt\, V_{l'}\, e^{i\boldsymbol{l}'\cdot[(\boldsymbol{x}-\boldsymbol{x}')-\boldsymbol{g}t]}\, i\boldsymbol{l}' \tag{5}$$

Using the definition of the δ_- functions [see (2.5.24), (4.1.5)] we get

$$\langle(\partial V/\partial \boldsymbol{x})\cdot(\partial \tilde{V}/\partial \boldsymbol{x})\rangle = \{(2\pi)^6/(L^3)^4\}\int d\boldsymbol{x} \sum_{ll'} V_l V_{l'} L^3 \delta^{\mathrm{Kr}}(\boldsymbol{l}+\boldsymbol{l}') \times i\boldsymbol{l}\pi\delta_-(\boldsymbol{l}\cdot\boldsymbol{g})\, i\boldsymbol{l}'$$

$$= \{\pi(2\pi)^6/(L^3)^2\}\sum_l |V_l|^2 i\boldsymbol{l}\delta(\boldsymbol{l}\cdot\boldsymbol{g})(-i\boldsymbol{l}) \qquad (6)$$

The principal part may be dropped because the integrand $|V_l|^2 l^2$ is even.

Substituting (4.4.6) into (4.4.1) we indeed recover (4.3.4). We see that the Boltzmann equation for weakly coupled systems corresponds to an averaging over space of the twice iterated force, taken along the *unperturbed trajectories.* The analogy with the quantum mechanical Born approximation is clear.

Let us now eliminate the integral over $\boldsymbol{l}$ which appears in (4.3.4).[1] We make use of the following identity, where $f(\boldsymbol{v})$ is an arbitrary function of the velocity

$$(\partial/\partial\boldsymbol{v})[\delta(\boldsymbol{l}\cdot\boldsymbol{g})f(\boldsymbol{v})] = f(\partial/\partial\boldsymbol{v})\delta(\boldsymbol{l}\cdot\boldsymbol{g})+\delta(\boldsymbol{l}\cdot\boldsymbol{g})(\partial f/\partial\boldsymbol{v})$$
$$= -f(\partial/\partial\boldsymbol{v}')\cdot\delta(\boldsymbol{l}\cdot\boldsymbol{g})+\delta(\boldsymbol{l}\cdot\boldsymbol{g})(\partial f/\partial\boldsymbol{v}) \qquad (7)$$

and integrate by parts. In this way we first get

$$\partial\varphi/\partial t = 8\pi^4\lambda^2 C\int d\boldsymbol{l}\int d\boldsymbol{v}'|V_l|^2\delta(\boldsymbol{l}\cdot\boldsymbol{g})\cdot\{\varphi'(\boldsymbol{l}\cdot(\partial/\partial\boldsymbol{v}))^2\varphi-\varphi(\boldsymbol{l}\cdot(\partial/\partial\boldsymbol{v}'))^2\varphi'\} \qquad (8)$$

In order to integrate over $\boldsymbol{l}$ let us use a coordinate system in which $\boldsymbol{g}$ is the polar axis and then transform the result into a form which is independent of this special choice. This is, of course, equivalent to the evaluation of (4.4.8) in an arbitrary coordinate system, which would, however require more laborious calculations. In our coordinate system we have for the first term in (4.4.8)

$$\int d\boldsymbol{l}|V_l|^2\delta(\boldsymbol{l}\cdot\boldsymbol{g})[\boldsymbol{l}\cdot(\partial/\partial\boldsymbol{v})]^2\varphi$$
$$= \int_0^\infty dl\, l^4|V_l|^2\int_0^\pi d\theta\,\sin\theta\int_0^{2\pi} d\Phi\,\delta(lg\cos\theta)$$
$$\times\{\sin^2\theta\sin^2\Phi(\partial^2\varphi/\partial v_x^2)+\sin^2\theta\cos^2\Phi(\partial^2\varphi/\partial v_y^2)$$
$$+\cos^2\theta(\partial^2\varphi/\partial v_z^2)+\text{mixed derivatives}\} \qquad (9)$$

[1] We follow here, and in subsequent paragraphs, the methods used by Prigogine and Balescu, 1957.

By integration over Φ the coefficients of the mixed derivatives vanish and we are left with

$$\pi \int_0^\infty dl\, l^4 |V_l|^2 \int_{-1}^{+1} d\xi\, \delta(lg\,\xi)$$
$$\times\{(1-\xi^2)[(\partial^2\varphi/\partial v_x^2)+(\partial^2\varphi/\partial v_y^2)]+\xi^2(\partial^2\varphi/\partial v_z^2)\}$$
$$= (\pi/g)[(\partial^2\varphi/\partial v_x^2)+(\partial^2\varphi/\partial v_y^2)] \int_0^\infty dl\, l^3 |V(l)|^2 \quad (10)$$

It is only through the quantity

$$B = \int_0^\infty dl\, l^3 |V(l)|^2 \quad (11)$$

that the intermolecular potential appears in the Boltzmann equation for weakly coupled gases. We shall discuss this factor in more detail below.

To obtain an expression which is independent of the coordinate system we observe that we may write

$$\partial/\partial v_z = (\boldsymbol{g}/g)\cdot(\partial/\partial \boldsymbol{v}) \quad (12)$$

Therefore, (4.4.10) takes the invariant form[1]

$$B(\pi/g)\{\nabla_v^2-[(\boldsymbol{g}/g)\cdot(\partial/\partial\boldsymbol{v})]^2\}\varphi =$$
$$B(\pi/g)\{\nabla_v^2-(1/g^2)[(\boldsymbol{g}\cdot\partial/\partial\boldsymbol{v})^2-\boldsymbol{g}(\partial/\partial\boldsymbol{v})]\}\varphi \quad (13)$$

We proceed in the same way with the second part of (4.4.8) and obtain

$$\partial\varphi/\partial t = 8\pi^5 C\lambda^2 B \int d\boldsymbol{v}'(1/g)\{\varphi'[\nabla_v^2-(1/g^2)(\boldsymbol{g}\cdot\partial/\partial\boldsymbol{v})^2+(1/g^2)(\boldsymbol{g}\cdot\partial/\partial\boldsymbol{v})]\varphi$$
$$-\varphi[\nabla_{v'}^2-(1/g^2)(\boldsymbol{g}\cdot\partial/\partial\boldsymbol{v}')^2-(1/g^2)(\boldsymbol{g}\cdot\partial/\partial\boldsymbol{v}')]\varphi'\} \quad (14)$$

In order to obtain more explicit results we shall now consider the special case of Brownian motion which corresponds to a linear problem for the distribution function.

We would like to mention that alternative or equivalent derivations of the Boltzmann equation (4.4.14) may be found in the literature (Landau, 1936; Rosenbluth, MacDonald, and

[1] The notation $[(\boldsymbol{g}/g)\cdot(\partial/\partial\boldsymbol{v})]^2\varphi$ always means $[(\boldsymbol{g}/g)\cdot(\partial/\partial\boldsymbol{v})]\times[(\boldsymbol{g}/g)\cdot(\partial/\partial\boldsymbol{v})]\varphi$ in which the derivatives act on everything to their right.

Judd, 1957). However, all these derivations start with the Fokker-Planck equation or with the usual Boltzmann equation for short-range forces and then introduce the assumption of weak forces a posteriori. As our main interest lies precisely in the derivation of such stochastic equations from mechanics, an account of these methods is not of interest here.

5. Brownian Motion

Let us suppose that a particle of velocity $\boldsymbol{v}$ moves in a medium which has already reached its thermal equilibrium. This corresponds to introducing into (4.4.14) the Maxwellian distribution function

$$\varphi' = (4/\sqrt{\pi})(m/2kT)^{3/2}\, e^{-mv'^2/2kT} \tag{1}$$

Performing the differentiation with respect to $\boldsymbol{v}'$ we first obtain

$$\begin{aligned}\partial\varphi/\partial t = {} & 32\pi^3 C/m^2(m\pi/2kT)^{3/2}\lambda^2 B \int d\boldsymbol{v}'\, e^{-mv'^2/2kT}(1/g) \\ & \times\{\nabla_v^2-(1/g^2)(\boldsymbol{g}\cdot(\partial/\partial\boldsymbol{v})^2+(1/g^2)(\boldsymbol{g}\cdot\partial/\partial\boldsymbol{v})+4(m/2kT) \\ & +4(m/2kT)^2(1/g^2)[(\boldsymbol{g}\cdot\boldsymbol{v})^2-g^2v^2]\}\varphi\end{aligned} \tag{2}$$

We have made use of the identity

$$(\boldsymbol{g}\cdot\boldsymbol{v}')^2-g^2v'^2 = (\boldsymbol{g}\cdot\boldsymbol{v})^2-g^2v^2 \tag{3}$$

which follows directly from the definition of $\boldsymbol{g}$ as the relative velocity.

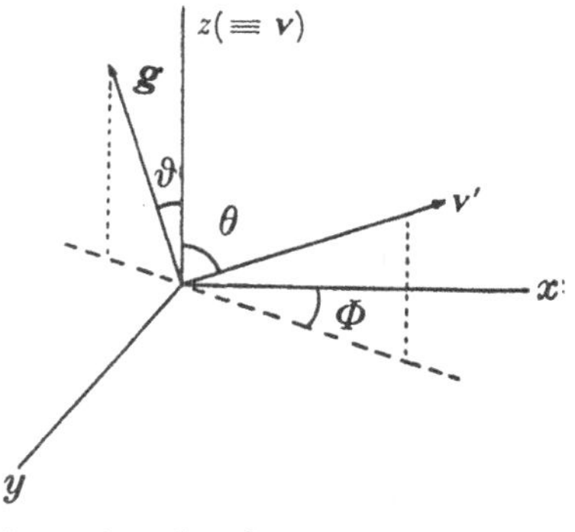

Fig. 4.5.1 Coordinate system for the evaluation of the integral in (4.5.2).

We now have to perform the integration over $\boldsymbol{v}'$. We use a coordinate system in which $\boldsymbol{v}$ is the polar axis (see Fig. 4.5.1). Let us consider the last term of (4.5.2) in more detail. We have (see Fig. 4.5.1)

$$(1/g^2)[(\boldsymbol{g}\cdot\boldsymbol{v})^2-g^2v^2] = v^2(\cos^2\vartheta-1) = -v^2\sin^2\vartheta = -v^2(v'^2/g^2)\sin^2\theta \tag{4}$$

Therefore

$$I = \int d\boldsymbol{v}'\, e^{-mv'^2/2kT}\,(1/g^3)[(\boldsymbol{g}\cdot\boldsymbol{v})^2-g^2v^2] = -2\pi v^2\int_0^\infty dv'\,v'^4 e^{-mv'^2/2kT}\int_0^\pi d\theta\,\sin\theta\,\sin^2\theta/g^3 \tag{5}$$

Instead of integrating over θ we take g as a new integration variable (see Fig. 4.5.2)

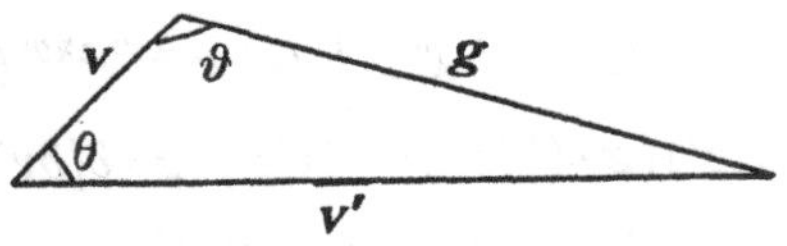

Fig. 4.5.2 See formula (4.5.6).

with

$$g^2 = v^2+v'^2-2vv'\cos\theta \tag{6}$$

$$gdg = vv'\sin\theta\, d\theta\,. \tag{7}$$

By substituting into (4.5.5) we obtain, therefore

$$I = (1/4)\int_0^\infty dv'\,v'^2 e^{-mv'^2/2kT}(1/vv')\int_{|v-v'|}^{v+v'} dg[g^2-2(v'^2+v^2)+g^{-2}(v^2-v'^2)^2]$$
$$= -(4/3)v^2[(1/v^3)\int_0^v dv'\,v'^4 e^{-mv'^2/2kT}+\int_v^\infty dv'\,v'\,e^{-mv'^2/2kT}] \tag{8}$$

These two integrals, as well as the other integrals which appear in (4.5.2), are related to the error function

$$\Phi(u) = (2/\sqrt{\pi})\int_0^u e^{-x^2}dx \tag{9}$$

For convenience we give the explicit expressions for the integrals which appear in this calculation as well as in the calculation of the other terms of (4.5.2)

$$\int_v^\infty dv'\, v'\, e^{-\gamma^2 v'^2} = (1/2\gamma^2)\,\Phi'(u) \tag{10}$$

$$\int_0^v dv'\, v'^2\, e^{-\gamma^2 v'^2} = (1/2\gamma^3)[\Phi(u)-u\Phi'] \tag{11}$$

$$\int_v^\infty dv'\, v'^3\, e^{-\gamma^2 v'^2} = (1/2\gamma^4)(1+u^2)\,\Phi' \tag{12}$$

$$\int_0^v dv'\, v'^4\, e^{-\gamma^2 v'^2} = (1/4\gamma^5)(3\Phi-3u\Phi'-2u^3\Phi') \tag{13}$$

with

$$\boldsymbol{u} = (m/2\boldsymbol{k}T)^{1/2}\,\boldsymbol{v} = \gamma \boldsymbol{v} \tag{14}$$

and

$$\Phi' = d\Phi/du \tag{15}$$

We therefore have

$$\begin{aligned} I &= -(4/3)\,v^2[(1/v^3)(1/4\gamma^5)(3\Phi-3u\Phi'-2u^3\Phi')+(1/2\gamma^2)\Phi'] \\ &= -(1/u\gamma^4)(\Phi-u\Phi') \end{aligned} \tag{16}$$

We now add the contributions of the different terms of (4.5.2) and use the dimensionless quantities (4.5.14) and

$$\tau = [32\pi^3(C/m^2)\,\lambda^2(m\pi/2\boldsymbol{k}T)^{3/2}B]\,t \tag{17}$$

We then obtain the basic equation for Brownian motion

$$\begin{aligned} \partial\varphi(\boldsymbol{u},\tau)/\partial\tau = \{&a(u)\nabla_u^2+b(u)[(\boldsymbol{u}\cdot\partial/\partial\boldsymbol{u})^2-\boldsymbol{u}\cdot\partial/\partial\boldsymbol{u}] \\ &+4\Phi'(u)\}\varphi(\boldsymbol{u},\tau) \end{aligned} \tag{18}$$

with

$$a(u) = u^{-1}\mathscr{H}(u), \qquad b(u) = u^{-1}\,da/du \tag{19}$$

$$\mathscr{H}(u) = \tfrac{1}{2}u^{-2}[u\Phi'(u)+(2u^2-1)\Phi(u)] \tag{20}$$

The function $\mathscr{H}(u)$ was introduced by Chandrasekhar (1942, 1943, 1945) in his pioneering work on weakly coupled gases (see also § 6).

The expression (4.5.17) shows that the relaxation time of the system will be of the order $\tau = 1$

$$t_r = \chi\, m^{1/2}(\boldsymbol{k}T)^{3/2}/\lambda^2 BC \qquad \text{with } \chi = [(8\sqrt{2})\pi^{9/2}]^{-1} \tag{21}$$

As could be expected, this relaxation time is inversely proportional to the concentration. Moreover it increases with temperature. This is natural because what matters here is the ratio of the perturbation energy to the unperturbed energy, which is proportional to $\boldsymbol{k}T$. Therefore for a given value of $\boldsymbol{k}T$ the relaxation time will decrease when the interaction increases, and for a given interaction the relaxation time will increase with temperature. This situation is very different from the usual case of strong, short-range interactions (for example hard spheres). There the ratio of potential energy to kinetic energy no longer appears in t_r (this ratio is infinite for hard spheres) and t_r is essentially determined by the number of collisions. This number will, for a given concentration, increase with temperature. Therefore, for hard spheres t_r decreases with temperature.

In the case of harmonic oscillators we also found that the relaxation time decreases as T increases [see (**3.3.36**)]. This, however, comes from the energy dependence of the interaction energy due to the anharmonic forces.

It is easy to transform (4.5.18) into polar coordinates u, θ, Φ.[1] One finds

$$\begin{aligned} \partial\varphi/\partial\tau = & \ (a+u\, da/du)\,\partial^2\varphi/\partial u^2 + (2a/u)\,\partial\varphi/\partial u \\ & + (a/u^2 \sin\theta)\,\partial/\partial\theta\,(\sin\theta\ \partial\varphi/\partial\theta) \\ & + (a/u^2 \sin^2\theta)\,\partial^2\varphi/\partial\Phi^2 + (8/\sqrt{\pi})e^{-u^2}\varphi \end{aligned} \tag{22}$$

One sees immediately that an isotropic distribution will remain isotropic.

On the other hand, even if we initially have an equilibrium distribution with respect to the absolute value of the velocity u that is a distribution function of the form

$$\varphi = u^2\, e^{-u^2} h(\theta, \Phi) \qquad \text{for } \tau = 0 \tag{23}$$

such a distribution will not be maintained in time because the variables u, θ, Φ do not separate in equation (4.5.22). However,

[1] The polar angle Φ should of course not be confused with the error function $\Phi(u)$ defined in (4.5.9).

if we substitute (4.5.23) into (4.5.22) we obtain

$$\begin{aligned}\partial h/\partial \tau &= a(u)\,[(1/\sin\theta)\,\partial/\partial\theta\,(\sin\theta\,\partial/\partial\theta) + (1/\sin^2\theta)\,\partial^2/\partial\Phi^2]\,h \\ &= a(u)\,\nabla^{(2)}_{\theta,\Phi} h \qquad (24)\end{aligned}$$

Therefore in the approximation in which $a(u)$ may be replaced in (4.5.24) by some average value (for example by the value of $a(u)$ for $u = 1$), the distribution (4.5.23) will indeed be a solution. Then h satisfies, as shown by (4.5.24), a simple diffusion equation with the usual Laplace operator in the polar angles θ, Φ.

Let us now write (4.5.18) in the form of the Fokker-Planck equation (2.9.5).

6. Fokker-Planck Equation and Dynamical Friction

We may write (4.5.18) in the form

$$\partial\varphi/\partial\tau = \sum_i \partial/\partial u_i[-(\overline{\Delta u_i/\Delta\tau}) + \tfrac{1}{2}\sum_j \partial/\partial u_j(\overline{\Delta u_i \Delta u_j})/(\Delta\tau)]\varphi \qquad (1)$$

with

$$(\overline{\Delta \mathbf{u}/\Delta\tau}) = -16 g(u)(\mathbf{u}/u) \qquad (2)$$

and

$$\overline{\Delta u_i \Delta u_j/\Delta\tau} = 12[g(u) - \tfrac{1}{3}\Phi(u)](u_i u_j/u^3) + a(u)\,\delta_{ij} \qquad (3)$$

where $a(u)$ is defined by (4.5.19) and $g(u)$ by

$$g(u) = (1/2u^2)[\Phi(u) - u\Phi'(u)] \qquad (4)$$

The simplest procedure for testing (4.6.2), (4.6.3) is to substitute back into (4.6.1) and to verify that we get (4.5.18).

Formula (4.6.2) expresses the fact that the fluctuating forces that the medium exerts on the particle give rise to a *systematic friction effect.* Indeed the definition (4.6.4) implies that $g(u)$ is positive. We therefore have

$$\overline{\Delta \mathbf{u}}/\Delta\tau = -\eta \mathbf{u} \qquad (5)$$

with

$$\eta = 16\, g(u)/u > 0 \qquad (6)$$

η is called the dynamical friction coefficient. This effect has been discovered by Chandrasekhar (1943, 1945). It has the interesting consequence that if even once, because of fluctuations, a particle has a large velocity, its interaction with the other particles will decelerate it.

It is easy to see, expanding $\Phi(u)$ and $u\Phi'(u)$ in a Taylor series around $u = 0$, that

$$g(u) \sim u, \qquad \eta = \text{independent of } u, \qquad \text{for } u \to 0 \tag{7}$$

On the contrary, for $u \to \infty$,

$$g(u) = (1/2u^2)\Phi(\infty) = 1/2u^2, \qquad \eta \sim 1/u^3 \tag{8}$$

Therefore the dynamical coefficient of friction, η, has the form represented in Fig. 4.6.1.

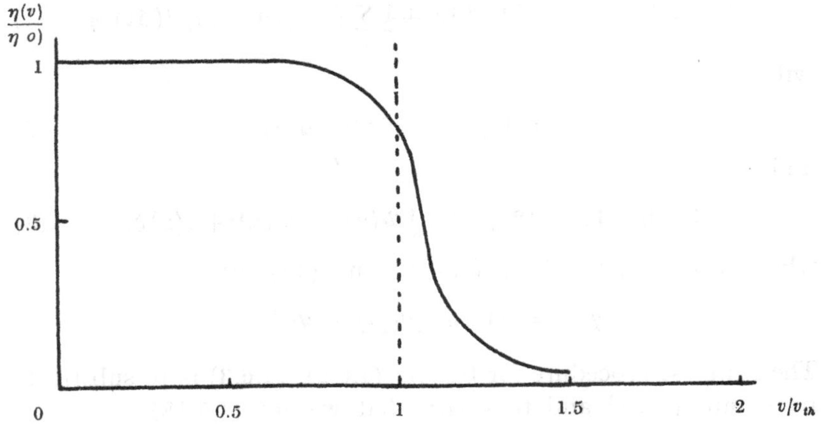

Fig. 4.6.1 Dynamical coefficient of friction, η, versus velocity (devided by the average thermal velocity).

As a consequence of this, a very rapid particle will experience little friction, and will keep its initial condition for a long time. Again, just the opposite behavior is to be expected for strong, short-range forces, where a very rapid particle may be expected to suffer more collisions, the effect of each being largely independent of the velocity. Such a particle will be brought to equilibrium more rapidly than a slow particle.

For weakly coupled systems the vanishing of the coefficient of friction for large values of u is especially significant when there are external forces. Such forces (for example, an electric field) will accelerate the particles, while only for velocities which are not too large can these accelerations be compensated by collisions. Therefore time-independent distribution functions cannot be expected in such situations (see especially H. Dreicer, 1958).

A similar behavior of the rate of approach to equilibrium as a function of the energy (or the velocity) has been found in other cases. For example, Bak, Henin, and Goche (1959) have studied the approach to equilibrium of a particle coupled to lattice vibrations. The coupling is supposed to be weak. Two limiting

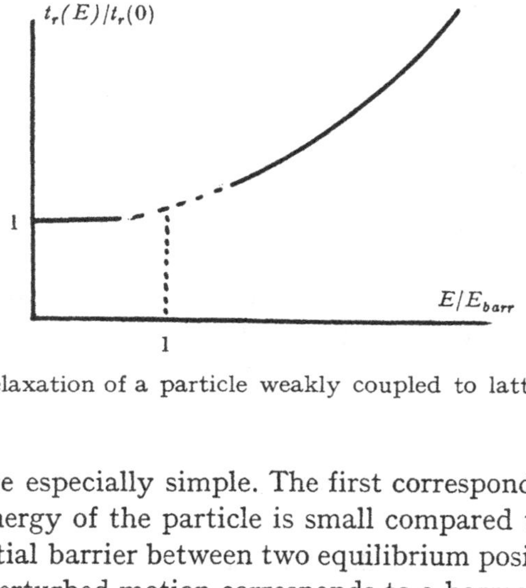

Fig. 4.6.2 Relaxation of a particle weakly coupled to lattice vibrations.

situations are especially simple. The first corresponds to the case where the energy of the particle is small compared to the height of the potential barrier between two equilibrium positions. In this case the unperturbed motion corresponds to a harmonic oscillator and the theory of Chapter 2 may be applied. In the second case the energy of the particle is large with respect to the potential barrier and we have a free translation to which we may apply the theory of this chapter. The relaxation time is represented in

Fig. 4.6.2. Again the relaxation time increases as the energy of the particle increases.[1]

7. Electrostatic Interactions — The Divergence Problem

We have seen that the precise form of the molecular interaction $V(r)$ appears only through the quantity B defined in (4.4.11) in the theory of weakly coupled gases. A change in the intermolecular forces therefore affects only the definition (4.5.17) of the reduced time.

Let us now consider the kind of intermolecular forces to which we may apply the theory of weakly coupled gases. It is quite clear that all strong, short-range forces have to be excluded, because for them the strength of the interaction λ is not small. Strictly speaking there exist no known intermolecular forces to which we could rigorously apply the theory of weakly coupled gases. Indeed the convergence of the integral (4.4.11) for $l \to 0$ implies that

$$V_l \sim 1/l^{2-\varepsilon} \qquad \text{for } l \to 0 \qquad (\varepsilon > 0) \tag{1}$$

while its convergence for $l \to \infty$ implies

$$V_l \sim 1/l^{2+\varepsilon} \qquad \text{for } l \to \infty \qquad (\varepsilon > 0) \tag{2}$$

In other words for $l \to 0$, V_l has to remain finite or at least not go to infinity more rapidly than l^{-2}. On the other hand for $l \to \infty$, V_l has to vanish more rapidly than l^{-2}. No such intermolecular forces are known.[2] Let us, for example, consider an electrostatic (or gravitational) interaction.

We have for the corresponding Fourier coefficient (see 4.1.2)

[1] For the relation of this problem to the theory of transport processes in solids see Prigogine and Bak (1959).

[2] For example, the potential $V_l = \gamma$ for $l_1 < l < l_2$, and $V_l = 0$ outside this range, would satisfy these conditions and give a convergent value to B; but the corresponding potential $V(r)$ obtained by a Fourier transform [see (4.1.2)] is proportional to $(\gamma/2)$ $(l_1 \sin rl_1 - l_2 \sin rl_2)$ and gives an oscillatory interaction of infinite range.

$$\begin{aligned} V_l &= (1/8\pi^3) \int d\boldsymbol{r}\, e^{-i\boldsymbol{l}\boldsymbol{r}} V(r) \\ &= (1/8\pi^3)\, e^2\, 2\pi \iint dr\, d\cos\theta\, (1/r) r^2 e^{-ilr\cos\theta} \\ &= e^2/l^2\, 2\pi^2 \end{aligned} \tag{3}$$

Therefore

$$B = (1/4\pi^4)\, e^4 \int_0^\infty dl\, l^{-1} \tag{4}$$

and we have a logarithmic divergence both for small and large values of l. The divergence for large values of the wave number l corresponds to a divergence for short distances. Its meaning is very simple: for short distances the interaction becomes too strong to be treated by a weakly coupled theory. The long-distance divergence (that is $l \to 0$) is more interesting.

This divergence also appears in all theories that deal with the equilibrium properties of particles interacting through long-range forces (see for example Mayer, 1950; Montroll and Ward, 1958; Gell-Mann and Brueckner, 1957) or with the waves which propagate in such a medium (the theory of plasma oscillations, see Bohm and Pines, 1953; Bohm, 1958; Sawada, 1957; Sawada, Brueckner, Fukuda and Brout, 1957; Brout, 1957). It may be explained intuitively as follows; when $l \to 0$, V_l goes to infinity as shown by (4.7.3), *whatever the value of* e^2. For large distances a simple perturbation treatment cannot be applied and the structure of the medium at such distances becomes very different from that of free, non-interacting particles. Using the concepts of the classical theory of strong electrolytes due to Debye and Hückel (see H. Falkenhagen, 1932) one can say that the modification of structure is due to an ionic cloud of opposite charge which screens the charge of each particle. The interaction between two charges is then described by the *screened Debye interaction*

$$V(r) = (e^2/r)\, e^{-\kappa r} \tag{5}$$

where $1/\kappa$ is the Debye length related to the ionic concentrations C_i, to their charges e_i and to the temperature by

$$1/\kappa^2 = \boldsymbol{k}T/4\pi \sum_i C_i e_i^2 \tag{6}$$

We see that only for

$$r \ll 1/\kappa \tag{7}$$

is the usual Coulomb law valid. For distances of the order of $1/\kappa$ or larger, the interaction tends to zero exponentially because of the screening.

We see, therefore, that the treatment of electrostatic interactions by the theory of weakly coupled gases is limited: (a) at short distances by a characteristic length a which we may take to be of the order of magnitude of a molecular diameter; (b) at long distances by the Debye length $1/\kappa$. One may, however introduce the following semi-empirical modifications into the theory: Let us introduce into (4.7.4) two cutoffs, one for small l — of the order κ and one for large l — of the order $1/a$. We then have

$$B = (1/4\pi^4)e^4 \int_\kappa^{1/a} dl\, l^{-1} = (1/4\pi^4)e^4 \ln (a\kappa)^{-1} \tag{8}$$

In this way the relaxation time (4.5.21) becomes

$$t_r = m^{1/2}(kT)^{3/2}/(2\sqrt{2\pi})\, Ce^4 \ln (a\kappa)^{-1} \tag{9}$$

A good account of the theory of transport processes based on such a model may be found in Spitzer's book (1956) on fully ionized gases.

It is interesting to compare (4.7.9) to the relaxation time for hard spheres, which is of order

$$t_r = 1/\pi C\bar{v}a^2 = 1/\pi C(\sqrt{kT/m})a^2 \tag{10}$$

Neglecting the slowly varying logarithm in (4.7.9), the ratio of the Debye relaxation time t_{D} to the hard-sphere relaxation time (4.7.10) is

$$t_{\mathrm{D}}/t_{\mathrm{HS}} \sim (kT)^2(a^2/e^4)$$

This is the square of the ratio of the thermal energy to the electrostatic energy evaluated at a distance equal to the radius of the hard sphere. Therefore, for usual temperatures we have

$$t_{\mathrm{D}}/t_{\mathrm{HS}} \ll 1 \tag{11}$$

Under such conditions the short-range forces will play no role in the approach to equilibrium and the theory of weakly coupled gases becomes very useful for the evaluation of relaxation times and transport coefficients. However, the way in which we have corrected the theory to suppress the long-distance divergence is by no means satisfactory. The relaxation time (4.7.9) is now of the form

$$t_r = 1/\lambda^2 \ln \lambda \tag{12}$$

and not

$$t_r = 1/\lambda^2 \tag{13}$$

as explicitly assumed in the derivation of the theory. We shall therefore come back to this problem in detail (see Chapter 9) and show how a more precise treatment permits us to obtain a consistent theory of the approach to equilibrium of particles interacting through long-range forces, in which the divergence at large separations is automatically eliminated.

As we have already noticed the basic divergence problems that are met in the approach to equilibrium are the same as those which appear in the theory of equilibrium properties. In both cases these problems are solved within a perturbation treatment by a summation of relevant contributions. We shall study two examples of this procedure in detail — one in Chapter 6 for strong, short-range forces and the other in Chapter 9 for collective effects.

CHAPTER 5

Approach to Equilibrium in Weakly Coupled Gases

1. Introduction

In this brief chapter we shall extend to weakly coupled gases the results we obtained for anharmonic oscillators in Chapter 3, §2. We want to show that inhomogeneities which exist initially will disappear in time as the result of interactions. The proof is slightly more involved than in the case of oscillators because in the case of a gas the "frequencies" $\boldsymbol{k} \cdot \boldsymbol{v}$ depend on the velocity, while in the case of a harmonic oscillator the frequencies are constant. The theorems we shall give here will be widely generalized in Chapters 11 and 12.

Let us first consider the H-theorem. Its proof for weakly coupled gases is identical to the one for anharmonic oscillators in Chapter 2, §8. It is in fact valid for all weakly coupled systems.

2. H-Theorem

It is shown in (2.7.10) that the operator which appears in the master equation (4.2.3) is self adjoint. Moreover (we use the notations $\boldsymbol{D}_{jn} = \partial/\partial \boldsymbol{v}_j - \partial/\partial \boldsymbol{v}_n$ and $\boldsymbol{g}_{jn} = \boldsymbol{v}_j - \boldsymbol{v}_n$) taking account of (4.2.3) we have

$$\begin{aligned}
&(d/dt) \int d\boldsymbol{v}\, \rho_{\{0\}} \ln \rho_{\{0\}} \\
&= (\lambda^2 8\pi^4/L^3) \sum_{jn} \int d\boldsymbol{l}\, |V_l|^2 \int d\boldsymbol{v} (1 + \ln \rho_{\{0\}}) \\
&\quad \times \boldsymbol{l} \cdot \boldsymbol{D}_{jn}\, \delta(\boldsymbol{l} \cdot \boldsymbol{g}_{jn})\, \boldsymbol{l} \cdot \boldsymbol{D}_{jn}\, \rho_{\{0\}} \\
&= -(\lambda^2 8\pi^4/L^3) \sum_{jn} \int d\boldsymbol{l}\, |V_l|^2 \int d\boldsymbol{v} (1/\rho_{\{0\}})\, \delta(\boldsymbol{l} \cdot \boldsymbol{g}_{jn}) [\boldsymbol{l} \cdot \boldsymbol{D}_{jn}\, \rho_{\{0\}}]^2 \leqq 0
\end{aligned} \tag{1}$$

As in Chapter 2, §8, this implies that

$$\rho_{\{0\}} \underset{t \to \infty}{\to} f(H_0) \tag{2}$$

The function $f(H_0)$ is again determined by the initial condition and is otherwise arbitrary.

We may however obtain a much more precise form of the H-theorem if we start from the Boltzmann equation for weakly coupled systems (4.3.4) instead of from the master equation. We then obtain instead of (5.2.1).

$$
\begin{aligned}
&(d/dt) \int d\mathbf{v}\, \varphi(\mathbf{v}) \ln \varphi(\mathbf{v}) \\
&= (8\pi^4 \lambda^2/m^2)\, C \int d\boldsymbol{l}\, |V_l|^2 \int d\mathbf{v}\, d\mathbf{v}' \ln \varphi(\mathbf{v})\, \boldsymbol{l}\cdot\boldsymbol{D}\, \delta(\boldsymbol{l}\cdot\boldsymbol{g}) \\
&\quad \times \boldsymbol{l}\cdot\boldsymbol{D}\, \varphi(\mathbf{v})\, \varphi(\mathbf{v}') \\
&= \tfrac{1}{2}(8\pi^4 \lambda^2 C/m^2) \int d\boldsymbol{l}\, |V_l|^2 \int d\mathbf{v}\, d\mathbf{v}' \ln \varphi(\mathbf{v})\, \varphi(\mathbf{v}')\, \boldsymbol{l}\cdot\boldsymbol{D}\, \delta(\boldsymbol{l}\cdot\boldsymbol{g}) \\
&\quad \times \boldsymbol{l}\cdot\boldsymbol{D}\, \varphi(\mathbf{v})\, \varphi(\mathbf{v}') \\
&= -\tfrac{1}{2}(8\pi^4 \lambda^2 C/m^2) \int d\boldsymbol{l}\, |V_l|^2 \int d\mathbf{v}\, d\mathbf{v}' [\delta(\boldsymbol{l}\cdot\boldsymbol{g})/\varphi(\mathbf{v})\, \varphi(\mathbf{v}')] \\
&\quad \times [\boldsymbol{l}\cdot\boldsymbol{D}\, \varphi(\mathbf{v})\, \varphi(\mathbf{v}')]^2 \leqq 0
\end{aligned}
\tag{3}
$$

The time independent state will be reached when the velocity distribution is such that the expression

$$\boldsymbol{l}\cdot[(\partial \ln \varphi/\partial \mathbf{v}) - (\partial \ln \varphi'/\partial \mathbf{v}')] = 0 \tag{4}$$

vanishes for all values of $\boldsymbol{l}$ for which the condition

$$\boldsymbol{l}\cdot\boldsymbol{g} = \boldsymbol{l}\cdot(\mathbf{v} - \mathbf{v}') = 0 \tag{5}$$

is satisfied. This implies that φ is of the form

$$\ln \varphi(\mathbf{v}) = a + \boldsymbol{b}\cdot\mathbf{v} + cv^2 \tag{6}$$

The simplest procedure (see also Chapter 14, §2 for a similar problem) is to expand $\ln \varphi(\mathbf{v})$ in a Taylor series in v_x, v_y, v_z, and to verify that all contributions other then (5.2.6) do not vanish identically when (5.2.5) is satisfied. For example, a contribution in v^4 to $\ln \varphi$ would give in (5.2.4)

$$\boldsymbol{l}\cdot(\mathbf{v}v^2 - \mathbf{v}'v'^2) \tag{7}$$

This expression does not indeed vanish identically when (5.2.5) is satisfied. The coefficients a, $\boldsymbol{b}$, and c can as usual (see for example Chapman and Cowling, 1939) be expressed in terms of the normalization condition [see (7.1.10)], the average velocity (which is here supposed to be zero) and the temperature. In this way we obtain the well-known expression for the Maxwell velocity distribution

$$\varphi(\mathbf{v}) = (m/2\pi kT)^{3/2} \exp(-mv^2/2kT) \tag{8}$$

We see therefore that the "molecular chaos" assumption (4.3.3) permits us to show that the velocity distribution for a single molecule tends to the same asymptotic form whatever the distribution $f(H_0)$ for the system as a whole. We may say that the other molecules provide a kind of "thermostat" for the evolution of $\varphi(\mathbf{v})$. It is there fore natural that $\varphi(\mathbf{v})$ tends to the canonical distribution (5.2.8).

3. Disappearance of Inhomogeneities

We shall now consider the "inhomogeneous" Brownian motion that we studied for oscillators in Chapter 3, §1. We consider distribution functions of the form

$$\rho(\mathbf{x}_n \mathbf{v}_n;\ \mathbf{v}_1 \cdots \mathbf{v}_{n-1}, \mathbf{v}_{n+1} \cdots \mathbf{v}_N) \tag{1}$$

which depends on the single coordinate $\mathbf{x}_n$. Therefore

$$\rho_{k_n\{k'\}} = 0 \qquad \text{unless} \quad \{k'\} = 0 \tag{2}$$

Instead of $\rho_{k_n\{0\}}$ we shall use the notation $\rho(n_k)$. We want to show that [see (3.2.3)]

$$\rho(n_k) \underset{t\to\infty}{\rightarrow} 0 \tag{3}$$

The initial inhomogeneity disappears as the result of interactions between the particles.

As in Chapter 3, §1 one may show that the basic process which determines the evolution of $\rho(n_k)$ is given in Fig. 5.3.1. Particle

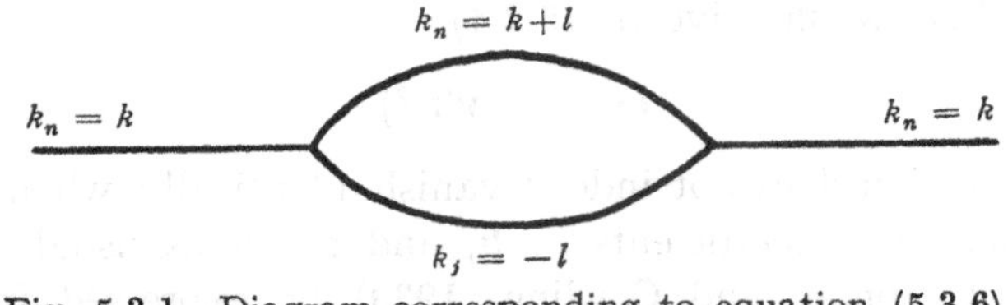

Fig. 5.3.1 Diagram corresponding to equation (5.3.6).

n propagates till it interacts with particle j. The cycle represents as usual a short lived correlation, corresponding to a collision. To the two equations (3.1.5) and (3.1.6) correspond the equations [see also (4.1.9)]

$$\partial \rho(n_k)/\partial t = \lambda \sum_j \int d\boldsymbol{l}\, e^{i\boldsymbol{k}\cdot\boldsymbol{v}_n t}\, V_l\, i\boldsymbol{l}\cdot\boldsymbol{D}_{jn}\, e^{-i[\boldsymbol{l}\cdot\boldsymbol{g}_{nj}+\boldsymbol{k}\cdot\boldsymbol{v}_n]t}\, \rho(n_{k+l}, j_{-l}) \quad (4)$$

$$\partial \rho(n_k+l, j_{-l})/\partial t = \lambda(2\pi/L)^3\, e^{i[\boldsymbol{l}\cdot\boldsymbol{g}_{nj}+\boldsymbol{k}\cdot\boldsymbol{v}_n]t}\, V_l\, i\boldsymbol{l}\cdot\boldsymbol{D}_{jn}\, e^{-i\boldsymbol{k}\cdot\boldsymbol{v}_n t}\, \rho(n_k) \quad (5)$$

By eliminating $\rho(n_k+l, j_{-l})$ we obtain the closed equation [see (3.1.7)]

$$(\partial/\partial t)\, \rho(n_k) = \lambda^2(2\pi/L)^3\, e^{i\boldsymbol{k}\cdot\boldsymbol{v}_n t} \sum_j \int d\boldsymbol{l}\, |V_l|^2\, i\boldsymbol{l}\cdot\boldsymbol{D}_{jn}\, e^{-i[\boldsymbol{l}\cdot\boldsymbol{g}_{nj}+\boldsymbol{k}\cdot\boldsymbol{v}_n]t}$$
$$\times \int_0^t dt_1\, e^{i[\boldsymbol{l}\cdot\boldsymbol{g}_{nj}+\boldsymbol{k}\cdot\boldsymbol{v}_n]t_1}\, i\boldsymbol{l}\cdot\boldsymbol{D}_{jn}\, e^{-i\boldsymbol{k}\cdot\boldsymbol{v}_n t_1}\, \rho(n_k, t_1) \quad (6)$$

The difference from (3.1.7) is that we have to be more careful with the position of the time dependent exponentials because they are velocity dependent and do not commute with the differential operator $\boldsymbol{D}_{jn} = \partial/\partial \boldsymbol{v}_j - \partial/\partial \boldsymbol{v}_n$. However we have the commutation rules

$$e^{i\boldsymbol{k}\cdot\boldsymbol{v}_n t}\, \boldsymbol{D}_{jn} = (\boldsymbol{D}_{jn} + i\boldsymbol{k}t)\, e^{i\boldsymbol{k}\cdot\boldsymbol{v}_n t} \quad (7)$$

$$\boldsymbol{D}_{jn}\, e^{-i\boldsymbol{k}\cdot\boldsymbol{v}_n t} = e^{-i\boldsymbol{k}\cdot\boldsymbol{v}_n t}\, (\boldsymbol{D}_{jn} + i\boldsymbol{k}t) \quad (8)$$

Therefore (5.3.6) may also be written

$$(\partial/\partial t)\, \rho(n_k) = \lambda^2(2\pi/L)^3 \sum_j \int d\boldsymbol{l}\, |V_l|^2\, i\boldsymbol{l}\cdot(\boldsymbol{D}_{jn} + i\boldsymbol{k}t)\, e^{-i\boldsymbol{l}\cdot\boldsymbol{g}_{nj}t}$$
$$\times \int_0^t dt_1\, e^{i\boldsymbol{l}\cdot\boldsymbol{g}_{nj}t_1}\, i\boldsymbol{l}\cdot(\boldsymbol{D}_{jn} + i\boldsymbol{k}t_1)\, \rho(n_k, t_1) \quad (9)$$

To this formula we may apply the methods of Chapter 2, §5 for asymptotic time integration [see especially (2.5.16)–(2.5.19)]. We have for t sufficiently large,

$$\int d\alpha f(\alpha)\, e^{-i\alpha t} \int_0^t dt_1\, e^{i\alpha t_1} = \int d\alpha f(\alpha)\, (e^{-i\alpha t} - 1)/-i\alpha$$
$$= \int d\alpha f(\alpha)\, \pi\delta_-(\alpha) \quad (10)$$

Similarly

$$\int d\alpha f(\alpha)\, e^{-i\alpha t} \int dt_1\, e^{i\alpha t_1} t_1 = \int d\alpha f(\alpha) [(t/i\alpha) - (1 - e^{-i\alpha t})/(i\alpha)^2]$$

$$= \int d\alpha f(\alpha)\, t \pi \delta_-(\alpha) \tag{11}$$

Therefore (5.3.9) becomes

$$(\partial/\partial t)\, \rho(n_k) = \lambda^2 (2\pi/L)^3 \sum_j \int d\boldsymbol{l}\, |V_l|^2 i\boldsymbol{l} \cdot (\boldsymbol{D}_{jn} + i\boldsymbol{k}t)\, \pi \delta_-(\boldsymbol{l} \cdot \boldsymbol{g}_{nj})$$

$$\times\, i\boldsymbol{l} \cdot (\boldsymbol{D}_{jn} + i\boldsymbol{k}t)\, \rho(n_k) \tag{12}$$

We write

$$\delta_-(\boldsymbol{l} \cdot \boldsymbol{g}_{nj}) = e^{i\boldsymbol{k} \cdot \boldsymbol{v}_n t}\, \delta_-(\boldsymbol{l} \cdot \boldsymbol{g}_{nj})\, e^{-i\boldsymbol{k} \cdot \boldsymbol{v}_n t} \tag{13}$$

and commute "back" the two exponentials to the left and the right. Using (5.3.7) and (5.3.8) again we obtain

$$(\partial/\partial t)\, \rho(n_k) = \lambda^2 (2\pi/L)^3 e^{i\boldsymbol{k} \cdot \boldsymbol{v}_n t} \sum_j \int d\boldsymbol{l}\, |V_l|^2 i\boldsymbol{l} \cdot \boldsymbol{D}_{jn} \pi \delta_-(\boldsymbol{l} \cdot \boldsymbol{g}_{nj})$$

$$\times\, i\boldsymbol{l} \cdot \boldsymbol{D}_{jn}\, e^{-i\boldsymbol{k} \cdot \boldsymbol{v}_n t}\, \rho(n_k) \tag{14}$$

The principal part of δ_- gives no contribution to the integral because it is odd in $\boldsymbol{l}$.

The two time dependent exponentials correspond to free propagation (the lines at the left and at the right in Fig. 5.3.1). Between them we have the collision operator as it appears in the equation for ρ_0 [see (4.2.3)].

We now want to show that (5.3.3) is satisfied for t sufficiently large. The simplest is to include the whole time dependence in $\rho(n_k)$. We shall therefore write

$$P(n_k, t) = e^{-i\boldsymbol{k} \cdot \boldsymbol{v}_n t}\, \rho(n_k, t) \tag{15}$$

In later chapters we shall often include the whole time dependence in the Fourier coefficients, but without using a special notation. We then obtain from (5.3.14)

$$\partial P(n_k)/\partial t + i\boldsymbol{k} \cdot \boldsymbol{v}_n P(n_k) = O_0 P(n_k) \tag{16}$$

where O_0 is the collision operator which appears in the homogeneous Fokker-Planck equation

$$O_0 = (2\pi/L)^3 \sum_j \int d\boldsymbol{l}\, |V_l|^2 i\boldsymbol{l}\cdot\boldsymbol{D}_{jn} \pi\delta(\boldsymbol{l}\cdot\boldsymbol{g}_{nj})\, i\boldsymbol{l}\cdot\boldsymbol{D}_{jn} \tag{17}$$

We shall also write (5.3.16) in the even more condensed form

$$\partial P(n_k)/\partial t = O_k P(n_k) \tag{18}$$

with

$$O_k = O_0 - i\boldsymbol{k}\cdot\boldsymbol{v}_n \tag{19}$$

Let us call φ_m and μ_m the eigenfunctions and the eigenvalues of the operator O_0. We know from the H-theorem that the eigenvalues are all negative or zero [see also (3.2.3)]. We also call $\varphi_m^{(k)}$, $\mu_m^{(k)}$ the eigenfunctions and eigenvalues of the operator O_k

$$O_k \varphi_m^{(k)} = \mu_m^{(k)} \varphi_m^{(k)} \tag{20}$$

We have by definition

$$\begin{aligned} \mu_m^{(k)} &= \int d\boldsymbol{v}\, \varphi_m^{*(k)}(\boldsymbol{v})\, O_k \varphi_m^{(k)}(\boldsymbol{v}) \\ &= \int d\boldsymbol{v}\, \varphi_m^{*(k)}(\boldsymbol{v})\, O_0 \varphi_m^{(k)}(\boldsymbol{v}) - \int d\boldsymbol{v}\, \varphi_m^{*(k)}(\boldsymbol{v})\, i\boldsymbol{k}\cdot\boldsymbol{v}_n \varphi_m^{(k)}(\boldsymbol{v}) \end{aligned} \tag{21}$$

As we have seen O_0 is an Hermitian (self adjoint) operator. The first term in (5.3.21) is real. On the contrary, $i\boldsymbol{k}\cdot\boldsymbol{v}_n$ is an antihermitian operator and the second term is imaginary. Therefore the real part of $\mu_m^{(k)}$ is given by the first term of (5.3.21),

$$\text{Re}\ \ \mu_m^{(k)} = \int d\boldsymbol{v}\, \varphi_m^{*(k)}(\boldsymbol{v})\, O_0 \varphi_m^{(k)}(\boldsymbol{v}) \tag{22}$$

We may now develop $\varphi_m^{(k)}$ in terms of the eigenfunctions of O_0.

$$\varphi_m^{(k)} = \sum_p C_{mp} \varphi_p \tag{23}$$

In this way we obtain

$$\text{Re}\ \ \mu_m^{(k)} = \sum_p |C_{mp}|^2 \mu_p \leqq 0 \tag{24}$$

But the equality sign would require

$$C_{mp} = \delta_{m0} \tag{25}$$

This is clearly impossible. Therefore

$$\text{Re} \quad \mu_m^{(k)} < 0 \tag{26}$$

in agreement with (5.3.3) [1]

4. Discussion

While the result we derived in §3 is the same as for interacting harmonic oscillators, the "forgetting" mechanism involved is deeply different. The characteristic features of the case of harmonic oscillators are: (a) the unperturbed frequencies are constants; (b) the perturbation energy (2.1.26) is action dependent.

It is the second feature which introduces the supplementary contribution M_l^2 into the collision term [see (3.3.12)].

In the case of gases we have on the contrary: (a) the unperturbed frequencies are action (or velocity) dependent; (b) the perturbation energy is (generally) velocity independent.

The action dependence of the frequencies introduces a kind of coupling between the flow term (the term $i\boldsymbol{k}\cdot\boldsymbol{v}_n P$ in 5.3.16) and the collision term, which finally drives the initial inhomogeneity to zero.

For velocity dependent forces between particles both mechanisms participate. An example would be the case of ionized gases if one retains the $(v/c)^2$ corrections due to relativistic effects and retardation in the transmission of interactions. This case has been studied (Prigogine and Leaf, 1959) but will not be included here.

On the contrary if both mechanisms are lacking, the initial inhomogeneity would not disappear. For example for an action independent perturbation energy between harmonic oscillators the fundamental solution $f(x\alpha t \mid x_0\alpha_0 t_0)$ [see (3.3.35)] becomes for $t \to \infty$

$$f(x\alpha t \mid x_0 \alpha_0 t_0) \sim e^{-x}\delta(\alpha - \alpha_0 - \omega t) \tag{1}$$

instead of the correct equilibrium value e^{-x}. In (5.4.1) the initial phase correlation persists.

[1] This proof is due to P. Résibois.

CHAPTER 6

Scattering Theory and Short-Range Forces

1. Two-Body Scattering Theory

As we have emphasized in Chapter 4, the case of weakly coupled gases corresponds to a rather exceptional situation. Real intermolecular potentials contain a strong, short-range repulsive part. We now want to take such forces into account. As a first step we shall treat the two-body scattering problem in our formalism. We consider a beam of particles diffused by a center of force situated at $\boldsymbol{r} = 0$. The beam is described in terms of the density in phase space $\rho(\boldsymbol{x}, \boldsymbol{p}, t)$. The scattering problem is perhaps one of the simplest physical situations which can be appropriately described in terms of a statistical ensemble. Indeed the basic problem here is not the determination of the motion of a single particle in the field of force but the effect of the scattering center on the whole beam formed by a large number of particles which are characterized by some distribution in phase space. Moreover two-body scattering is really the "elementary mechanism" of change in the evolution of N-body systems. For this reason the methods we shall develop now will be widely used in the later chapters (see especially Chapters 8, 12, 14).

In the present chapter we shall adopt an "asymptotic" point of view. We shall only be interested in the relation which exists between ρ at large times and its initial value $\rho(\boldsymbol{x}, \boldsymbol{p}, t = 0)$[1] (see Fig. 6.1.1).

This corresponds to what is known as the Heisenberg S-matrix point of view in quantum mechanics. The duration of the collision process may then be neglected in comparison with the times one is interested in. This approximation is enough to obtain the evolution of a sufficiently dilute gas, in which the duration of

[1] Instead of $t = 0$ it is sometimes useful to consider $t = -\infty$.

a collision is negligible with respect to the time between two collisions. But as we shall see later (cf. especially Chapters 8 and 11), this is no longer so in the general theory, which is valid for arbitrary conditions. There we need a much more accurate description of the collision process.

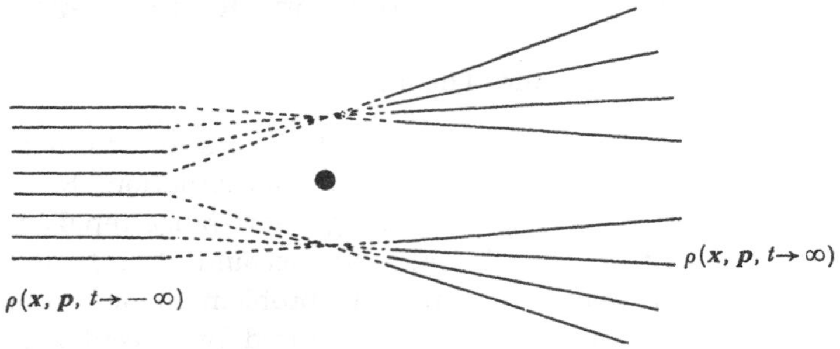

Fig. 6.1.1 Description of a scattering experiment in terms of the distribution function.

For a two-body scattering problem with central forces the Hamiltonian is simply [see (4.1.1)]

$$H = p^2/2m + \lambda V(|\boldsymbol{x}|) \tag{1}$$

We again expand V in the Fourier series [see (4.1.2)]

$$V(|\boldsymbol{x}|) = (2\pi/L)^3 \sum_l V_l \, e^{i\boldsymbol{l}\cdot\boldsymbol{x}} \tag{2}$$

The Liouville equation (4.1.9) is now (we take $m = 1$)

$$(\partial/\partial t)\, \rho_k = \lambda (2\pi/L)^3 \sum_l e^{i\boldsymbol{k}\cdot\boldsymbol{v}t} \, V_l \, i\boldsymbol{l} \cdot (\partial/\partial \boldsymbol{v}) e^{-i(\boldsymbol{k}-\boldsymbol{l})\cdot\boldsymbol{v}t} \rho_{k-l} \tag{3}$$

We may notice that the law of conservation of wave vectors (4.1.8) does not apply here, because the potential energy V is not invariant with respect to translation (we have an "external" force).

The equation (6.1.3) expresses the laws of mechanics in $(\boldsymbol{k}, \boldsymbol{v})$ space instead of the usual description in phase space $(\boldsymbol{x}, \boldsymbol{v})$ (see also Chapter 1, §10). The Fourier coefficient V_l of the potential induces "transitions" between the Fourier coefficients ρ_{k-l} and ρ_k of the phase density. To each Fourier coefficient corresponds a periodic space dependence determined by its wave vector. We have the situation represented in a schematic way in Fig. 6.1.2.

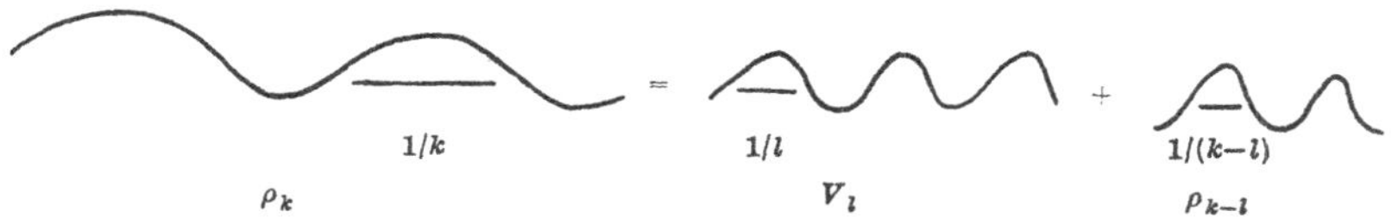

Fig. 6.1.2 Schematic representation of equation (6.1.3).

To each Fourier coefficient ρ_k corresponds a plane wave in phase space. Out of such plane waves we may form wave packets which we shall study in detail in Chapter 8, §1.

In most scattering problems it is not the complete distribution function ρ which is of interest but only the velocity distribution ρ_0. We assume moreover that before the interaction comes into play the system was homogeneous (see Bohm, 1951 for an excellent discussion of similar initial conditions in quantum mechanics). We therefore take the initial condition

$$\rho_0(\mathbf{v},\, t = 0) \neq 0, \qquad \rho_k(\mathbf{v},\, t = 0) = 0 \qquad (k \neq 0) \tag{3'}$$

and solve (6.1.3) by iteration. We then have

$$\begin{aligned}
\rho_0(t) &= \rho_0(0) - \lambda^2 (2\pi/L)^6 \sum_{l} \int_0^t dt_1 \int_0^{t_1} dt_2\, V_l\, i\boldsymbol{l} \cdot (\partial/\partial \mathbf{v})\, e^{i\boldsymbol{l}\cdot \mathbf{v}(t_1-t_2)} \\
&\times V_{-l}\, i\boldsymbol{l} \cdot (\partial/\partial \mathbf{v}) \rho_0(0) - \lambda^3 (2\pi/L)^9 \sum_{ll'} \int_0^t dt_1 \int_0^{t_1} dt_2 \int_0^{t_2} dt_3\, V_l\, i\boldsymbol{l} (\partial/\partial \mathbf{v}) \\
&\times e^{i\boldsymbol{l}\cdot \mathbf{v}(t_1-t_2)}\, V_{|l'-l|}\, i\,(\boldsymbol{l}'-\boldsymbol{l}) \cdot (\partial/\partial \mathbf{v})\, e^{i\boldsymbol{l}'\cdot \mathbf{v}(t_2-t_3)}\, V_{-l'}\, i\boldsymbol{l}' \cdot (\partial/\partial \mathbf{v})\, \rho_0(0) \\
&- \lambda^4 (2\pi/L)^{12} \sum_{ll'l''} \int_0^t dt_1 \int_0^{t_1} dt_2 \int_0^{t_2} dt_3 \int_0^{t_3} dt_4\, V_l\, i\boldsymbol{l} \cdot (\partial/\partial \mathbf{v}) e^{i\boldsymbol{l}\cdot \mathbf{v}(t_1-t_2)} \\
&\times V_{|l'-l|}\, i(\boldsymbol{l}'-\boldsymbol{l}) \cdot (\partial/\partial \mathbf{v})\, e^{i\boldsymbol{l}'\cdot \mathbf{v}(t_2-t_3)}\, V_{|l''-l'|}\, i(\boldsymbol{l}''-\boldsymbol{l}') \cdot (\partial/\partial \mathbf{v}) \\
&\times e^{i\boldsymbol{l}''\cdot \mathbf{v}(t_3-t_4)}\, V_{l''}\, i\boldsymbol{l}'' \cdot (\partial/\partial \mathbf{v}) \rho_0(0) + \ldots
\end{aligned} \tag{4}$$

We shall now give an asymptotic evaluation appropriate to the problem of scattering in which we are interested. We have to take account of the following remarks:

(1) We are interested in times much longer than the duration

of the interaction. Let us call l^* a characteristic wave length corresponding to V_l. Then $1/l^*$ is a characteristic length which measures the range of the potential. Therefore, the duration of the collision is

$$t_{\text{int}} \approx 1/\boldsymbol{l}^* \cdot \boldsymbol{v}$$

and we are interested in times such that

$$t \gg 1/\boldsymbol{l}^* \cdot \boldsymbol{v} \tag{5}$$

The time t is therefore much longer than the time necessary for a molecule to cross the distance $1/l^*$ with a velocity $\boldsymbol{v}$.

(2) Because we are considering a large system, we may use relation (4.2.4) for transforming the sums over $\boldsymbol{l}$ into integrals.

(3) The time t and the length L are chosen in such a way that multiple collisions or "recollisions" with the scattering center can be neglected. For example, we may take t to be of the order of L/v. This corresponds to the time necessary to cross the whole system with a velocity $\boldsymbol{v}$.

Then

$$t/L \approx 1/v \tag{6}$$

is finite, but

$$t/L^3 \approx 1/vL^2 \tag{7}$$

is very small. Therefore, terms in $(t/L^3)^2$ which, as we shall see, correspond to a recollision, may be neglected.

Proceeding now as in Chapter 2, § 5, the second-order term in (6.1.4) may immediately be written

$$-\lambda^2 t(2\pi/L)^3 \int d\boldsymbol{l} |V_l|^2 i\boldsymbol{l} \cdot (\partial/\partial \boldsymbol{v}) \pi\delta_+(\boldsymbol{l} \cdot \boldsymbol{v}) i\boldsymbol{l} \cdot (\partial/\partial \boldsymbol{v}) \rho_0(0) \tag{8}$$

We have a similar expression for the third-order term. In the case of the fourth-order term we have to consider two possibilities:

(α) $\boldsymbol{l}' \neq 0$, in which case we have, as in (6.1.8),

$$\begin{aligned} -\lambda^4 t(2\pi/L)^3 \int d\boldsymbol{l}\, d\boldsymbol{l}'\, d\boldsymbol{l}''\, V_l i\boldsymbol{l} \cdot (\partial/\partial \boldsymbol{v}) \pi\delta_+(\boldsymbol{l} \cdot \boldsymbol{v}) V_{|l'-l|}\, i(\boldsymbol{l}'-\boldsymbol{l}) \cdot (\partial/\partial \boldsymbol{v}) \\ \times \pi\delta_+(\boldsymbol{l}' \cdot \boldsymbol{v}) V_{|l''-l'|}\, i(\boldsymbol{l}''-\boldsymbol{l}') \cdot (\partial/\partial \boldsymbol{v}) \pi\delta_+(\boldsymbol{l}'' \cdot \boldsymbol{v}) V_{-l''} i\boldsymbol{l}'' \cdot (\partial/\partial \boldsymbol{v}) \rho_0(0) \end{aligned} \tag{9}$$

(β) $\boldsymbol{l}' = 0$, where we have

$$-\lambda^4(2\pi/L)^6 \int d\boldsymbol{l}\, d\boldsymbol{l}'' \int_0^t dt_1 \int_0^{t_1} dt_2 \int_0^{t_2} dt_3 \int_0^{t_3} dt_4 \,[V_l i\boldsymbol{l}\cdot(\partial/\partial\boldsymbol{v})\, e^{i\boldsymbol{l}\cdot\boldsymbol{v}(t_1-t_2)}$$
$$\times V_{-l} i\boldsymbol{l}\cdot(\partial/\partial\boldsymbol{v})] \times [V_{l''} i\boldsymbol{l}''\cdot(\partial/\partial\boldsymbol{v})\, e^{i\boldsymbol{l}''\cdot\boldsymbol{v}(t_3-t_4)}\, V_{-l''} i\boldsymbol{l}''\cdot(\partial/\partial\boldsymbol{v})]\,\rho_0(0) \tag{10}$$

We now have a sequence of two disconnected transitions (each represented by one of the bracketed terms). The first transition goes from 0 to $\boldsymbol{l}''$ and back to 0, while the second starts again from zero, goes to $\boldsymbol{l}$, and returns to 0. This situation is exactly the same as that in Chapter 2, § 6, and thus we obtain from (6.1.10) a contribution of the order

$$\lambda^4 t^2/L^6 \tag{11}$$

Such a term corresponds to two successive independent transitions, i.e., to two successive collisions of the particles of the beam with the scattering center. In accordance with (6.1.7) such terms will be neglected and we are left with (6.1.9).

Proceeding in the same way to all orders we have, therefore,

$$\begin{aligned}\rho_0(t) = \rho_0(0) - t(2\pi/L)^3 \Big[\lambda^2 \int d\boldsymbol{l}\, V_l i\boldsymbol{l}\cdot(\partial/\partial\boldsymbol{v})\pi\delta_+(\boldsymbol{l}\cdot\boldsymbol{v})V_{-l} i\boldsymbol{l}\cdot(\partial/\partial\boldsymbol{v})\\
+\lambda^3 \int d\boldsymbol{l}\, d\boldsymbol{l}'\, V_l i\boldsymbol{l}\cdot(\partial/\partial\boldsymbol{v})\pi\delta_+(\boldsymbol{l}\boldsymbol{v})V_{|l'-l|} i(\boldsymbol{l}'-\boldsymbol{l})\cdot(\partial/\partial\boldsymbol{v})\pi\\
\times\delta_+(\boldsymbol{l}'\boldsymbol{v})V_{-l'} i\boldsymbol{l}'\cdot(\partial/\partial\boldsymbol{v}) \qquad (12)\\
+\lambda^4 \int d\boldsymbol{l}\, d\boldsymbol{l}'\, d\boldsymbol{l}''\, V_l i\boldsymbol{l}\cdot(\partial/\partial\boldsymbol{v})\pi\delta_+(\boldsymbol{l}\cdot\boldsymbol{v})V_{|l'-l|} i(\boldsymbol{l}'-\boldsymbol{l})\cdot(\partial/\partial\boldsymbol{v})\\
\times\pi\delta_+(\boldsymbol{l}'\cdot\boldsymbol{v})V_{|l''-l'|} i(\boldsymbol{l}''-\boldsymbol{l}')\cdot(\partial/\partial\boldsymbol{v})\pi\delta_+(\boldsymbol{l}''\cdot\boldsymbol{v})V_{-l''} i\boldsymbol{l}''\cdot(\partial/\partial\boldsymbol{v})\\
+\cdots\Big]\,\rho_0(0)\end{aligned}$$

As could be expected, the velocity distribution function $\rho_0(\boldsymbol{v}, t)$ contains two parts: one corresponds to the incident beam, the other to the scattered beam. The number of scattered particles increases linearly with time because of our homogeneous initial conditions (6.1.3′).

Therefore, the scattering is represented by operators in $\partial/\partial\boldsymbol{v}$ of higher and higher order applied to the initial distribution $\rho_0(0)$. If we retain only the second-order operator we would have the situation corresponding to weak coupling. In such a case the velocity distribution at time t would differ by an infinitesimal amount from that at $t = 0$. It is only by adding the effect of all terms in (6.1.12) that we obtain a finite change in the velocity distribution, as is necessary for a "strong" collision.[1]

The structure of the series (6.1.12) is very simple. The term in λ^n corresponds to n successive interactions. To each of them corresponds an operator of the form $i\boldsymbol{l}V_l \cdot (\partial/\partial \boldsymbol{v})$, which expresses the modification of the velocity distribution through the force $i\boldsymbol{l}V_l$ which acts on the particle. Between the interactions the particle propagates freely, and this propagation is expressed by the function $\delta_+(\boldsymbol{l} \cdot \boldsymbol{v})$. We shall analyze the physical meaning of this propagator in § 2.

Using the terminology of quantum mechanics we could say that the successive terms in (6.1.12) correspond to successive Born approximations (see for example Bohm, 1951).

It is possible to write (6.1.12) into a much more compact form. Indeed, let us write

$$\rho_0(t) = \rho_0(0) - \lambda^2 t (2\pi/L)^3 \int d\boldsymbol{l}\, V_l i\boldsymbol{l} \cdot (\partial/\partial \boldsymbol{v}) \pi \delta_+(\boldsymbol{l} \cdot \boldsymbol{v}) \mathscr{g}_{l,0}(\boldsymbol{v}) \rho_0(0) \quad (13)$$

where the operator $\mathscr{g}_{l,0}(\boldsymbol{v})$ satisfies the integral equation

$$\mathscr{g}_{l,0}(\boldsymbol{v}) = iV_{-l}\boldsymbol{l} \cdot (\partial/\partial \boldsymbol{v}) + \lambda \int d\boldsymbol{l}'\, V_{|l'-l|} i(\boldsymbol{l}' - \boldsymbol{l}) \cdot (\partial/\partial \boldsymbol{v}) \pi \delta_+(\boldsymbol{l}' \cdot \boldsymbol{v}) \mathscr{g}_{l,0} \quad (14)$$

If we expand $\mathscr{g}_{l,0}$ in powers of λ by iteration using (6.1.14) and substitute into (6.1.13) we recover (6.1.12). The integral equation (6.1.14) is the analogue in classical mechanics of the integral equations obeyed by the quantum mechanical scattering matrix. We have used a simple iteration procedure to derive (6.1.13), (6.1.14) (Prigogine and Henin, 1957, 1958). We might mention

[1] A difference operator can be expressed as the sum of successive differential operators. For example

$$f(x+a) = f(x) + a(\partial f(x)/\partial x) + (a^2/2!)(\partial^2 f(x)/\partial x^2) + \cdots = e^{a(\partial/\partial x)} f(x)$$

We may say that each of the differential operators corresponds to an infinitesimal change of x while the sum gives rise to the finite change from x to $x + a$. In the case of weak coupling a is proportional to λ and we have, therefore, to keep only the lowest non-vanishing derivative.

that there exists a more elegant method (Résibois, 1959), using a procedure similar to that of Lippman and Schwinger in quantum mechanics (Lippman and Schwinger, 1949).

2. Propagators

While the basic equations for scattering are most readily established by using as variables the wave vectors $\boldsymbol{l}$ and velocities $\boldsymbol{v}$, their physical meaning becomes much clearer when one transforms the equations (6.1.12) or (6.1.13) and (6.1.14) back into phase space $\boldsymbol{x}$, $\boldsymbol{v}$ by means of Fourier transforms (Résibois, 1959).

An essential role will be played by the Fourier transform of the propagator $\delta_+(\boldsymbol{l}\cdot\boldsymbol{v})$. We shall denote it by $G(\boldsymbol{x}, \boldsymbol{v})$, or more briefly $G(\boldsymbol{x})$. We have, by definition,[1]

$$G(\boldsymbol{x}, \boldsymbol{v}) = (1/8\pi^3)\int d\boldsymbol{l}\,\pi\delta_+(\boldsymbol{l}\cdot\boldsymbol{v})e^{-i\boldsymbol{l}\cdot\boldsymbol{x}} \tag{1}$$

and inversely

$$\pi\delta_+(\boldsymbol{l}\cdot\boldsymbol{v}) = \int d\boldsymbol{x}\,G(\boldsymbol{x}, \boldsymbol{v})e^{i\boldsymbol{l}\cdot\boldsymbol{x}} \tag{2}$$

We shall call G the *free particle propagator in phase space.* Using the explicit form of δ_+ given in (2.5.25) and (2.5.26) we get

$$\begin{aligned} G(\boldsymbol{x}, \boldsymbol{v}) &= (1/8\pi^3)\int_0^\infty d\tau \int d\boldsymbol{l}\,e^{i\boldsymbol{l}\cdot\boldsymbol{v}\tau}e^{-i\boldsymbol{l}\cdot\boldsymbol{x}} \\ &= \int_0^\infty d\tau\,\delta(\boldsymbol{x}-\boldsymbol{v}\tau) \end{aligned} \tag{3}$$

where we use the obvious abbreviation

$$\delta(\boldsymbol{x}-\boldsymbol{v}\tau) = \delta(x-v_x\tau)\delta(y-v_y\tau)\delta(z-v_z\tau) \tag{4}$$

Let us study (6.2.3) for a particular value of the velocity $\boldsymbol{v}$.

[1] Let us stress the fact that the expression $\pi\delta_+(\boldsymbol{l}\cdot\boldsymbol{v})$ for the free particle propagator is related to the definition (6.1.2) of the Fourier transform of $V(x)$. If we use

$$V(|\boldsymbol{x}|) = (2\pi/L)^3 \sum_{l} V_l e^{-i\boldsymbol{l}\cdot\boldsymbol{x}}$$

instead, the free particle propagator is $\pi\delta_-(\boldsymbol{l}\cdot\boldsymbol{v})$. In both cases the propagator in phase space is given by (6.2.3).

In subsequent chapters we shall use δ_- or δ_+, whichever appears to be more convenient.

It becomes especially simple in a coordinate system in which $\mathbf{v}$ is taken as the polar axis. Then

$$\begin{aligned} G(\mathbf{x}-\mathbf{x}', \mathbf{v}) &= \int_0^\infty d\tau\, \delta(x-x')\delta(y-y')\delta(z-z'-v\tau) \\ &= (1/v)\eta(z-z')\delta(x-x')\delta(y-y') \end{aligned} \tag{5}$$

where η is the step function defined by

$$\begin{aligned} \eta(z) &= 1 \qquad z > 0 \\ \eta(z) &= 0 \qquad z < 0 \end{aligned} \tag{6}$$

The relation between this step function η and the Dirac δ function is

$$d\eta(z)/dz = \delta(z) \tag{7}$$

Formula (6.2.5) states that for a positive velocity $\mathbf{v}$ directed along the z axis, the particle "propagates" in a straight line ($x = x'$, $y = y'$) in the direction of increasing z ($z > z'$). Therefore, the only points $\mathbf{x}'$ which are connected to $\mathbf{x}$ through (6.2.5) are the points which correspond to *earlier* times on the trajectory. This expresses the causality conditions in classical mechanics. We shall come back to it in § 3. The function G plays somewhat the same role here as the *retarded potentials* in electromagnetism or the propagators of quantum field theory (see for example Schweber, Bethe, and de Hoffmann, 1956).

It is also easy to see that G (or more precisely iG) is the Green's function corresponding to the unperturbed Liouville operator L_0.[1]

Indeed the first equality (6.2.3) shows that

$$\begin{aligned} \mathbf{v}[\partial G(\mathbf{x}-\mathbf{x}', \mathbf{v})/\partial \mathbf{x}] &= -\int_0^\infty d\tau(\partial/\partial\tau) \int d\mathbf{l}\, e^{i\mathbf{l}\cdot\mathbf{v}\tau - i\mathbf{l}\cdot(\mathbf{x}-\mathbf{x}')} \\ &= \delta(\mathbf{x}-\mathbf{x}') \end{aligned} \tag{8}$$

Taking account of (1.5.3) this can also be written

$$L_0 iG(\mathbf{x}-\mathbf{x}', \mathbf{v}) = \delta(\mathbf{x}-\mathbf{x}') \tag{9}$$

Therefore, the solution of the inhomogeneous equation

$$L_0 f(\mathbf{x}, \mathbf{v}) = F(\mathbf{x}, \mathbf{v}) \tag{10}$$

[1] For a detailed study of the Green's function of the Liouville operators, see Andrews, 1960; Balescu, 1962.

can always be expressed in the closed form

$$f(\boldsymbol{x}, \boldsymbol{v}) = \int d\boldsymbol{x}' iG(\boldsymbol{x}-\boldsymbol{x}', \boldsymbol{v}) F(\boldsymbol{v}, \boldsymbol{x}') \tag{11}$$

To verify (6.2.11) it is sufficient to apply the operator L_0 to both sides and to take account of (6.2.9).

It should be noticed that in (6.2.8) we have neglected a contribution coming from very long distances $\boldsymbol{x}-\boldsymbol{x}'$. Indeed, strictly speaking, we have

$$-\int_0^\infty d\tau(\partial/\partial\tau) \int d\boldsymbol{l}\, e^{i\boldsymbol{l}\cdot\boldsymbol{v}\tau - i\boldsymbol{l}\cdot(\boldsymbol{x}-\boldsymbol{x}')} = \delta(\boldsymbol{x}-\boldsymbol{x}') - \delta(\boldsymbol{x}-\boldsymbol{x}'-\infty) \tag{12}$$

This second contribution can, however, always be neglected as long as the Green's function G is used in conjunction with functions $F(\boldsymbol{v}, \boldsymbol{x}')$ [see (6.2.11)] which go to zero sufficiently quickly when $\boldsymbol{x}' \to \infty$. This will always be the case[1] in the applications we shall study.

$G(\boldsymbol{x}, \boldsymbol{v})$ as given by (6.2.3) corresponds to free motion. In the presence of a force $\boldsymbol{F}(\boldsymbol{x})$ the corresponding expression is

$$G(\boldsymbol{x}\boldsymbol{v} | \boldsymbol{v}'\boldsymbol{x}') = \int_0^\infty d\tau\, \delta\left[\boldsymbol{x}-\boldsymbol{x}' - \int_0^\tau \boldsymbol{v}(\tau_1)\, d\tau_1\right] \delta\left[\boldsymbol{v}-\boldsymbol{v}' - \int_0^\tau \boldsymbol{F}(\boldsymbol{x}(\tau_1))\, d\tau_1\right] \tag{13}$$

For a vanishing force $\boldsymbol{F} = 0$, the velocity $\boldsymbol{v}(\tau_1)$ reduces to a constant and we come back to (6.2.3) (for more details see Andrews, 1960; Balescu, 1962).

3. Scattering in Phase Space

Using (6.2.1), (6.2.2) we may easily transform (6.1.12) back into phase space. We also have to express the Fourier components V_l in terms of the force $\partial V/\partial \boldsymbol{x}$ by [see (6.1.2)]

$$i\boldsymbol{l} V_l = (1/8\pi^3) \int d\boldsymbol{x} (\partial V/\partial \boldsymbol{x}) e^{-i\boldsymbol{l}\cdot\boldsymbol{x}} \tag{1}$$

Let us consider a typical term in (6.1.12). We have

[1] One could also use an adiabatic switching process, as in quantum mechanics (see for example, Schweber, Bethe, and de Hoffmann, 1956).

$$\lambda^3 \int d\boldsymbol{l}\, d\boldsymbol{l}'\, V_l i\boldsymbol{l} \cdot (\partial/\partial \boldsymbol{v}) \pi\delta_+(\boldsymbol{l}\cdot\boldsymbol{v}) V_{|\boldsymbol{l}'-\boldsymbol{l}|} i(\boldsymbol{l}'-\boldsymbol{l}) \cdot (\partial/\partial\boldsymbol{v})$$
$$\times\, \pi\delta_+(\boldsymbol{l}'\cdot\boldsymbol{v}) V_{-l'} i\boldsymbol{l}' \cdot (\partial/\partial\boldsymbol{v})\rho_0(0)$$
$$= -[\lambda^3/(8\pi^3)^3] \int d\boldsymbol{l}\, d\boldsymbol{l}' \int d\boldsymbol{x}_1 (\partial V/\partial \boldsymbol{x}_1) e^{-i\boldsymbol{l}\cdot\boldsymbol{x}_1} (\partial/\partial\boldsymbol{v}) \int d\boldsymbol{x}_2\, G(\boldsymbol{x}_2) e^{i\boldsymbol{l}\cdot\boldsymbol{x}_2}$$
$$\times \int d\boldsymbol{x}_3 (\partial V/\partial\boldsymbol{x}_3) e^{-i(\boldsymbol{l}'-\boldsymbol{l})\cdot\boldsymbol{x}_3} (\partial/\partial\boldsymbol{v}) \int d\boldsymbol{x}_4\, G(\boldsymbol{x}_4) e^{i\boldsymbol{l}'\cdot\boldsymbol{x}_4}$$
$$\times \int d\boldsymbol{x}_5 (\partial V/\partial\boldsymbol{x}_5) e^{i\boldsymbol{l}'\cdot\boldsymbol{x}_5} (\partial/\partial\boldsymbol{v})\rho_0(0) \tag{2}$$

The integrations over $\boldsymbol{l}$ and $\boldsymbol{l}'$ can be performed immediately, and give simply $\delta(\boldsymbol{x}_1-\boldsymbol{x}_2-\boldsymbol{x}_3)\delta(\boldsymbol{x}_3-\boldsymbol{x}_4-\boldsymbol{x}_5)$. This then permits us to integrate over $\boldsymbol{x}_2$ and $\boldsymbol{x}_4$ as well and we are left with

$$-(\lambda^3/8\pi^3) \int d\boldsymbol{x}\, d\boldsymbol{x}'\, d\boldsymbol{x}'' (\partial V/\partial\boldsymbol{x}) \cdot (\partial/\partial\boldsymbol{v}) G(\boldsymbol{x}-\boldsymbol{x}') (\partial V/\partial\boldsymbol{x}') \cdot (\partial/\partial\boldsymbol{v})$$
$$\times\, G(\boldsymbol{x}'-\boldsymbol{x}'') (\partial V/\partial\boldsymbol{x}'') \cdot (\partial/\partial\boldsymbol{v})\rho_0(0) \tag{3}$$

The interpretation of such a term is very simple. It gives a contribution to the scattering, due to the interactions which occur in succession at three points $\boldsymbol{x}''$, $\boldsymbol{x}'$ and $\boldsymbol{x}$. Because these points are arbitrary, we have to perform an integration over $\boldsymbol{x}''$, $\boldsymbol{x}'$ and $\boldsymbol{x}$. Between the interactions, the motion is determined by the free particle propagator G (see Fig. 6.3.1). This formulation is very similar to Feynman's path-integral formulation in quantum mechanics (Feynman, 1949; see also for example Schweber, Bethe and de Hoffmann, 1956).

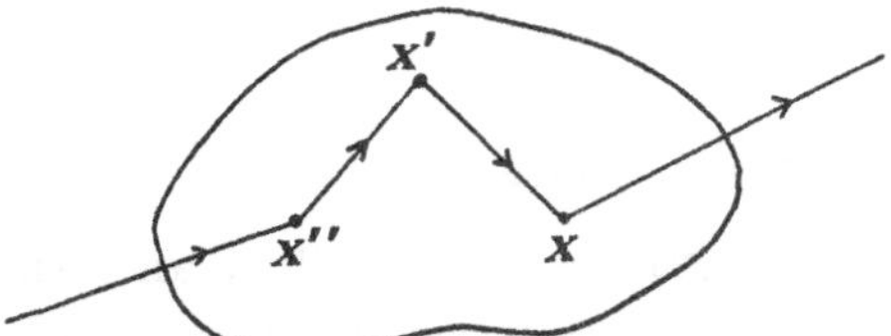

Fig. 6.3.1 Example of third-order scattering [see (6.3.3)].

The importance of incorporating the causality condition into the propagator G is obvious. Indeed the phase distribution at say point $\boldsymbol{x}'$ (see Fig. 6.3.1) can only depend on the value of the distribution function at points which correspond to earlier times on the trajectory.

It is interesting to compare the scattering process as described in the variables $\boldsymbol{l}$, $\boldsymbol{v}$ to that in the variables $\boldsymbol{x}$, $\boldsymbol{v}$.

Each interaction alters the value of the wave vector $\boldsymbol{l}$. On the contrary between the interactions we have the propagator $\delta_+ (\boldsymbol{l} \cdot \boldsymbol{v})$ to which corresponds a well defined value of $\boldsymbol{l}$. Now in the $\boldsymbol{x}$ description, each interaction occurs at a well-defined point. These points are connected by the propagator $G(\boldsymbol{x}'-\boldsymbol{x}'')$. This is a direct consequence of the already noted fact (see Chapter 1, §10) that δL is non-diagonal in the representation of the eigenfunctions of L_0, while it is diagonal is the coordinate representation (it is simply a function of the coordinates). The inverse is true for L_0, which is non-diagonal in the coordinate representation [see (1.10.3), and Chapter 7, §6) and diagonal in the $\boldsymbol{l}$ representation.

All other terms in (6.1.12) may be transformed in the same way. We obtain

$$\begin{aligned}\rho_0(t) = \rho_0(0) + (t/L^3) \Big[\lambda^2 \int d\boldsymbol{x}\, d\boldsymbol{x}' (\partial V/\partial \boldsymbol{x}) \cdot (\partial/\partial \boldsymbol{v}) \\ \times G(\boldsymbol{x}-\boldsymbol{x}')(\partial V/\partial \boldsymbol{x}') \cdot (\partial/\partial \boldsymbol{v}) + \lambda^3 \int d\boldsymbol{x}\, d\boldsymbol{x}'\, d\boldsymbol{x}'' \\ \times (\partial V/\partial \boldsymbol{x}) \cdot (\partial/\partial \boldsymbol{v}) G(\boldsymbol{x}-\boldsymbol{x}')(\partial V/\partial \boldsymbol{x}') \cdot (\partial/\partial \boldsymbol{v}) G(\boldsymbol{x}'-\boldsymbol{x}'') \\ \times (\partial V/\partial \boldsymbol{x}'') \cdot (\partial/\partial \boldsymbol{v}) + \cdots \Big] \rho_0(0)\end{aligned} \tag{4}$$

As we did for (6.1.12), we may write (6.3.4) in a more compact form. Indeed (6.3.4) is equivalent to the set of two equations

$$\rho_0(t) = \rho_0(0) + (t/L^3) \int d\boldsymbol{x} (\partial V/\partial \boldsymbol{x}) \cdot (\partial/\partial \boldsymbol{v}) f(\boldsymbol{x}, \boldsymbol{v}) \tag{5}$$

$$f(\boldsymbol{x}, \boldsymbol{v}) = \rho_0(0) + \lambda \int d\boldsymbol{x}' G(\boldsymbol{x}-\boldsymbol{x}', \boldsymbol{v})(\partial V/\partial \boldsymbol{x}') \cdot (\partial/\partial \boldsymbol{v}) f(\boldsymbol{x}', \boldsymbol{v}) \tag{6}$$

If we expand f in powers of λ by iteration in (6.3.6) and substitute into (6.3.5) we recover (6.3.4) (note that $\int d\boldsymbol{x} (\partial V/\partial \boldsymbol{x}) \times (\partial/\partial \boldsymbol{v}) \rho_0(\boldsymbol{v}, t=0) = 0$ because ρ_0 is independent of $\boldsymbol{x}$).

The function $f(\boldsymbol{x}, \boldsymbol{v})$ is simply the time-independent solution of the Liouville equation

$$\boldsymbol{v}(\partial f/\partial \boldsymbol{x}) - (\partial V/\partial \boldsymbol{x}) \cdot (\partial f/\partial \boldsymbol{v}) = 0 \qquad \text{or} \qquad Lf = 0 \tag{7}$$

satisfying the boundary condition that the incident flow is given by $\rho_0(\mathbf{v}, t = 0)$. Indeed the fundamental property (6.2.8) of the propagator G gives us

$$\begin{aligned} \mathbf{v} \cdot (\partial f/\partial \mathbf{x}) &= \mathbf{v} \cdot (\partial/\partial \mathbf{x}) \int d\mathbf{x}' G(\mathbf{x}-\mathbf{x}', \mathbf{v}) (\partial V/\partial \mathbf{x}') \cdot (\partial/\partial \mathbf{v}) f(\mathbf{x}', \mathbf{v}) \\ &= \int d\mathbf{x}' \delta(\mathbf{x}-\mathbf{x}') (\partial V/\partial \mathbf{x}') \cdot (\partial/\partial \mathbf{v}) f(\mathbf{x}', \mathbf{v}) \qquad (8) \\ &= (\partial V/\partial \mathbf{x}) \cdot (\partial/\partial \mathbf{v}) f(\mathbf{x}, \mathbf{v}) \end{aligned}$$

in agreement with (6.3.7).

The integral equation (6.3.6) takes an especially simple form if the expression (6.2.5) for the propagator is used. We then have

$$f(x, y, z, \mathbf{v}) = \rho_0(0) + (1/v) \int_{z'<z} dz' (\partial V/\partial \mathbf{x}')_{xyz'} (\partial/\partial \mathbf{v}) f(x, y, z', \mathbf{v}) \qquad (9)$$

The scattering formulation we have developed in this chapter provides an alternative approach to the evaluation of cross sections. While this approach is of no interest for simple situations for which there exist well-known classical methods (see for example Goldstein 1953; Chapman and Cowling, 1939; Hirschfelder, Curtiss, and Bird, 1954) it is of interest for more complicated situations (threebody cross section, influence of internal degrees of freedom, see for example Light, 1961). We shall not go into a description of such calculations but show that by eliminating f between (6.3.5) and (6.3.6) one obtains the familiar expression which appears in the Boltzmann transport equation (see *loc. cit.*) as the result of a two-body collision.

Let us make $z \to \infty$ in (6.3.9). We then obtain

$$\underset{\mathbf{r}\cdot\mathbf{v}\to\infty}{f(\mathbf{r}, \mathbf{v})} = \rho_0(\mathbf{v}, 0) + (1/v) \int dz' (\partial V/\partial \mathbf{r}')_{xyz'} (\partial/\partial \mathbf{v}) f(x, y, z', \mathbf{v}) \qquad (10)$$

We use this equation to write (6.3.5) in the form

$$\begin{aligned} \rho_0(\mathbf{v}, t) - \rho_0(\mathbf{v}, 0) &= (t/L^3) \int dx\,dy \int dz' (\partial V/\partial \mathbf{r}')_{xyz'} (\partial/\partial \mathbf{v}) f(x, y, z', \mathbf{v}) \\ &= (t/L^3) \int dx\,dy\, v[\underset{\mathbf{r}\cdot\mathbf{v}\to\infty}{f(\mathbf{r}, \mathbf{v})} - \rho_0(\mathbf{v}, 0)] \qquad (11) \end{aligned}$$

In (6.3.11) the only values of the stationary distribution f which appear are those calculated asymptotically very far from the scattering center.

We now use polar coordinates b, φ in the plane (x, y) perpendicular to the direction of the velocity

$$dx\,dy = b\,db\,d\varphi$$

b is called the "collision parameter." Moreover it is shown in textbooks on the kinetic theory of gases (see reference p. 124) that the relation of b to the angle of deflection θ is given by (see Fig. 6.3.2)

$$b\,db\,d\varphi = \sigma(\theta, \varphi)d\Omega \tag{12}$$

where σ is the cross section and $d\Omega$ an element of solid angle.

We may also express $f_{\mathbf{r}\cdot\mathbf{v}\to\infty}(\mathbf{r}, \mathbf{v})$ in (6.3.11) in terms of $\rho_0(\mathbf{v}, t = 0)$. In order to do so we shall not solve the integral equation (6.3.6) explicitly, but shall use the classical device of expressing the fact that f is conserved along the trajectory of the motion. Indeed f is equal to the density $\rho_0(\mathbf{v}', 0)$ before the interaction, for some value of $\mathbf{v}'$ of the velocity:

$$f(\mathbf{r}, \mathbf{v}) = \rho_0(\mathbf{v}', t = 0) \tag{13}$$

To obtain $\mathbf{v}'$ we have to trace back the trajectories of the particles which have velocity $\mathbf{v}$ at point $\mathbf{r}$ (see Fig. 6.3.2).

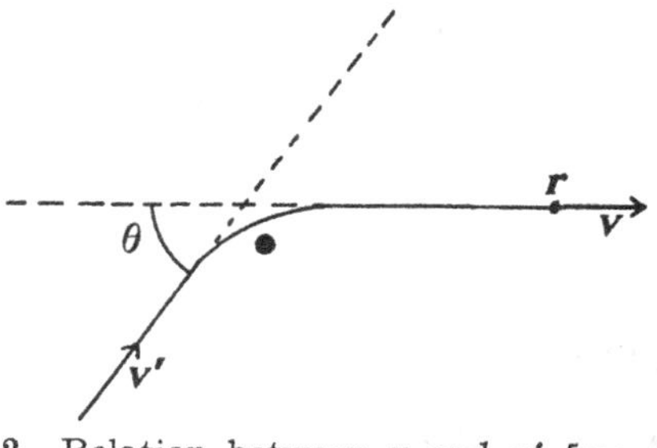

Fig. 6.3.2 Relation between $\mathbf{v}$ and $\mathbf{v}'$ [see (6.3.12)].

Using (6.3.12) and (6.3.13) we write (6.3.11) in the form

$$\rho_0(\mathbf{v}, t) - \rho_0(\mathbf{v}, 0) = t(1/L)^3 \int d\Omega\sigma v[\rho_0(\mathbf{v}', t = 0) - \rho_0(\mathbf{v}, t = 0)] \tag{14}$$

This is the familar expression from the kinetic theory of gases. Indeed we have on the right-hand side the difference between the initial distribution functions taken for the initial and the final values of the velocities.

We have shown in this way that equation (6.1.12) which is obtained from the Liouville equation, is equivalent to (6.3.14). Therefore the effect of all successive differential operators in (6.1.12) is equivalent to the difference operator in (6.3.14).

4. Equilibrium Distribution and Scattering Theory

Before we discuss the N-body problem, let us consider a simple application of the scattering formalism we have developed. Our interest in the method we shall outline is due to the fact that it is very similar to the one we shall develop in Chapter 12 in connection with the general problem of the approach to equilibrium.[1]

Let us consider an incident flow corresponding to a Maxwell distribution (we always take $m = 1$)

$$\rho_0(0) = (\tfrac{1}{2}\pi kT)^{3/2} e^{-v^2/2kT} \tag{1}$$

We shall show that the stationary phase distribution $f(\mathbf{x}, \mathbf{v})$ which satisfies the integral equation (6.3.6) is the equilibrium Boltzmann distribution

$$f(\mathbf{x}, \mathbf{v}) = (\tfrac{1}{2}\pi kT)^{3/2} e^{-(1/kT)[(v^2/2)+\lambda V(x)]} \tag{2}$$

Indeed we obtain by iteration of (6.3.6) the series

$$\begin{aligned}
f(\mathbf{x}, \mathbf{v}) &= (\tfrac{1}{2}\pi kT)^{3/2} e^{-v^2/2kT} \\
&\quad + \lambda(\tfrac{1}{2}\pi kT)^{3/2} \int d\mathbf{x}'\, G(\mathbf{x}-\mathbf{x}')(\partial V/\partial \mathbf{x}') \cdot (\partial/\partial \mathbf{v})\, e^{-v^2/2kT} \\
&\quad + \lambda^2(\tfrac{1}{2}\pi kT)^{3/2} \int d\mathbf{x}'\, d\mathbf{x}''\, G(\mathbf{x}-\mathbf{x}')(\partial V/\partial \mathbf{x}') \cdot (\partial/\partial \mathbf{v}) G(\mathbf{x}'-\mathbf{x}'') \\
&\quad \times (\partial V/\partial \mathbf{x}'') \cdot (\partial/\partial \mathbf{v})\, e^{-v^2/2kT} \\
&\quad + \lambda^3 \cdots
\end{aligned} \tag{3}$$

Let us consider the λ^2 term in detail. We have, using the basic

[1] This example is due to P. Résibois.

relation (6.2.8) twice,

$$
\begin{aligned}
-(1/kT)\int d\mathbf{x}'\,d\mathbf{x}''\,G(\mathbf{x}-\mathbf{x}')\,(\partial V/\partial \mathbf{x}')\cdot(\partial/\partial \mathbf{v}) \\
\times\, G(\mathbf{x}'-\mathbf{x}'')\,\mathbf{v}\,(\partial V/\partial \mathbf{x}'')\,e^{-v^2/2kT} \\
= (1/kT)\int d\mathbf{x}'\,d\mathbf{x}''\,G(\mathbf{x}-\mathbf{x}')\,(\partial V/\partial \mathbf{x}')\cdot(\partial/\partial \mathbf{v}) \\
\times\, V(x'')\,\mathbf{v}\cdot(\partial/\partial \mathbf{x}'')G(\mathbf{x}'-\mathbf{x}'')\,e^{-v^2/2kT} \\
= (1/kT)\int d\mathbf{x}'\,d\mathbf{x}''\,G(\mathbf{x}-\mathbf{x}')\,(\partial V/\partial \mathbf{x}')\cdot(\partial/\partial \mathbf{v}) \\
\times\, V(x'')\,\delta(\mathbf{x}'-\mathbf{x}'')\,e^{-v^2/2kT} \\
= -\tfrac{1}{2}(1/kT)^2\int d\mathbf{x}'\,G(\mathbf{x}-\mathbf{x}')\,(\partial V^2/\partial \mathbf{x}')\cdot \mathbf{v}\;e^{-v^2/2kT} \\
= \tfrac{1}{2}(1/kT)^2\,V^2(x)\,e^{-v^2/2kT} \qquad (4)
\end{aligned}
$$

This is precisely the second order term as obtained from (6.4.2) when expanded in powers of λ. The proof is identical for all other orders in λ.

This result permits us to consider the equilibrium spatial distribution from what may be called a "dynamical" point of view. It is the expression for a stationary state which is built up as a result of the deflections of the particles of the beam by the force center. In this way there appears a deep relationship between the Boltzmann factor and the Maxwell velocity distribution. With a generalization of this argument we shall show (in Chapter 12) that all results of equilibrium statistical mechanics which correspond to properties involving a finite number of degrees of freedom may be obtained as the long time effect of the application of the dynamical interaction to the equilibrium velocity distribution. We shall in this way obtain a complete "dynamical" derivation of equilibrium statistical mechanics.

5. *N*-Body Problem in the Limit of Low Concentrations

Let us now go back to the N-body problem. In contrast to the case of weakly coupled gases we no longer assume that the interactions are weak; we allow arbitrarily strong forces that produce large deflections during a single collision. On the other hand, we now suppose that the concentration C is very small.

This is the usual model used in a discussion of the transport properties of a dilute gas.

We shall again, as in Chapter 2, proceed by iteration. First we solve the Liouville equation for ρ_0 formally, assuming, as before, the initial condition (4.2.1), and then we retain a well-defined class of terms. The terms which we have to retain appear clearly if we first consider the Boltzmann equation for weakly coupled gases (4.3.4).

We may indeed repeat the argument we used for weakly coupled systems in Chapter 2, § 3. For low concentrations the relaxation time will be inversely proportional to C [see (4.7.10)]

$$t_r \sim 1/C$$

We shall therefore be interested in times such that

$$Ct \sim 1$$

and we have to keep all terms of the form $(Ct)^n$ irrespective of their order in λ. On the other hand contributions of the form $C^m(Ct)^n$, $m \geqq 1$ are concentration-dependent corrections to the relaxation time and may be neglected for dilute systems.

It should also be noted that the right-hand side of the master equation (4.2.3) ,from which we derived the Boltzmann equation (4.3.4), is proportional to $(\lambda^2/L^3) \sum_{jn}(*)_{jn} = N\lambda^2C$. The factor of N drops out in the transition from (4.2.3) to (4.4.8). In general all factors of N which appear in this paragraph drop out when one calculates intensive properties depending on a finite number of degrees of freedom. We shall discuss this problem in detail in Chapter 7, § 8.

To obtain the correct form of the Boltzmann equation we have to retain all powers of NCt, again irrespective of their order in λ but neglect terms of order $C^m(NCt)^n$, $m \geqq 1$. This can easily be done (Prigogine and Henin, 1958). We write down the formal solution for ρ_0, which will be given by exactly the same equation as (6.1.4); the only ditference is that now it refers to the N-body problem characterized by the Hamiltonian (4.1.1). Let us consider the terms of lowest order in a systematic way before we write the general expression.

Second-Order Terms. These will be exactly the same as for weakly coupled gases [see also (6.1.4), (6.1.8)]

$$-\lambda^2(2\pi/L)^6 \sum_{jn} \sum_{l} \int_0^t dt_1 \int_0^{t_1} dt_2 V_l i\boldsymbol{l} \cdot \boldsymbol{D}_{jn} e^{i\boldsymbol{l}\cdot\boldsymbol{g}_{jn}(t_1-t_2)} V_{-l} i\boldsymbol{l} \cdot \boldsymbol{D}_{jn} \rho_0$$
$$= -\lambda^2 t(2\pi/L)^3 \sum_{jn} \int d\boldsymbol{l} V_l i\boldsymbol{l} \cdot \boldsymbol{D}_{jn} \pi\delta_+(\boldsymbol{l}\cdot\boldsymbol{g}_{jn}) V_{-l} i\boldsymbol{l} \cdot \boldsymbol{D}_{jn} \rho_0 \quad (1)$$

From now on, we shall use the notation

$$\boldsymbol{D}_{jn} = (\partial/\partial \boldsymbol{v}_j) - (\partial/\partial \boldsymbol{v}_n) \quad (2)$$

Using the diagrams introduced in Chapter 4, §1, this would again correspond to Fig. 4.2.3 which we reproduce here.

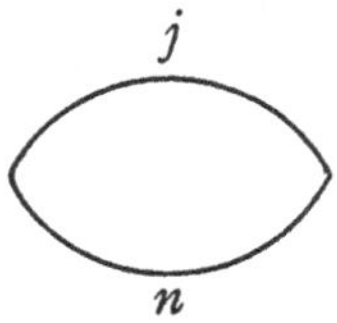

Fig. 6.5.1 Diagram corresponding to (6.5.1).

The contribution (6.5.1) is of the order

$$t(1/L^3) \sum_{jn} (*) \sim NtC \quad (3)$$

and, of course, has to be retained.

Third-Order Terms. Basically there are two ways in which the particles may interact in three steps to go from ρ_0 to ρ_0. We may have a two-particle interaction of exactly the same kind as in the two-body scattering problem [see (6.1.4) and (6.1.12)]

$$= -\lambda^3(2\pi/L)^9 \sum_{ll'} \sum_{jn} \int_0^t dt_1 \int_0^{t_1} dt_2 \int_0^{t_2} dt_3 V_l i\boldsymbol{l} \cdot \boldsymbol{D}_{jn} e^{i\boldsymbol{l}\cdot\boldsymbol{g}_{jn}(t_1-t_2)}$$
$$\times V_{|l'-l|} i(\boldsymbol{l}'-\boldsymbol{l}) \cdot \boldsymbol{D}_{jn} e^{i\boldsymbol{l}'\cdot\boldsymbol{g}_{jn}(t_2-t_3)} V_{-l'} i\boldsymbol{l}' \cdot \boldsymbol{D}_{jn} \rho_0$$
$$= -\lambda^3 t(2\pi/L)^3 \sum_{jn} \int d\boldsymbol{l} \int d\boldsymbol{l}' V_l i\boldsymbol{l} \cdot \boldsymbol{D}_{jn} \pi\delta_+(\boldsymbol{l}\cdot\boldsymbol{g}_{jn}) V_{|l'-l|} i(\boldsymbol{l}'-\boldsymbol{l}) \cdot \boldsymbol{D}_{jn}$$
$$\times \pi\delta_+(\boldsymbol{l}'\cdot\boldsymbol{g}_{jn}) V_{-l'} i\boldsymbol{l}' \cdot \boldsymbol{D}_{jn} \rho_0 \quad (4)$$

This is again a term of order NtC [see (6.5.3)]. The diagram representation of (6.5.4) is given in Fig. 6.5.2.

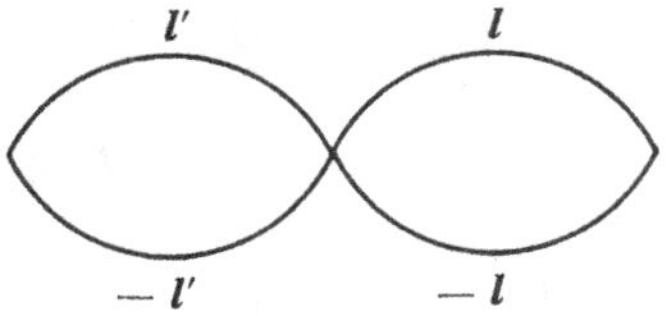

Fig. 6.5.2 Two-body diagram of order λ^3 corresponding to (6.5.4).

The diagram of Fig. 6.5.2 is a three vertex diagram ($\sim \lambda^3$) in which the two particles j and n start from the state $\boldsymbol{k}_j = \boldsymbol{k}_n = 0$, go over to $\boldsymbol{l}$ and $-\boldsymbol{l}$, then by a second interaction to $\boldsymbol{l}'$ and $-\boldsymbol{l}'$, and, at the third interaction, back to $\boldsymbol{k}_j = \boldsymbol{k}_n = 0$.

An alternative possibility for achieving a third-order transition from ρ_0 to ρ_0 is through a *three-body interaction*. Indeed let us consider the diagram of Fig. 6.5.3.

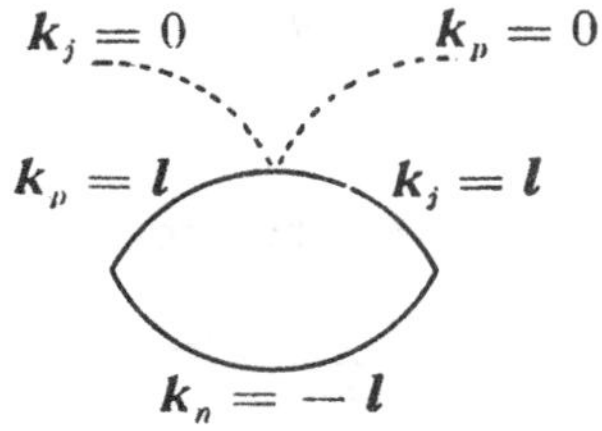

Fig. 6.5.3 Three-body diagram of order λ^3.

In Fig. 6.5.3. two particles n, j start from the state $\boldsymbol{k}_j = \boldsymbol{k}_n = 0$ and go to $\boldsymbol{l}$ and $-\boldsymbol{l}$. But now particle j interacts at the second vertex with a particle p for which $\boldsymbol{k}_p = 0$, to give $\boldsymbol{k}_j = 0$ and $\boldsymbol{k}_p = \boldsymbol{l}$. Such a process is in agreement with the law of wave-vector conservation (4.1.8). We shall come back to such interactions in much greater detail in Chapters 7 and 9. The particles p and n finally interact at the third vertex and go to the state $\boldsymbol{k}_p = \boldsymbol{k}_n = 0$. Therefore, both the initial and the final states indeed correspond to all $\{\boldsymbol{k}\} = 0$. In order to have a simpler graphical representation we shall use the diagram of Fig. 6.5.4 instead of Fig. 6.5.3.

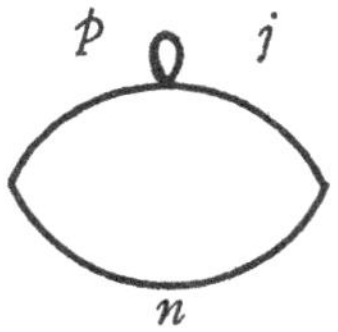

Fig. 6.5.4 Three-body diagram identical to that of Fig. 6.5.3.

Generally, we shall indicate only the labels of the particles in the figures (and not the value of the wave vectors). The small circle in Fig. 6.5.4 represents the wave vector "exchange" between j and p, which we just described.

It is now easy to estimate the order of such three-body contributions. Indeed we gain a summation over a supplementary particle p and lose a summation over $\boldsymbol{l}$. (Instead of the two summations over $\boldsymbol{l}$, $\boldsymbol{l}'$ in Fig. 6.5.2, we now have only one as is necessary for wave-vector conservation.)

On the other hand, our usual arguments show (see Chapter 2, §§ 5, 6) that the time integration will bring in a factor of t as in (6.5.4) (see also Chapter 8). Therefore, we now have a contribution of order

$$t(\lambda^3/L^9)\sum_{l}\sum_{jnp} \sim (t\lambda^3/L^6)\sum_{jnp}\int d\boldsymbol{l} \sim C(NCt) \tag{5}$$

instead of NCt.

The contribution (6.5.5) has, therefore, to be neglected. This is exactly the result we would have expected on intuitive grounds. *In the low density limit we must retain only two-body diagrams.*

Fourth-Order Terms. As in our discussion of the two-body problem in § 1, we have to distinguish between connected and disconnected diagrams. Using a terminology we have already introduced (Chapter 2, § 6), we could also say that we have to distinguish between diagrams which are formed by one or two diagonal fragments. Let us first consider connected diagrams containing a single diagonal fragment. We may again have a two-particle interaction of the same type as in (6.1.9). The corresponding diagram is represented in Fig. 6.5.5.

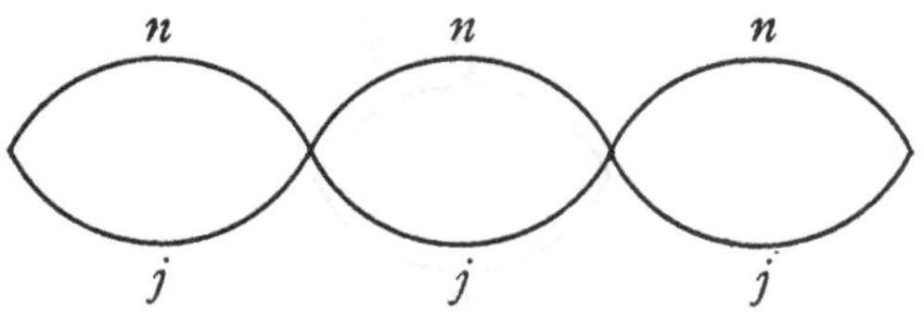

Fig. 6.5.5 Two-body diagram of order λ^4.

We also have three- or even four-body diagrams of the type represented in Figs. 6.5.6 and 6.5.7.

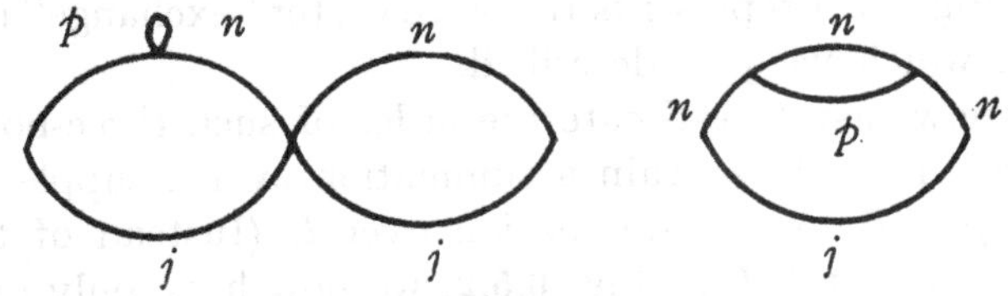

Fig. 6.5.6 Three-body diagrams of order λ^4.

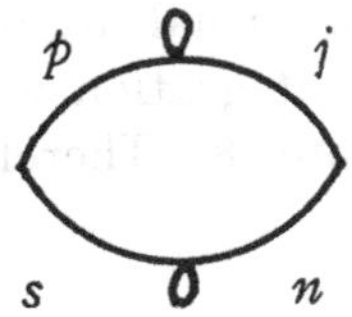

Fig. 6.5.7 Four-body diagram of order λ^4.

For the same reasons as for the diagrams for the third-order term, these diagrams must be neglected in the low density limit. Let us next consider the contribution of two diagonal fragments. To (6.1.10) now corresponds

$$\lambda^4(2\pi/L)^{12}\sum_{ll'}\sum_{jn}\sum_{pr}\int_0^t dt_1\cdots\int_0^{t_3} dt_4[V_l\, i\boldsymbol{l}\cdot\boldsymbol{D}_{jn}\, e^{i\boldsymbol{l}\cdot\boldsymbol{g}_{jn}(t_1-t_2)}\, V_{-l}\, i\boldsymbol{l}\cdot\boldsymbol{D}_{jn}] \times [V_{l''}\, i\boldsymbol{l}''\cdot\boldsymbol{D}_{pr}\, e^{i\boldsymbol{l}\cdot\boldsymbol{g}_{pr}(t_3-t_4)}\, V_{-l''}\, i\boldsymbol{l}''\cdot\boldsymbol{D}_{pr}]\,\rho_0 \qquad (6)$$

The time integrations can be performed [see (6.1.11)], and we obtain

$$(t^2/2)\lambda^4(2\pi/L)^6\left[\sum_{jn}\int d\boldsymbol{l}\, V_l\, i\boldsymbol{l}\cdot\boldsymbol{D}_{jn}\,\pi\delta_+(\boldsymbol{l}\cdot\boldsymbol{g}_{jn})V_{-l}\, i\boldsymbol{l}\cdot\boldsymbol{D}_{jn}\right] \times\left[\sum_{pr}\int d\boldsymbol{l}''\, V_{l''}\, i\boldsymbol{l}''\cdot\boldsymbol{D}_{pr}\,\pi\delta_+(\boldsymbol{l}\cdot\boldsymbol{g}_{pr})V_{-l''}\, i\boldsymbol{l}''\cdot\boldsymbol{D}_{pr}\right]\rho_0 \qquad (7)$$

This corresponds to two *independent* two-body interactions and is represented by the diagrams of Fig. 6.5.8.

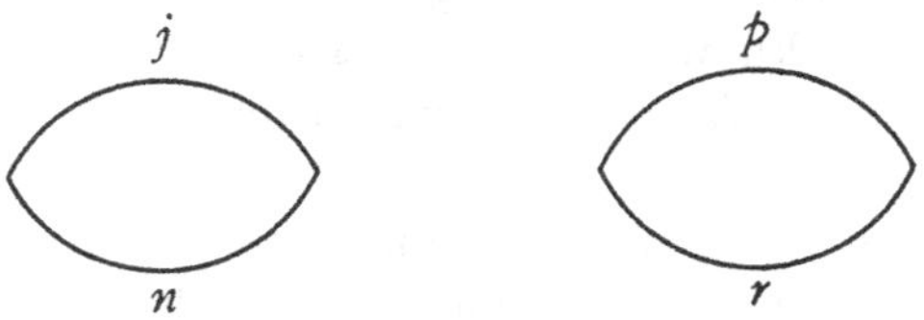

Fig. 6.5.8 Two two-body interactions corresponding to (6.5.7).

It is clear that this term must be retained (it was retained for weakly coupled gases.) It is indeed of order

$$(tNC)^2 \tag{8}$$

Finally we have to consider disconnected "overlapping" diagrams as represented in Fig. 6.5.9 (see also Fig. 2.6.2); such diagrams are formed by a single diagonal fragment.

Fig. 6.5.9 Overlapping two-body diagrams.

But exactly as in Chapter 2, § 6 we then lose a factor of t and instead of (6.5.8), we are left with

$$NC(tNC) \tag{9}$$

The factors of N drop out again in the calculation of properties depending on a finite number of degrees of freedom (see Chapter 7, § 8), and we have to neglect (6.5.9) in the low density limit.

6. Master Equation in the Limit of Low Concentrations

The detailed discussion we presented in § 5 can be summarized by the following rule:

In the limit of low concentrations the only diagrams we have to retain are two-body diagrams (Figs. 6.5.1, 6.5.2, and 6.5.5) *and disconnected products of such diagrams* (Fig. 6.5.8).

Moreover, we know that a *succession of n independent two-body diagrams will give us the time factor $t^n/n!$* (see Chapter 2, §§ 6, 7).

With this in mind we can immediately write the expression for $\rho_{\{0\}}$ at time t. In terms of diagrams we shall have [see (2.7.4)]

$$\rho_{\{0\}}(t) = \rho_{\{0\}}(0)$$

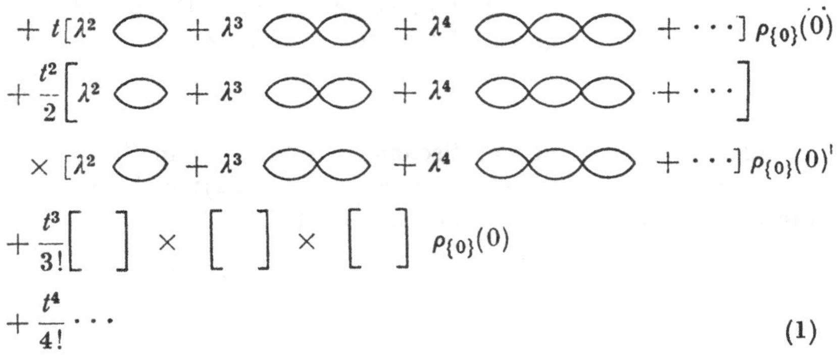

$$+ t[\lambda^2 \;\bigcirc\; + \lambda^3 \;\infty\; + \lambda^4 \;\infty\infty\; + \cdots] \rho_{\{0\}}(0)$$

$$+ \frac{t^2}{2}\Big[\lambda^2 \;\bigcirc\; + \lambda^3 \;\infty\; + \lambda^4 \;\infty\infty\; + \cdots \Big]$$

$$\times [\lambda^2 \;\bigcirc\; + \lambda^3 \;\infty\; + \lambda^4 \;\infty\infty\; + \cdots] \rho_{\{0\}}(0)'$$

$$+ \frac{t^3}{3!}\Big[\quad \Big] \times \Big[\quad \Big] \times \Big[\quad \Big] \rho_{\{0\}}(0)$$

$$+ \frac{t^4}{4!} \cdots \tag{1}$$

Let us introduce the operator

$$S = \lambda^2 \;\bigcirc\; + \lambda^3 \;\infty\; + \lambda^4 \;\infty\infty\; + \cdots \tag{2}$$

which we may call the scattering operator. We then have

$$\rho_{\{0\}}(t) = e^{tS} \rho_{\{0\}}(0) \tag{3}$$

and by differentiation with respect to time

$$\partial \rho_{\{0\}}/\partial t = S \rho_{\{0\}}(t) \tag{4}$$

This is the master equation in the low density limit. Using (6.5.1), (6.5.4) we have

$$S = \sum_{jn} S_{jn} \tag{5}$$

with

$$S_{jn} = -(2\pi/L)^3 \Big[\lambda^2 \int d\boldsymbol{l} V_l i\boldsymbol{l} \cdot \boldsymbol{D}_{jn} \pi\delta_+(\boldsymbol{l} \cdot \boldsymbol{g}_{jn}) V_{-l} i\boldsymbol{l} \cdot \boldsymbol{D}_{jn}$$
$$+ \lambda^3 \int d\boldsymbol{l} \int d\boldsymbol{l}' V_l i\boldsymbol{l} \cdot \boldsymbol{D}_{jn} \pi\delta_+(\boldsymbol{l} \cdot \boldsymbol{g}_{jn}) V_{|l'-l|} i(\boldsymbol{l}' - \boldsymbol{l}) \cdot \boldsymbol{D}_{jn}$$
$$\times \pi\delta_+(\boldsymbol{l}' \cdot \boldsymbol{g}_{jn}) V_{-l'} i\boldsymbol{l}' \cdot \boldsymbol{D}_{jn} + \lambda^4 \cdots \Big] \tag{6}$$

Comparing this to (6.1.12), we see that S_{jn} represents exactly the scattering operator corresponding to the two-body scattering between j and n.

Therefore, using the equivalence between (6.1.12) and (6.3.14) established in § 3, we may write (6.6.6) in the form

$$S_{jn} \rho_{\{0\}} = (1/L^3) \int d\Omega_{jn} \sigma(\Omega_{jn}, |g_{jn}|) |g_{jn}|$$
$$\times [\rho_{\{0\}}(\boldsymbol{v}_1 \cdots \boldsymbol{v}'_j \cdots \boldsymbol{v}'_n \cdots \boldsymbol{v}_l \cdots) - \rho_{\{0\}}(\boldsymbol{v}_1 \cdots \boldsymbol{v}_j \cdots \boldsymbol{v}_n \cdots \boldsymbol{v}_l)] \tag{7}$$

and the master equation (6.6.4) takes the explicit form

$$\partial \rho_{\{0\}}/\partial t = (1/L)^3 \sum_{jn} \int d\Omega_{jn} \sigma(\Omega_{jn}, |g_{jn}|) |g_{jn}| [\rho_{\{0\}}(\boldsymbol{v}_1 \cdots \boldsymbol{v}'_j \cdots \boldsymbol{v}'_n \cdots \boldsymbol{v}_l)$$
$$- \rho_{\{0\}}(\boldsymbol{v}_1 \cdots \boldsymbol{v}_j \cdots \boldsymbol{v}_n \cdots \boldsymbol{v}_l)] \tag{8}$$

The change in the velocity distribution function at low densities occurs as the result of all possible binary collisions. By a procedure identical to that in Chapter 4, § 3 we may, using the factorization condition (4.3.3), deduce from (6.6.8) the classical Boltzmann equation

$$\partial \varphi(\boldsymbol{v})/\partial t = C \int d\boldsymbol{v}_1 \int d\Omega \sigma g [\varphi(\boldsymbol{v}') \varphi(\boldsymbol{v}'_1) - \varphi(\boldsymbol{v}) \varphi(\boldsymbol{v}_1)] \tag{9}$$

This equation applies to a homogeneous gas. The case of an inhomogeneous gas will be studied in Chapter 10.

Our derivation of the master equation or of the Boltzmann equation provides us with a non-trivial application of the diagram technique. The important point is that only a single class of diagrams, corresponding to repeated interactions of the same pair of particles, contribute to the order of approximation we have considered.

The physical picture of the approach to equilibrium is the same both for weakly coupled systems and for strong forces in the limit of low concentrations. Perhaps the main point is that the duration of an interaction t_{int} is in both cases small with respect to the time interval between interactions t_r (which is of the order of the relaxation time). Therefore, each perturbation may be described as a "complete" collision giving a persistent effect proportional to time. The importance of this point has often been stressed (see, for example, Kirkwood and Ross, **1956**; Green, **1956**; Brout, **1956**). The whole evolution of ρ_0 may then be considered as a succession of complete, independent scattering processes.

When we go to higher concentrations the separation between the two time scales, t_{int} and t_r, becomes less absolute. As we shall see in Chapter 11 this gives to the evolution equations corresponding to an arbitrary concentration (or an arbitrary interaction strength λ) a very different form.

Let us finally notice that the scattering theory shows that any transport equation of the Boltzmann type can hold only in the limit

$$\begin{gathered} L^3 \to \infty, \qquad N \to \infty \\ N/L^3 = \text{finite} \end{gathered} \tag{10}$$

Indeed in order to have "t" factors at all we have to make $L^3 \to \infty$ (see § 1). But in order to keep a non-vanishing collision probability per molecule, we also have to make $N \to \infty$ in agreement with (6.6.10).

This limiting procedure will be analyzed in detail in Chapter 7. As we shall show there it plays an essential role in the general theory.

7. Scattering Theory and the N-Body Problem

The main difference between the scattering theory discussed in §§ 1–4 of this chapter and the N-body problem we discussed in §§ 5–6 is that in the first problem, a single collision process is considered; for the derivation of the transport equation an arbitrary number of collisions has to be considered. It is the *summation* over all those collisions in (6.6.1) which gives rise to the *exponential* character of the transformation (6.6.3).

There exist some "simplified" derivations of the transport equation which tend to obscure this issue. One first considers times such that

$$Ct \ll 1 \tag{1}$$

instead of $Ct \approx 1$. Then the series (6.6.1) may be limited to

$$\rho_{\{0\}}(t) = \rho_{\{0\}}(0) + tS\rho_{\{0\}}(0) + O[(Ct)^2] \tag{2}$$

and by differentiation with respect to time we have

$$(\partial \rho_{\{0\}}(t)/\partial t)_{t=0} = S\rho_{\{0\}}(t)_{t=0} \tag{3}$$

It is then argued that the time $t = 0$ can play no special role and it is therefore *assumed* that formula (6.7.3) is valid for all t. We then recover the master equation (6.6.4).

This derivation makes use of the assumption that after each time interval (6.7.1) conditions (2.3.3) or (4.2.1) which we assume at time $t = 0$, remain valid.

We have seen in Chapter 2, § 3 that these conditions express an assumption of random phases. This assumption also plays an important role in the derivation of the quantum mechanical transport equation (Pauli, 1928; Tolman, 1938; Van Hove, 1955; see also Chapter 13). We may therefore say that the simplified derivation makes repeated use of the random phase assumption. On the contrary in our derivation it is only assumed once at time $t = 0$. This is of course much more satisfactory. We want to note that the first derivation of a transport equation without the repeated use of the random phase approximation was achieved in the quantum mechanical case by Van Hove (1955).

It is true that the simplified derivation leads to the Boltzmann equation. But we believe that the value of a theory of non-equilibrium processes cannot be judged by the "derivation" of the Boltzmann equation. The main goal is to go beyond the classical results and to construct a general theory valid under much wider conditions. This will be the subject of the next chapters.

CHAPTER 7

Distribution Functions — The Diagram Representation

1. Distribution Functions

In the preceding chapters we have been mainly concerned with especially simple situations: weakly coupled systems (Chapters 2–5) and dilute gases (Chapter 6). We shall now deal with the general theory of non-equilibrium statistical mechanics. The main object of this chapter is to develop in a more detailed way the diagram representation we introduced before (Prigogine and Balescu, 1959). As in equilibrium problems it is useful to introduce reduced distribution functions f_s referring to s molecules. (For more details about distribution functions see especially de Boer, 1948–1949; Massignon, 1957.) We shall first present a summary of their properties and then discuss their relation to the Fourier expansion of the phase density in eigenfunctions of the unperturbed Liouville operator.

From the normalized probability distribution in phase space ρ we may deduce the probability $\rho_s(\boldsymbol{x}_1 \cdots \boldsymbol{x}_s, \boldsymbol{p}_1 \cdots \boldsymbol{p}_s)$ of finding, at a given time t, a set of s specific molecules $1, 2, \ldots, s$ with momenta $\boldsymbol{p}_1, \ldots, \boldsymbol{p}_s$ and coordinates $\boldsymbol{x}_1, \ldots, \boldsymbol{x}_s$

$$\rho_s(\boldsymbol{x}_1 \cdots \boldsymbol{x}_s, \boldsymbol{p}_1 \cdots \boldsymbol{p}_s, t) = \int (d\boldsymbol{x})^{N-s} (d\boldsymbol{p})^{N-s} \rho(\boldsymbol{x}_1 \cdots \boldsymbol{p}_N, t) \quad (1)$$

The relation between ρ_s and ρ_{s+1} is clearly

$$\rho_s = \int d\boldsymbol{x}_{s+1} d\boldsymbol{p}_{s+1} \rho_{s+1} \quad (2)$$

Distribution functions that refer to specified molecules are called *specific* distribution functions. We shall in general be more interested in the probability of finding s arbitrary molecules at positions $\boldsymbol{x}_1, \cdots, \boldsymbol{x}_s$ and momenta $\boldsymbol{p}_1, \ldots, \boldsymbol{p}_s$.

This probability, which we shall call f_s is found by multiplying ρ_s by the factor $N!/(N-s)!$ This is the number of possible ways in which a sequence of s molecules can be chosen out of N. Therefore

$$f_s(\boldsymbol{x}_1 \cdots \boldsymbol{x}_s, \boldsymbol{p}_1 \cdots \boldsymbol{p}_s) = N!/(N-s)!\,\rho_s(\boldsymbol{x}_1 \cdots \boldsymbol{x}_s, \boldsymbol{p}_1 \cdots \boldsymbol{p}_s) \qquad (3)$$

In consequence of the relation (7.1.2) between ρ_s and ρ_{s+1}, we now have

$$f_s = 1/(N-s) \int d\boldsymbol{x}_{s+1} d\boldsymbol{p}_{s+1} f_{s+1} \qquad (4)$$

and by iteration

$$\begin{aligned} f_s &= 1/(N-s)! \int (d\boldsymbol{x})^{N-s} (d\boldsymbol{p})^{N-s} f_N \\ &= N!/(N-s)! \int (d\boldsymbol{x})^{N-s} (d\boldsymbol{p})^{N-s} \rho \end{aligned} \qquad (5)$$

where ρ is the normalized phase distribution function (from 7.1.3 it follows that $f_N = N!\ \rho$). The reduced distribution functions f_s are sometimes called *generic* distribution functions to distinguish them from the specific distribution functions. As we shall use the generic distribution functions exclusively, we shall henceforth simply call the f_s the distribution functions.

From (7.1.5) we see that the normalization condition for the distribution function of s particles is

$$\int (d\boldsymbol{x})^s (d\boldsymbol{p})^s f_s = N!/(N-s)! \qquad (6)$$

Besides the distribution functions f_s in phase space we shall also use distribution functions φ_s in momentum space, n_s in coordinate space, and also mixed distribution functions, $f_{s,r}$, of s position vectors and r momenta:

$$\varphi_s = (N-s)!/N! \int (d\boldsymbol{x})^s f_s \qquad (7)$$

$$n_s = \int (d\boldsymbol{p})^s f_s \qquad (8)$$

$$f_{s,r} = 1/(N-s)! \int (d\boldsymbol{x})^{N-s} (d\boldsymbol{p})^{N-r} f_N \qquad (9)$$

The normalization in (7.1.7)–(7.1.8) is such that

$$\int (dp)^s \varphi_s = 1, \qquad \int (dx)^s n_s = N!/(N-s)!$$
$$\int (dx)^s (dp)^s f_{s,r} = N!/(N-s)! \tag{10}$$

Then n_1 is simply the number density, n_2/n_1^2 the radial distribution function (or pair correlation function), and φ_1 is the already familiar velocity distribution function. All averages over physically important functions of momenta and coordinates may be expressed in terms of the distribution function of a *small* number of particles (usually one or two). For example, using the definitions we have given, we obtain for the average energy [see (4.1.1)]

$$\bar{E} = \int (dx)^N (dp)^N \{\sum_j p_j^2/2m + \sum_{j<n} V_{jn}(|x_j - x_n|)\} \rho$$
$$= N[\int dp\,(p^2/2m)\,\varphi_1(p) + (L^3/N) \int dr\, n_2(r)\, V(|r|)]$$

where we have supposed that $n_2(x_1, x_2)$ depends only on $r = x_1 - x_2$.

One may also express more complicated quantities like the pressure tensor in terms of the distribution functions (see for example Massignon, 1957), but we shall not use these expressions here

It is easy to deduce the equations of change for the reduced distribution functions f_s from the Liouville equation. In this way one obtains a hierarchy of equations relating the time change of f_s to f_{s+1}. These equations were first derived by Yvon (1935, 1937) and rediscovered and widely used by Born and Green (1946–1947), Kirkwood (1946–1947), and Bogolioubov (1946). The first of these equations is (neglecting a term in $1/N$)

$$\partial f_1(x_\alpha, p_\alpha, t)/\partial t + p_\alpha/m\,(\partial f_1/\partial x_\alpha)$$
$$= \lambda \int dx_\beta\, dp_\beta\, (\partial V_{\alpha\beta}/\partial x_\alpha) \cdot (\partial/\partial p_\alpha) f_2(x_\alpha, x_\beta, p_\alpha, p_\beta, t) \tag{11}$$

The whole set of these equations (we shall call them the "Y–B–G–B–K" equations) is of course equivalent to the Liouville equation.

The definitions of the reduced distribution functions given above are valid for all values of the number N of particles, and of the volume L^3. However, as stated before (see Chapter 6, §6) we are mainly interested in the limit

$$\begin{matrix} N \to \infty \\ L^3 \to \infty \end{matrix} \qquad C = N/L^3 = \text{finite constant} \tag{12}$$

Now in such a limit the reduced distributions do not necessarily approach finite values, which may depend on C but are otherwise independent of N or L^3. To understand this point let us consider a system defined by the following distribution,

$$\begin{aligned} n_1(\boldsymbol{x}) &= n = \text{constant} \quad \boldsymbol{x} \ \text{inside} \ \omega \\ &= 0 \qquad\qquad\qquad\ \ \boldsymbol{x} \ \text{outside} \ \omega \end{aligned} \tag{13}$$

where ω is a fixed volume, *independent of* L^3. By the normalization condition (7.1.10) one has to take

$$n = N/\omega \tag{14}$$

Therefore the density n_1 inside ω is proportional to N. Moreover in the limit (7.1.12) the density goes to infinity in ω and remains zero elsewhere. If on the other hand one defines ω as a given fraction of the total volume L^3, say

$$\omega = aL^3 \tag{15}$$

then n_1 becomes independent of N and is finite in the whole system.

It should be noted that the distribution functions f_s have the same limiting behavior as n_s because the integrations over momenta do not introduce any new volume factors [see (7.1.8)].

If the reduced distribution functions do not tend to well-defined limit (7.1.12), then no intensive variables (pressure, energy density, ...) could be defined in the limit of a large system. We want to exclude such situations explicitly and shall consider only situations for which the reduced distribution functions have well-defined limits in the sense of (7.1.12). We shall show in § 3 that this has important consequences for the Fourier expansion of $\rho(\boldsymbol{x}_1 \cdots \boldsymbol{x}_N, \boldsymbol{p}_1 \cdots \boldsymbol{p}_N, t)$ in terms of the eigenfunctions of the unperturbed Liouville operator.

One may directly understand why it is necessary to exclude situations in which the density would go to infinity. Particles belonging to such a region would have a greater and greater number of first neighbours. The relaxation time of such a system would go to zero in the limit (7.1.12) and we can, therefore, no longer expect a Boltzmann kind of equation to exist at all.

The assumption that reduced distribution functions exist plays an essential role in our theory and we shall analyze it more in detail in § 3. This assumption has to be made at the initial time $t = 0$ only. It can indeed be easily shown (see §8) that it will automatically be preserved by the equations of motion.

2. Fourier Expansion and Distribution Functions

Let us study the Fourier expansion (1.5.20) in detail. We shall use the following notation: $\rho(j_k, l_{k'}, m_{k''})$ will indicate a Fourier coefficient such that in the ensemble, to particle j "corresponds" the wave vector $\boldsymbol{k}$, to l the wave vector $\boldsymbol{k}'$, and to m the wave vector $\boldsymbol{k}''$.[1] For convenience the whole time dependence is included in the Fourier coefficients. We have

$$\begin{aligned}\rho = L^{-3N}[\rho_0 &+ \Omega^{-1}\sum_k{}' \sum_j \rho(j_k)e^{i\boldsymbol{k}\cdot\boldsymbol{x}_j} + \Omega^{-1}\sum_k{}'\sum_{jl} \\ &\times \rho(j_k, l_{-k})\, e^{i\boldsymbol{k}\cdot(\boldsymbol{x}_j-\boldsymbol{x}_l)} + \Omega^{-2}\sum_{kk'}{}' \rho(j_k, l_{k'})\, e^{i(\boldsymbol{k}\cdot\boldsymbol{x}_j+\boldsymbol{k}'\cdot\boldsymbol{x}_l)} \\ &+ \Omega^{-2}\sum_{kk'}{}'\sum_{jlr} \rho(j_k, l_{k'}, r_{-k-k'})\, e^{i[\boldsymbol{k}\cdot\boldsymbol{x}_j+\boldsymbol{k}'\cdot\boldsymbol{x}_l-(\boldsymbol{k}+\boldsymbol{k}')\cdot\boldsymbol{x}_r]} \\ &+ \cdots] \end{aligned} \tag{1}$$

with

$$\Omega = L^3/(2\pi)^3$$

The prime in the summation over wave vectors means that the value $\boldsymbol{k} = 0$ always has to be omitted.

Similarly in a term like

$$\Omega^{-2}\sum_{kk'}{}'\sum_{jl} \rho(j_k, l_{k'})\, e^{i(\boldsymbol{k}\cdot\boldsymbol{x}_j+\boldsymbol{k}'\cdot\boldsymbol{x}_l)}$$

[1] The momentum dependence will generally not be indicated by special notation. According to the context for example $\rho(j_k)$ may mean $\rho(j_k, \boldsymbol{p}_1, \ldots, \boldsymbol{p}_N)$ as in (7.2.1) or $\rho(j_k, \boldsymbol{p}_j)$, in which an integration over all momenta except $\boldsymbol{p}_j$ has been performed.

the contribution $\boldsymbol{k}' = -\boldsymbol{k}$ is excluded because it has already been accounted for in a preceding term. In comparison with (1.5.20) the only modification we have introduced is that we have now inserted a volume factor Ω^{-r} where r is the number of independent, non-vanishing wave vectors.[1]

In the limit $\Omega \to \infty$ the sums over the wave vectors in (7.2.1) may be replaced by integrals [see (4.2.4)].[2]

We then have

$$\begin{aligned}\rho = L^{-3N}[\rho_0 &+ \sum_j \int d\boldsymbol{k}\, \rho(j_{\boldsymbol{k}})\, e^{i\boldsymbol{k}\cdot\boldsymbol{x}_j} \\ &+ \sum_{jl} \int d\boldsymbol{k}\, \rho(j_{\boldsymbol{k}}, l_{-\boldsymbol{k}})\, e^{i\boldsymbol{k}\cdot(\boldsymbol{x}_j - \boldsymbol{x}_l)} \\ &+ \sum_{jl} \int\int d\boldsymbol{k}\, d\boldsymbol{k}'\, \rho(j_{\boldsymbol{k}}, l_{\boldsymbol{k}'})\, e^{i(\boldsymbol{k}\cdot\boldsymbol{x}_j + \boldsymbol{k}'\cdot\boldsymbol{x}_l)} \\ &+ \cdots] \end{aligned} \tag{2}$$

We shall now assume that in the limit (7.1.12) the Fourier coefficients ρ_0, $\rho(j_{\boldsymbol{k}})$, $\rho(j_{\boldsymbol{k}}, l_{-\boldsymbol{k}})$, ... do not depend explicitly on N or Ω (they may depend, of course, on the ratio N/Ω). We shall show in this chapter that this assumption ensures the existence and finiteness of all reduced distribution functions.

Let us consider a simple example. Using (7.1.5) and (7.2.2), we have (for simplicity of notation we shall not explicitly write the integration over momenta, which is irrelevant here)

$$\begin{aligned} f_1(\boldsymbol{x}_j) &= N!/(N-1)! \int (d\boldsymbol{x})^{N-1} \rho \\ &= N(L^3)^{-N}[(L^3)^{N-1}\rho_0 + (L^3)^{N-1} \int d\boldsymbol{k}\, \rho(j_{\boldsymbol{k}})\, e^{i\boldsymbol{k}\boldsymbol{x}_j}] \\ &= C[\rho_0 + \int d\boldsymbol{k}\, \rho(j_{\boldsymbol{k}})\, e^{i\boldsymbol{k}\boldsymbol{x}_j}] \end{aligned} \tag{3}$$

We see indeed that as long as ρ_0 and $\rho(j_{\boldsymbol{k}})$ do not depend explicitly on N or L alone (but only on their ratio) the same is true for

[1] This number is defined as the total number of non-vanishing wave vectors minus the number of relations of the type $\boldsymbol{k}_\alpha + \boldsymbol{k}_\beta + \cdots = 0$ which are imposed upon them.

[2] Strictly speaking all the integrals contain principal parts to take account of the restrictions in the summations over the $\boldsymbol{k}$'s.

$f_1(\boldsymbol{x}_j)$. Moreover we have

$$\rho(j_k) = 1/C \int d\boldsymbol{x}_j [f_1(\boldsymbol{x}_j) - \rho_0] e^{-i\boldsymbol{k}\boldsymbol{x}_j} \tag{4}$$

Therefore $\rho(j_k)$ is the Fourier transform of the deviation of the one-particle distribution function f_1 from its average value.

It is useful to distinguish between Fourier coefficients for which the sum of the wave vectors $\boldsymbol{k}+\boldsymbol{k}'+\cdots$ vanishes and Fourier coefficients for which $\boldsymbol{k}+\boldsymbol{k}'+\cdots \neq 0$. For *homogeneous* situations such that

$$\begin{aligned} &\rho(\boldsymbol{x}_1+\boldsymbol{a}, \ldots, \boldsymbol{x}_N+\boldsymbol{a}, \boldsymbol{p}_1, \boldsymbol{p}_2, \ldots, \boldsymbol{p}_N, t) \\ &\quad = \rho(\boldsymbol{x}_1 \cdots \boldsymbol{x}_N, \boldsymbol{p}_1, \boldsymbol{p}_2 \cdots \boldsymbol{p}_N, t) \end{aligned} \tag{5}$$

the expansion (7.2.1) shows us that

$$\rho(j_k, l_{k'}, \ldots) = 0 \qquad \text{for} \qquad \boldsymbol{k}+\boldsymbol{k}'+\cdots \neq 0 \tag{6}$$

Because of the fundamental law of the conservation of wave vectors (4.1.8), this condition will be preserved in time. Moreover (4.1.8) shows that the Fourier coefficients $\rho(j_k, l_{k'}, \ldots)$ with $\boldsymbol{k}+\boldsymbol{k}'+\cdots = 0$ and those with $\boldsymbol{k}+\boldsymbol{k}'+\cdots \neq 0$ transform in time in completely independent ways. Therefore even in an inhomogeneous system the rate of change with time of each one of these two types of Fourier coefficients is completely independent of that of the other. In § 5 we shall analyze the physical meaning of the Fourier coefficients which appear in the expansion of ρ for homogeneous systems. Inhomogeneous systems will be studied in Chapter 10.

Perhaps the most remarkable feature of (7.2.1) or (7.2.2) is that the velocity distribution ρ_0 is completely separated from the space dependent part represented by the other Fourier coefficients. It is this feature which will permit us to treat ρ_0 as we would the ground state in quantum mechanics (or the "vacuum" in quantum field theory) and the space dependent part as "excitations." As we have seen in example (7.2.3), (7.2.4) this separation expresses the existence and finiteness of reduced distribution functions in the limit (7.1.12).

It is therefore important to compare (7.2.2) to the standard expression for the Fourier integral of an arbitrary function.

3. Singularities in the Fourier Expansion

Let us compare (7.2.2) to the general expression of the Fourier integral of a function of the N variables $\mathbf{x}_1, \cdots, \mathbf{x}_N$

$$\rho(\mathbf{x}_1 \cdots \mathbf{x}_N, \mathbf{p}_1 \cdots \mathbf{p}_N) = \int (d\mathbf{k})^N \rho(\mathbf{k}_1 \cdots \mathbf{k}_N, \mathbf{p}_1 \cdots \mathbf{p}_N) \times \exp i \sum_j \mathbf{k}_j \cdot \mathbf{x}_j \tag{1}$$

In order to reduce this to the form (7.2.2) it is necessary to *assume* that the Fourier transform $\rho(\mathbf{k}_1 \cdots \mathbf{k}_N, \mathbf{p}_1 \cdots \mathbf{p}_N)$ is of the form

$$\begin{aligned}\rho(\mathbf{k}_1, \cdots \mathbf{k}_N, \mathbf{p}_1 \cdots \mathbf{p}_N) = L^{-3N} [&\rho_0(\mathbf{p}_1 \cdots \mathbf{p}_N) \prod \delta(\mathbf{k}_m) \\ &+ \sum_j \rho(j_k, \mathbf{p}_1 \cdots \mathbf{p}_N) \prod_{(j)} \delta(\mathbf{k}_m) \\ &+ \sum_{jl} \rho(j_k, l_{-k}, \mathbf{p}_1 \cdots \mathbf{p}_N) \prod_{(j,l)} \delta(\mathbf{k}_m) \\ &+ \cdots]\end{aligned} \tag{2}$$

Here, for example, $\prod_{(j)} \delta(\mathbf{k}_m)$ means the product

$$\delta(\mathbf{k}_1)\delta(\mathbf{k}_2) \cdots \delta(\mathbf{k}_{j-1})\delta(\mathbf{k}_{j+1}) \cdots$$

Indeed, if we introduce (7.3.2) into (7.3.1) we obtain (7.2.2). The Fourier transform $\rho(\mathbf{k}_1 \cdots, \mathbf{p}_1 \cdots)$ appears, therefore, as a sum of terms, each having a characteristic δ singularity.

This singularity condition is an alternative statement of our assumption that reduced distribution functions exist in the limit (7.1.12). It expresses a restriction on the class of distribution functions ρ to which our theory applies. This restriction appears to us to be more of a mathematical than of a physical nature. For molecules with realistic forces (repulsive at short distances) it is hard to imagine situations for which reduced distribution functions would not exist in the limit (7.1.12), at least in a coarse-grained sense.

Let us consider as an example the distribution of N points on the interval $(0, L)$, (Philippot, 1961). The one-particle distribution for given positions $\mathbf{x}_1^0, \cdots, \mathbf{x}_N^0$ of the N particles may be represented as a sum of δ functions.

$$\rho(x) = \sum_{k=1}^{N} \delta(x - x_k^0) \tag{3}$$

The mean distance between successive points is $d = L/N$. We go to the limit of an infinite system, $L \to \infty, N \to \infty, d = L/N$ remaining finite. We consider distributions for which there exists a characteristic length Δx

$$L \gg \Delta x \gg d \tag{4}$$

which permits us to define the function $g(x)$

$$\int_x^{x+\Delta x} \rho(x)dx = \Delta x[N/L+g(x)]+O(1) \tag{5}$$

This function remains finite for $N \to \infty$, $L \to \infty$, N/L finite. We may then call

$$\hat{\rho}(x) = N/L + g(x) \tag{6}$$

the *coarse-grained density.*

Such a density will exist provided each finite interval Δx sufficiently large contains a finite number of points of the order of $\Delta x/d$. Under very general conditions (see for example Lighthill, 1958) $g(x)$ will admit a Fourier transform $f(k)$ and we may write for $\tilde{\rho}(x)$

$$\tilde{\rho}(x) = \int [(N/L)\ \delta(k)+f(k)]\, e^{ikx}\, dk \tag{7}$$

The Fourier transform of $\tilde{\rho}(x)$

$$(N/L)\ \delta(k)+f(k) \tag{8}$$

is precisely of the form (7.3.2) with the characteristic δ singularity.

We may say therefore that (7.3.8) or (7.3.2) applies either to distribution functions defined for the ensemble or in the *coarse-grained sense,* to *single* systems.

The situation is different in quantum mechanics (for more details see Philippot, 1961). There the density distribution for even a single system does not in general have the highly singular character of (7.3.3) (sum of δ functions) and there exist classes of pure states in the quantum mechanical sense to which the singularity condition (7.3.2) may be applied without any supplementary coarse-graining.

The Fourier expansion (7.2.2) or the singularity condition

(7.3.2) is one of the two fundamental assumptions on which our method is based. The other assumption is the existence of a finite range of correlations (see Chapter 4, § 3, Chapter 11). In a sense we may say that in our theory condition (7.3.2) replaces the condition of random phases used in the classical theory of the approach to equilibrium (see also Appendix IV). However (7.3.2) is much more general.

4. Cluster Expansion of the Distribution Function

Let us consider the reduced functions f_s, φ_s, and $f_{s,r}$ which were defined in § 1. We shall always assume the following factorization properties:

$$\varphi_s(\boldsymbol{p}_1 \cdots \boldsymbol{p}_s, t) = \varphi_1(\boldsymbol{p}_1, t)\varphi_1(\boldsymbol{p}_2, t) \cdots \varphi_1(\boldsymbol{p}_s, t) \tag{1}$$

$$\begin{aligned} &f_{s,r}(\boldsymbol{x}_1 \cdots \boldsymbol{x}_s, \boldsymbol{p}_1 \cdots \boldsymbol{p}_r, t) \\ &= f_s(\boldsymbol{x}_1 \cdots \boldsymbol{x}_s, \boldsymbol{p}_1 \cdots \boldsymbol{p}_s, t)\varphi_1(\boldsymbol{p}_{s+1}, t) \cdots \varphi_1(\boldsymbol{p}_r, t) \qquad s < r \end{aligned} \tag{2}$$

We have already discussed the meaning of the factorization of velocity distribution functions in Chapter 4, § 3; it expresses our assumption that correlations have a finite range. The meaning of (7.4.2) is similar. The situation for n_s of f_s is not the same; these functions are generally *not* factorized and one may say that it is only for these functions that the problem of "molecular chaos" has a non-trivial meaning (see especially Chapter 10, § 3).

Let us introduce the following convenient abbreviations

$$\begin{aligned} F(1, 2, \ldots, s) &= C^{-s} f_s(\boldsymbol{x}_1 \cdots \boldsymbol{x}_s, \boldsymbol{p}_1 \cdots \boldsymbol{p}_s, t) \\ \Phi(\alpha) &= C^{-1}\varphi_1(\boldsymbol{p}_\alpha, t) \end{aligned} \tag{3}$$

Now we want to split $F(1, 2, \ldots, s)$ into a sum of terms depending on s positions, $s-1$ positions, . . ., 0 positions. We thus introduce functions $u(1 \cdots s) \equiv u(\boldsymbol{x}_1 \cdots \boldsymbol{x}_s, \boldsymbol{p}_1 \cdots \boldsymbol{p}_s, t)$ which we define as follows

$$\begin{aligned} F(1) &= u(1) + \Phi(1) \\ F(1, 2) &= u(1, 2) + u(1)\Phi(2) + u(2)\Phi(1) + \Phi(1)\Phi(2) \end{aligned}$$

$$
\begin{aligned}
F(1,2,3) = {} & u(1,2,3)+u(1,2)\,\Phi(3)+u(2,3)\,\Phi(1)+u(3,1)\,\Phi(2) \\
& +u(1)\,\Phi(2)\,\Phi(3)+u(2)\,\Phi(3)\,\Phi(1)+u(3)\,\Phi(1)\,\Phi(2) \\
& +\Phi(1)\,\Phi(2)\,\Phi(3) \\
F(1,2,3,4) = {} & u(1,2,3,4)+u(1,2,3)\,\Phi(4)+\cdots \\
& +u(1,2)\,\Phi(3)\,\Phi(4)+\cdots+u(1)\,\Phi(2)\,\Phi(3)\,\Phi(4)+\cdots \\
& +\Phi(1)\,\Phi(2)\,\Phi(3)\,\Phi(4) \qquad \text{etc.}
\end{aligned} \tag{4}
$$

The inversion of these formulae gives

$$
\begin{aligned}
u(1) = {} & F(1)-\Phi(1) \\
u(1,2) = {} & F(1,2)-F(1)\,\Phi(2)-F(2)\,\Phi(1)+\Phi(1)\Phi(2) \\
u(1,2,3) = {} & F(1,2,3)-F(1,2)\Phi(3)-F(1,3)\Phi(2)-F(2,3)\Phi(1) \\
& +F(1)\,\Phi(2)\,\Phi(3)+F(2)\,\Phi(3)\,\Phi(1)+F(3)\,\Phi(1)\,\Phi(2) \\
& -\Phi(1)\,\Phi(2)\,\Phi(3) \\
u(1,2,3,4) = {} & F(1,2,3,4)-F(1,2,3)\Phi(4) \\
& -\cdots+F(1,2)\,\Phi(3)\,\Phi(4)+\cdots \\
& -F(1)\Phi(2)\Phi(3)\Phi(4)-\cdots+\Phi(1)\Phi(2)\Phi(3)\Phi(4)\cdots \text{etc.}
\end{aligned} \tag{5}
$$

The functions $u(1\cdots s)$ have the following properties:

(a) they *factorize* if the group of particles $1, \ldots, s$ splits into two mutually independent subgroups. If these subgroups are formed by particles $1, 2, \cdots, r$, and $r+1, \cdots, s$ we have

$$F(1,2,\ldots,s) = F(1,2,\ldots,r)F(r+1,\ldots,s)$$

and we deduce from (7.4.5)

$$u(1,2,\ldots,s) = u(1,2,\ldots,r)\,u(r+1,\ldots,s) \tag{6}$$

In the special case in which all particles $1, 2, \ldots, s$ are statistically independent [i.e., $F(1,2,\ldots,s) = F(1)\cdots F(s)$], we have

$$u(1,2,\cdots s) = u(1)\,u(2)\cdots u(s) \tag{7}$$

(b) The function $u(1,2,\cdots,s)$ *vanishes* if particle $1, 2, \ldots, s$ split into two mutually independent subgroups and if, moreover, the system is homogeneous [i.e., $F(1) = \Phi(1)$]. For this reason $u(1)$ may be considered as a measure of the inhomogeneity of the system.

For example the second relation (7.4.5) gives us for two independent particles

$$\begin{aligned} u(1, 2) &= F(1)F(2) - F(1)\Phi(2) - F(2)\Phi(1) + \Phi(1)\Phi(2) \\ &= [u(1)+\Phi(1)][u(2)+\Phi(2)] \\ &\quad -[u(1)+\Phi(1)]\Phi(2) - [u(2)+\Phi(2)]\Phi(2) + \Phi(1)\Phi(2) \\ &= u(1)u(2) \end{aligned}$$

This expression vanishes for a homogeneous system.

Expansions (7.4.4) and (7.4.5) generalize to non-equilibrium systems the cluster expansions in the form first introduced by Uhlenbeck and Kahn (Uhlenbeck and Kahn, 1938; see also de Boer, 1948–1949). Non-equilibrium cluster expansions have been introduced by M. Green (1956) and R. Brout (1956) for homogeneous systems. There exists a close connection between the functions u and our Fourier expansion. Indeed using (7.4.1) through (7.4.3) we obtain in succession, from (7.2.2) by integration over all coordinates and momenta except for one, two, . . ., particles [see also (7.2.3)],

$$\Omega^{-1} \sum_k{}' \rho(1_k)\, e^{i\mathbf{k}\cdot\mathbf{x}_1} = F(1) - \Phi(1) \tag{8}$$

$$\Omega^{-2} \sum_{kk'}{}' [\rho(1_k, 2_{k'})\, e^{i(\mathbf{k}\cdot\mathbf{x}_1 + \mathbf{k}'\cdot\mathbf{x}_2)} + \Omega\, \delta^{\mathrm{Kr}}_{k+k'} \rho(1_k, 2_{-k})\, e^{i\mathbf{k}\cdot(\mathbf{x}_1 - \mathbf{x}_2)}] \\ = F(1, 2) - F(1)\Phi(2) - F(2)\Phi(1) + \Phi(1)\Phi(2) \tag{9}$$

$$\begin{aligned} &\Omega^{-3} \sum_{kk'k''}{}' [\rho(1_k, 2_{k'}, 3_{k''})\, e^{i(\mathbf{k}\cdot\mathbf{x}_1 + \mathbf{k}'\cdot\mathbf{x}_2 + \mathbf{k}''\cdot\mathbf{x}_3)} \\ &+ \Omega\, \delta^{\mathrm{Kr}}_{k+k'+k''} \rho(1_k, 2_{k'}, 3_{k''})\, e^{i(\mathbf{k}\cdot\mathbf{x}_1 + \mathbf{k}'\cdot\mathbf{x}_2 + \mathbf{k}''\cdot\mathbf{x}_3)} \\ &+ \Omega\, \delta^{\mathrm{Kr}}_{k'+k''} \rho(1_k, 2_{k'}, 3_{k''})\, e^{i(\mathbf{k}\cdot\mathbf{x}_1 + \mathbf{k}'\cdot\mathbf{x}_2 + \mathbf{k}''\cdot\mathbf{x}_3)} + \cdots] \\ &= F(1, 2, 3) - F(1, 2)\Phi(3) - F(1, 3)\Phi(2) - F(2, 3)\Phi(1) \\ &+ F(1)\Phi(2)\Phi(3) + F(2)\Phi(3)\Phi(1) + F(3)\Phi(1)\Phi(2) \\ &- \Phi(1)\Phi(2)\Phi(3) \end{aligned} \tag{10}$$

By comparison with (7.4.5) we obtain, therefore, the following important result: the functions $u(1, 2, \ldots, s)$ are the Fourier transforms of the sum of all coefficients $\rho(1_{k_1} \cdots s_{k_s})$ with non-vanishing wave vectors corresponding to particles $1, \ldots, s$. For example

$$\Omega^{-2} \sum_{kk'}' [\rho(1_k, 2_{k'}) e^{i(\boldsymbol{k}\cdot\boldsymbol{x}_1 + \boldsymbol{k}'\cdot\boldsymbol{x}_2)} + \Omega\, \delta^{\mathrm{kr}}_{k+k'}\, \rho(1_k, 2_{-k}) e^{i\boldsymbol{k}\cdot(\boldsymbol{x}_1 - \boldsymbol{x}_2)}] = u(1, 2) \tag{11}$$

Now in the limit of a large system, each of the sums in relations (7.4.8) through (7.4.10) goes over to an integral and the Ω factors cancel [see (4.2.4)]. Therefore, the assumption made in § 2 that in the limit (7.1.12) all Fourier coefficients depend on N or Ω only through the ratio N/Ω automatically ensures that the same is true for the functions $u(1, 2, \ldots, s)$ and, therefore, by (7.4.4), for all reduced distribution functions.

The relation between the functions u and the Fourier transforms of the sum of all coefficients $\rho(1_{k_1}, \ldots, s_{k_s})$ which we have established shows that this sum has a simple meaning in terms of the reduced distribution functions. Now we want to go further and analyze the meaning of each Fourier coefficient more closely.

5. Physical Meaning of the Fourier Coefficients

In this section we shall consider the case of homogeneous systems. (Inhomogeneous systems will be studied in Chapter 10.) In this case expressions (7.4.8) through (7.4.10) reduce to

$$u(1, 2) = \Omega^{-1} \sum_{kk'}' \delta^{\mathrm{Kr}}_{k+k'}\, \rho(1_k, 2_{k'})\, e^{i(\boldsymbol{k}\cdot\boldsymbol{x}_1 + \boldsymbol{k}'\cdot\boldsymbol{x}_2)} \tag{1}$$

$$u(1, 2, 3) = \Omega^{-2} \sum_{kk'k''}' \delta^{\mathrm{Kr}}_{k+k'+k''}\, \rho(1_k, 2_{k'}, 3_{k''})\, e^{i(\boldsymbol{k}\cdot\boldsymbol{x}_1 + \boldsymbol{k}'\cdot\boldsymbol{x}_2 + \boldsymbol{k}''\cdot\boldsymbol{x}_3)} \tag{2}$$

$$\begin{aligned}
u(1, 2, 3, 4) &= \Omega^{-3} \sum_{kk'k''k'''}' \delta^{\mathrm{Kr}}_{k+k'+k''+k'''}\, \rho(1_k, \ldots, 4_{k'''})\, e^{i(\boldsymbol{k}\cdot\boldsymbol{x}_1 + \cdots + \boldsymbol{k}'''\cdot\boldsymbol{x}_4)} \\
&+ \Omega^{-2} \sum_{kk'}' \sum_{k''k'''}' \delta^{\mathrm{Kr}}_{k+k'}\, \delta^{\mathrm{Kr}}_{k''+k'''}\, \rho(1_k, 2_{k'}, 3_{k''}, 4_{k'''})\, e^{i(\boldsymbol{k}\cdot\boldsymbol{x}_1 + \boldsymbol{k}'\cdot\boldsymbol{x}_2)}\, e^{i(\boldsymbol{k}''\cdot\boldsymbol{x}_3 + \boldsymbol{k}'''\cdot\boldsymbol{x}_4)} \\
&+ \Omega^{-2} \sum_{kk'}' \sum_{k''k'''}' \delta^{\mathrm{Kr}}_{k+k'}\, \delta^{\mathrm{Kr}}_{k''+k'''}\, \rho(1_k, 3_{k'}, 2_{k''}, 4_{k'''})\, e^{i(\boldsymbol{k}\cdot\boldsymbol{x}_1 + \boldsymbol{k}'\cdot\boldsymbol{x}_3)}\, e^{i(\boldsymbol{k}''\cdot\boldsymbol{x}_2 + \boldsymbol{k}'''\cdot\boldsymbol{x}_4)} \\
&+ \Omega^{-2} \sum_{kk'}' \sum_{k''k'''}' \delta^{\mathrm{Kr}}_{k+k'}\, \delta^{\mathrm{Kr}}_{k''+k'''}\, \rho(1_k, 4_{k'}, 2_{k''}, 3_{k'''})\, e^{i(\boldsymbol{k}\cdot\boldsymbol{x}_1 + \boldsymbol{k}'\cdot\boldsymbol{x}_4)}\, e^{i(\boldsymbol{k}''\cdot\boldsymbol{x}_2 + \boldsymbol{k}'''\cdot\boldsymbol{x}_3)}
\end{aligned} \tag{3}$$

The functions u have especially simple properties in homogeneous systems. As we have seen in § 4 they then represent correlations [remember that for homogeneous systems $u(1) = 0$ and that $u(1, 2, \ldots, s)$ vanish for independent particles]. We see, therefore, from (7.5.1) that $\rho(1_k, 2_{-k})$ is simply the Fourier transform of the binary correlation function $u(1, 2)$. Similarly the only contributions to $\rho(1_k, 2_{k'}, 3_{-k-k'})$ arise from situations in which *all* three particles 1, 2, 3 are correlated; for all other configurations $u(1, 2, 3)$ vanishes.

The different contributions in (7.5.3) have a similar meaning. Using (7.4.5) we may write, after a few calculations,

$$\begin{aligned} u(1, 2, 3, 4) = [&F(1, 2, 3, 4) - F(1, 2, 3)F(4) - \cdots - F(1, 2)F(3,4) - \cdots \\ &+ 3F(1, 2)\,F(3)\,F(4) + \cdots - 12F(1)\,F(2)\,F(3)\,F(4)] \\ &+ u(1, 2)\,u(3, 4) + u(1, 3)\,u(2, 4) + u(1, 4)\,u(2, 3) \qquad (4) \end{aligned}$$

It can be verified that the contribution in the brackets is different from zero only when *all four* particles are simultaneously correlated. By comparison with (7.5.3) and (7.5.1) we see that $\rho(1_k, 2_{k'}, 3_{k''}, 4_{k'''})\,\delta^{\mathrm{Kr}}(\boldsymbol{k}+\boldsymbol{k}'+\boldsymbol{k}''+\boldsymbol{k}''')$

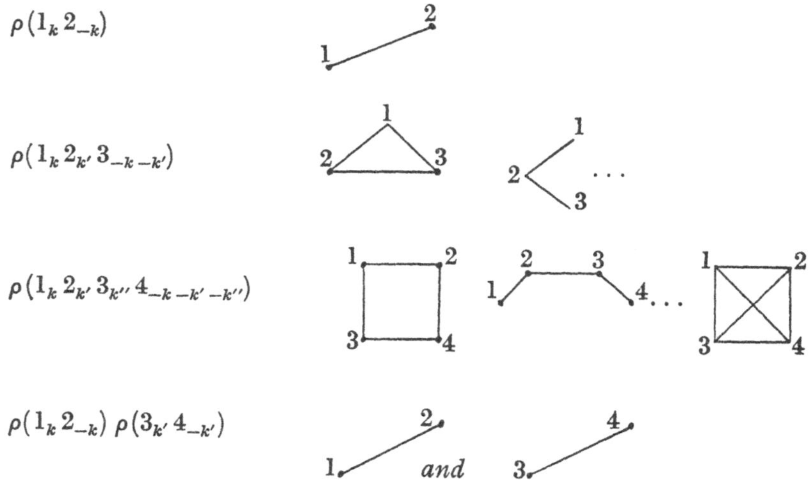

Fig. 7.5.1 The physical meaning of some Fourier coefficients in terms of correlations.

comes from situations in which all four particles are correlated. On the contrary the other terms in (7.5.3) correspond to two binary correlations. Moreover we have the factorization theorem

$$\rho(1_k, 2_{k'}, 3_{k''}, 4_{k'''})\,\delta^{\mathrm{Kr}}(\boldsymbol{k}+\boldsymbol{k}')\,\delta^{\mathrm{Kr}}(\boldsymbol{k}''+\boldsymbol{k}''')$$
$$= \rho(1_k, 2_{k'})\,\delta^{\mathrm{Kr}}(\boldsymbol{k}+\boldsymbol{k}') \times \rho(3_{k''}, 4_{k'''})\,\delta^{\mathrm{Kr}}(\boldsymbol{k}''+\boldsymbol{k}''') \quad (5)$$

The usual notation in equilibrium statistical mechanics (see for example Mayer and Mayer, 1940) is a line connecting particles which are correlated.[1]

Figure 7.5.1 gives the physical meaning of some Fourier coefficients in terms of the Mayer diagrams.

We see, therefore, that the Fourier coefficients of the distribution function ρ are directly related to the "state" of the system when expressed in terms of Mayer's diagrams.

We shall now study the change of this state in time and introduce a diagram technique which permits us to visualize the way the Mayer diagrams are modified through the effect of molecular interactions.

6. Diagrams

We have already studied the Liouville equation for interacting particles in Chapter 4, § 1. The only modification we have to introduce now comes from the inclusion of the volume factors in (7.2.1). Let us call r the number of independent wave vectors (see note p. 143) of $\rho_{\{k\}}$ in (4.1.9), and r' the corresponding number for $\rho_{k_j-l,\, k_n+l,\, \{k'\}}$. We shall now have [for the definition of $\boldsymbol{D}_{jn}$ see (6.5.2); remember also $\Omega = L^3/8\pi^3$]:

$$(1/\Omega^r)(\partial\rho_{\{k\}}/\partial t) = (\lambda/\Omega)\sum_{jn}\sum_{l} 1/\Omega^{r'}\, e^{i\Sigma \boldsymbol{k}_m\cdot \boldsymbol{v}_m t}\, V_l\, i\boldsymbol{l}\cdot\boldsymbol{D}_{jn}$$
$$\times \exp\{-i[\textstyle\sum \boldsymbol{k}_m\cdot\boldsymbol{v}_m - \boldsymbol{l}\cdot(\boldsymbol{v}_j-\boldsymbol{v}_n)]t\}\,\rho_{k_j-l,\, k_n+l,\, \{k'\}} \quad (1)$$

As in Chapter 4, § 2 and in Chapter 6, §§ 5 and 6 we shall use a

[1] Because this graphical representation has been widely used by J. E. Mayer in his work on equilibrium statistical mechanics, we shall call such diagrams the "*Mayer diagrams.*"

diagram representation in which only lines corresponding to $\boldsymbol{k} \neq 0$ are represented.[1]

Taking account of the conservation law for wave vectors (4.1.8) we have the following six possibilities:

$$
\begin{array}{llll}
\text{(a)} & \boldsymbol{k}_j, \boldsymbol{k}'_n & \neq 0 & \boldsymbol{k}'_j, \boldsymbol{k}_n = 0 \\
\text{(b)} & \boldsymbol{k}'_j, \boldsymbol{k}'_n & \neq 0 & \boldsymbol{k}_j, \boldsymbol{k}_n = 0 \\
\text{(c)} & \boldsymbol{k}_j, \boldsymbol{k}_n & \neq 0 & \boldsymbol{k}'_j, \boldsymbol{k}'_n = 0 \\
\text{(d)} & \boldsymbol{k}'_j, \boldsymbol{k}'_n, \boldsymbol{k}_j & \neq 0 & \boldsymbol{k}_n \quad = 0 \\
\text{(e)} & \boldsymbol{k}'_j, \boldsymbol{k}_j, \boldsymbol{k}_n & \neq 0 & \boldsymbol{k}'_n \quad = 0 \\
\text{(f)} & \boldsymbol{k}'_j, \boldsymbol{k}'_n, \boldsymbol{k}_j, \boldsymbol{k}_n & \neq 0 &
\end{array}
\tag{2}
$$

and the situations obtained by permutation of j and n. The case $\boldsymbol{k}'_j = \boldsymbol{k}'_n = \boldsymbol{k}_j = 0$, $\boldsymbol{k}_n \neq 0$ is forbidden by (4.1.8).

As in Chapter 6, § 4 we use the diagram

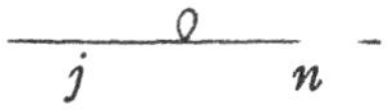

for the process

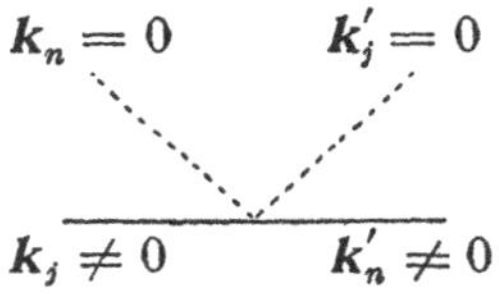

We have drawn the six basic diagrams corresponding to (7.6.2) in Fig. 7.6.1; we have already met most of these diagrams. Indeed (b) and (c) form together the basic diagram for weakly coupled systems represented in Fig. 4.2.3, while (a) and (f) have been discussed in Chapter 6, § 5.

[1] The very possibility of such a diagram representation is based on the fact that for the large systems we are studying here, ρ_0 (the "correlation-vacuum") gives a contribution to ρ of the same order as the *sum* of all other Fourier coefficients that represent correlations and inhomogeneities (see §§ 2—3 of this chapter).

As in Chapter 2, § 4, we may introduce the following classification of diagrams:

(a) diagrams corresponding to the *destruction of correlations*: the number of lines is smaller at the left than at the right of the vertex [(b) and (d) of Fig. 7.6.1]. For example in diagram (b) we have at the right a correlation between particles j and n (see Fig. 7.5.1) which is destroyed by the vertex.

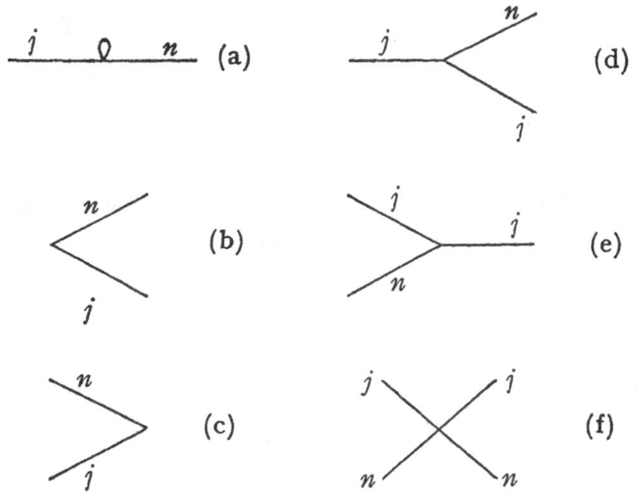

Fig. 7.6.1 The six one-vertex diagrams corresponding to (7.6.2).

(b) diagrams corresponding to the *creation of correlations* the number of lines is larger at the left [(c) and (e) of Fig. 7.6.1].

(c) diagrams corresponding to the *propagation of correlations*: the number of lines is not modified by the vertex [(a) and (f) of Fig. 7.6.1].

It is interesting to write equation (7.6.1) for the first Fourier coefficients in terms of these diagrams. We have already done this in the case of anharmonic oscillators in (3.1.3) and (3.1.4). In this way we obtain Figs. 7.6.2 and 7.6.3.

In most problems (see § 8) we shall be concerned with the evolution of a certain number of "fixed" particles. We shall

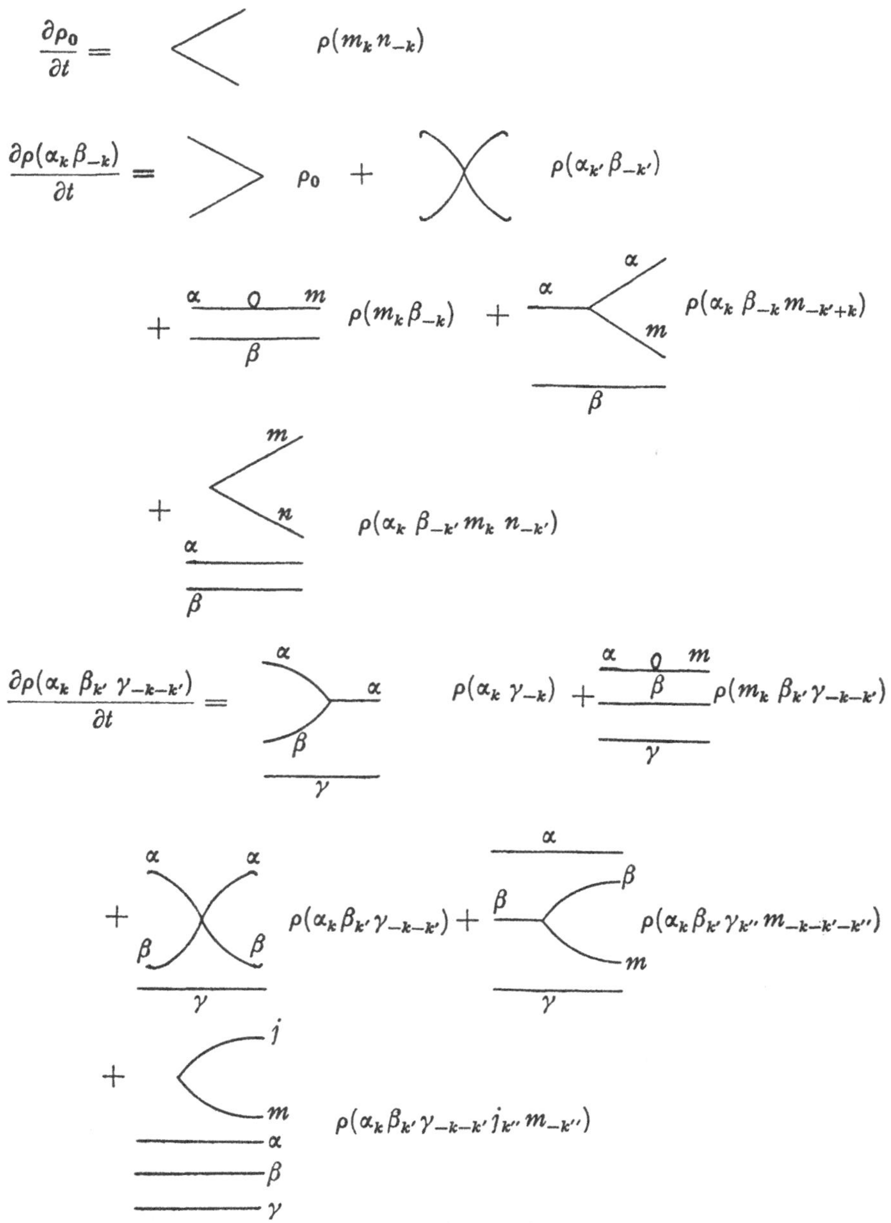

Fig. 7.6.2 Liouville equation for simple classes of Fourier coefficients. (In all these coefficients the sum of the wave vectors vanishes.)

generally use Greek letters for them, while we shall use roman letters for "dummy" particles, over which we have to perform a summation in (7.6.1). This notation is used in Figs. 7.6.2 through 7.6.4.

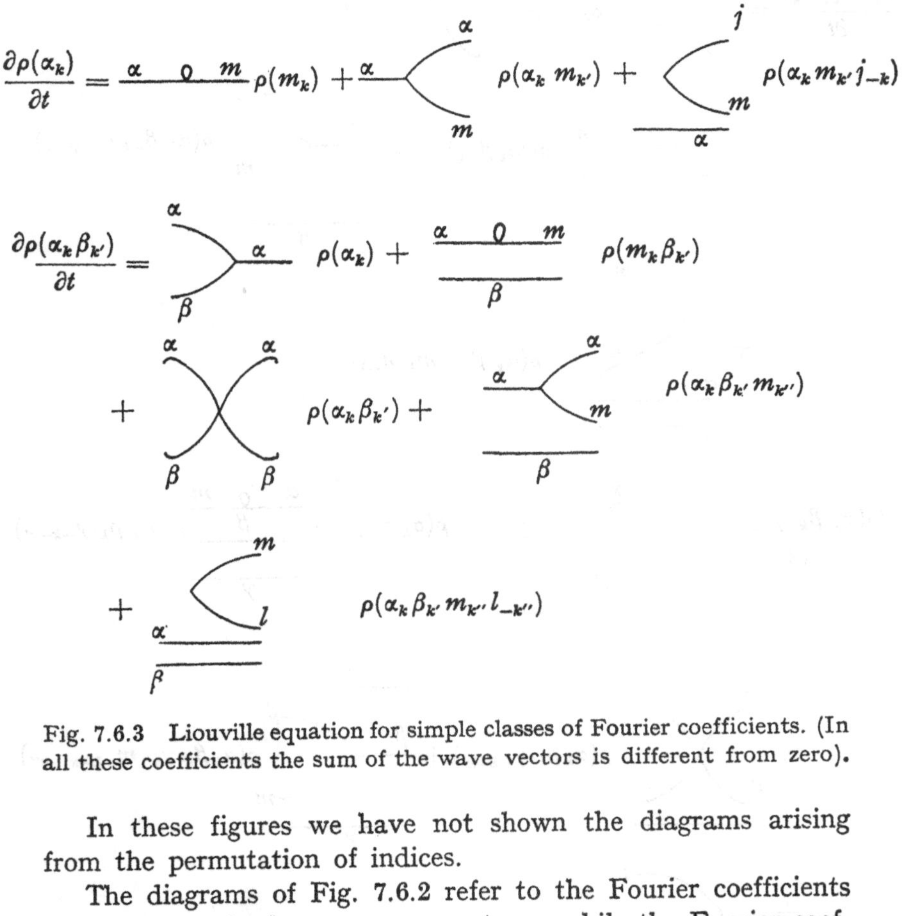

Fig. 7.6.3 Liouville equation for simple classes of Fourier coefficients. (In all these coefficients the sum of the wave vectors is different from zero).

In these figures we have not shown the diagrams arising from the permutation of indices.

The diagrams of Fig. 7.6.2 refer to the Fourier coefficients which appear for homogeneous systems, while the Fourier coefficients in Fig. 7.6.3 refer to inhomogeneous situations [see (7.2.4)].

We have seen in § 5 how the Fourier coefficients are related to correlations. For example the second equation in Fig. 7.6.2 refers in terms of correlations to the following processes involving

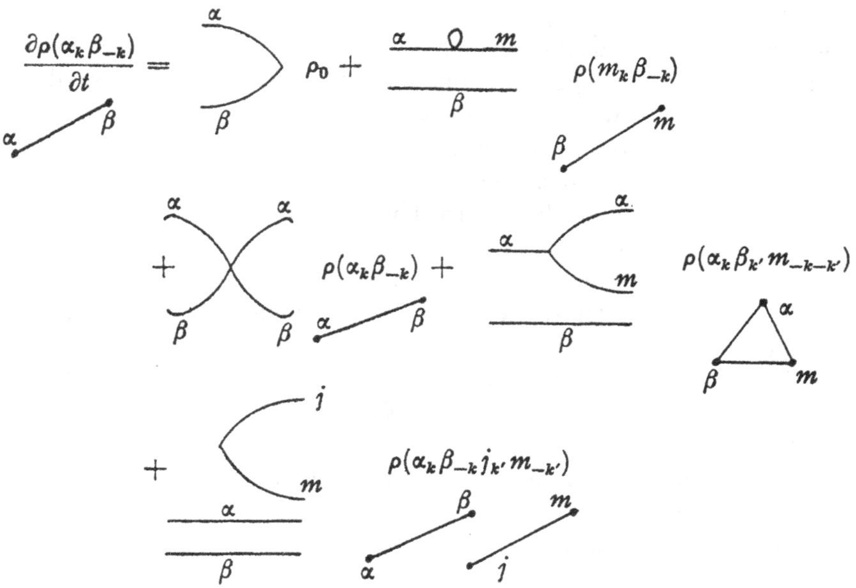

Fig. 7.6.4 Second equation of Fig. 7.6.2 with references to corresponding correlations (see § 5).

the "fixed" particles α and β (see Fig. 7.6.4): the creation of a binary correlation, the "transfer" of the correlation from an arbitrary molecule m to α, the destruction of a three-body correlation, or of one of two independent two-body correlations. Now we clearly see the physical meaning of our method: It is a method for describing the *dynamics of correlations* (and inhomogeneities) through the effect of interactions. It is a mechanics in which the objects are the correlations, the Mayer diagrams.

It may also be interesting to observe that the equations obtained by integrating over the momenta of all particles in Figs. 7.6.2 and 7.6.3, except those of a given set, could also be obtained from the hierarchy of equations (7.1.11). We then have to start from the Fourier expansions (7.5.1), (7.5.2), (7.5.3), . . . instead from the Liouville equation itself.

The essential point however is to use as independent variables the wave vectors $\boldsymbol{k}_1, \ldots, \boldsymbol{k}_N$ and the momenta $\boldsymbol{p}_1, \ldots, \boldsymbol{p}_N$ instead of the coordinates $q_1, \ldots q_N$ and the momenta. The wave vectors are the variables which appear because we expressed ρ in terms of the eigenfunctions of the unperturbed Liouville operator L_0. As we have already noticed in Chapter 1, § 10, in this representation δL becomes a *non-diagonal operator* and we may speak of "*transitions*" between $\boldsymbol{k}$ and $\boldsymbol{k}'$ such that $\langle \boldsymbol{k} | \delta L | \boldsymbol{k}' \rangle$ is different from zero. They are precisely the transitions represented by our diagram technique. On the contrary, in the coordinate representation, the operator L_0 is non-diagonal; [$\exp(itL_0) = \exp(t\boldsymbol{v}\,(\partial/\partial \boldsymbol{x}))$ is a displacement operator in coordinate space] while δL is diagonal (it is an ordinary function of the coordinates). Therefore in the coordinate representation there is nothing like a "transition" induced by δL (see also Chapter 6, § 3).

The most remarkable point here is probably the use of different representations in classical mechanics, quite in the same spirit as in quantum mechanics.

7. Concentration Dependence of Diagrams

Let us now show that the structure of the diagrams uniquely fixes the N and L^3 dependence of the corresponding contributions to the Liouville equation. We shall consider two of the diagrams of Fig. 7.6.1 in detail. We again use Greek letters for fixed particles and roman letters for dummy particles.

Diagram (*a*). Conservation of wave vector together with $\boldsymbol{k}'_\alpha = \boldsymbol{k}_j = 0$ implies [see for example (4.1.6)]

$$\boldsymbol{k}_\alpha = \boldsymbol{l} = \boldsymbol{k}'_j \tag{1}$$

Therefore such a diagram gives to (7.6.1) a contribution of order ($r = r' = 1$)

$$\lambda L^{-3} \sum_j = \lambda N L^{-3} = \lambda C \tag{2}$$

Indeed a summation over $\boldsymbol{l}$ is lost because of (7.7.1), as well as a summation over one particle (say n) because α is a *fixed* particle.

Diagram (*b*). We have here $r = 0$, $r' = 1$; therefore, the order of this contribution is

$$\lambda L^{-6} \sum_{jn} \sum_{l} = \lambda L^{-3} \sum_{jn} = \lambda NC \tag{3}$$

We may proceed exactly in the same way for the other diagrams. In every case we have to take into account the number of lines at the right and at the left of the vertex, the eventual existence of relations of the form $\boldsymbol{k}_j' + \boldsymbol{k}_n' + \cdots = 0$, and the number of "roman" letters over which a summation is performed.

The result may be expressed in the following way: For each of the six basic diagrams we introduce the "*index*" T which is given in Table 7.7.1. This index is uniquely associated with the type of connection at the vertex, and is independent of the nature ("Greek" or "roman") of the particle indices of the lines.

Table 7.7.1 Index T [see formula (7.7.4)].

	Vertex	T		*Vertex*	T
(a)		-1	(d)	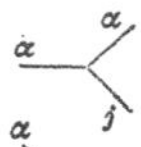	-1
(b)		-1	(e)	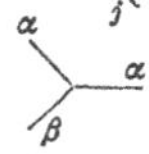	0
(c)		0	(f)	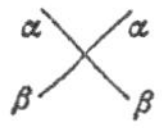	0

Let us also call R the number of distinct "roman" letters in each diagram (see Table 7.7.1). The order of the diagram is then given by

$$\lambda N^R L^{3T} \tag{4}$$

For example we have for diagram (a)

$$R = 1, \qquad T = -1$$

and for (b)

$$R = 2, \qquad T = -1$$

in agreement with (7.7.2) and (7.7.3).

The result (7.7.4) is very simple: Whenever we have a particle at the right of the vertex which does not appear at the left and over which we may freely sum, we have a factor $N/L^3 = C$.

This is the situation for diagrams (a) and (d). In (b) we have two "roman" particles at the right and this gives the factor NC. However, as we shall see in § 8 the extra factor of N disappears when we go to the reduced distribution functions. Therefore, diagrams (a), (b), and (d) essentially introduce a factor of C, while (c), (e), and (f) do not.

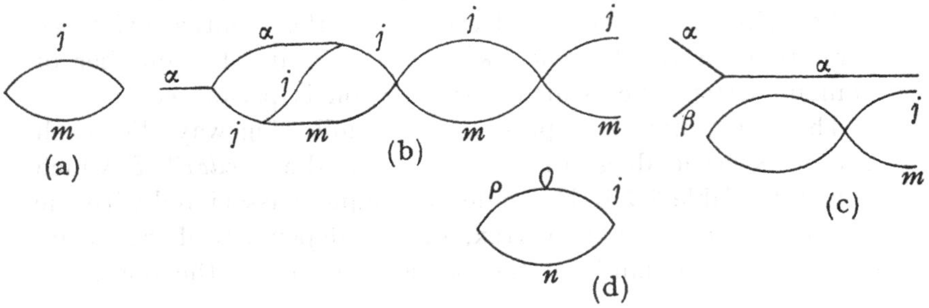

Fig. 7.7.1 Examples of diagrams.

These rules may be extended to the combination of diagrams obtained by iteration. The asymptotic order of magnitude of a diagram δ containing n vertices of indices $T_1, T_2, T_3, \ldots, T_n$ and a total number R of distinct roman letters is given by

$$\lambda^n N^R L^{3\Sigma T_i} \tag{5}$$

For example in Fig. 7.7.1 the order of magnitude of each of the diagrams (a) through (d) is, respectively

(a): $\lambda^2 N^2 L^{-3} = \lambda NC$ (c): $\lambda^3 N^2 L^{-3} = \lambda^3 NC$

(b): $\lambda^5 N^2 L^{-6} = \lambda^5 C^2$ (d): $\lambda^3 N^3 L^{-6} = \lambda^3 NC^2$

The order of diagram (a) of Fig. 7.7.1 is of course in agreement with our result (4.2.3). Similarly (d) coincides with (6.5.5).

There is, however, one case to which the rule (7.7.5) cannot

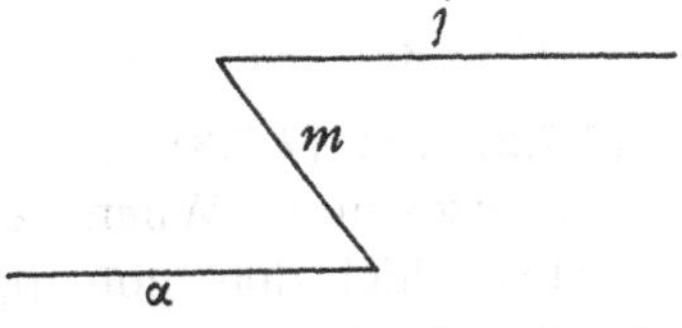

Fig. 7.7.2 Diagram to which the rule given by (7.7.5) does not apply.

be applied because the putting "together" of two diagrams introduces a supplementary restriction.

Let us indeed consider the diagram of Fig. 7.7.2. The application of (7.7.5) would give us

$$\lambda^2 L^{-3} N^2 = \lambda^2 NC \tag{6}$$

while two successive applications of (7.6.1) gives us

$$(\lambda^2/L^6) \sum_{mj} = \lambda^2 C^2 \tag{7}$$

Indeed in order to "connect" diagrams (b) and (c) of Fig. 7.6.1 one has to impose the supplementary restrictions

$$\boldsymbol{k}_j = -\boldsymbol{k}_m = \boldsymbol{k}_\alpha$$

Therefore, there is no summation left over $\boldsymbol{l}$ in (7.7.2). This "transfer" of the wave vector is the only case where (7.7.5) is not applicable.

8. Reduced Distribution Functions and Diagrams

In all problems of physical significance one is finally interested in the reduced distribution functions depending on a given finite number of velocities. Important simplifications then arise. If for example in the second equation of Fig. 7.6.2 we integrate over all velocities except $\mathbf{v}_\alpha$, $\mathbf{v}_\beta$ the contribution of the last diagram vanishes because of the very form of (4.1.7), which contains the operator $(\partial/\partial \mathbf{v}_j - \partial/\partial \mathbf{v}_m)$. Indeed an expression like

$$\int d\mathbf{v}_j \cdot (\partial/\partial \mathbf{v}_j)\, F(\mathbf{v}_j)\, \rho$$

is a surface integral taken for values of $|\mathbf{v}_j| \to \infty$. But ρ and its derivatives are always supposed to vanish at infinity.

We may state the following rules: *The only diagrams which give a non-vanishing contribution to the Fourier components of the reduced distribution function $f_s(\mathbf{x}_1 \cdots \mathbf{x}_s, \mathbf{v}_1 \cdots \mathbf{v}_s, t)$ are those in which every vertex has at least one of the following properties*: (a) There is a fixed line corresponding to one of the particles 1, . . ., s starting or ending at that vertex; (b) One of the two indices of the vertex appears on a line situated at the left of that vertex. Indeed, as each vertex with indices j, l corresponds to an operator $(\partial/\partial \mathbf{v}_j - \partial/\partial \mathbf{v}_l)$ acting on everything to its right, the only ways in which the contribution will not vanish are that one does *not* have to integrate over $\mathbf{v}_j$ or $\mathbf{v}_l$ or that this operator is preceded (at its left) by a function of $\mathbf{v}_j$ or $\mathbf{v}_l$.

For example in Fig. 7.8.1 we represent two diagrams that contribute to $f_1(\mathbf{x}_\alpha, \mathbf{v}_\alpha, t)$. It is useful to have a graphical representation which stresses the fact that the components of a diagram have a particle *in common*. This is indicated in Fig. 7.8.1 by a dotted line; such diagrams will be called *semi-connected*.

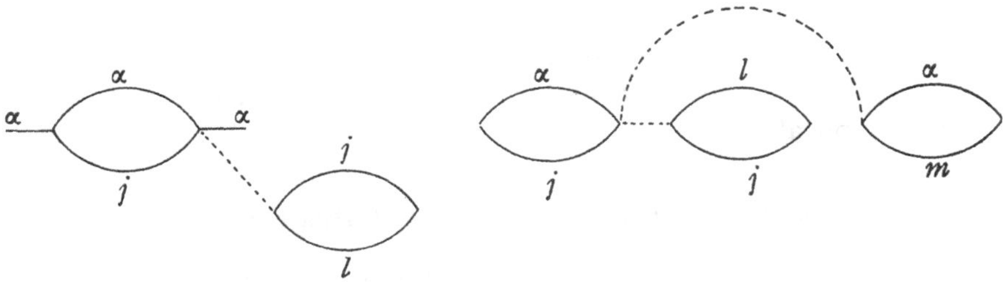

Fig. 7.8.1 Semi-connected diagrams that contribute to $f_1(\mathbf{x}_\alpha, \mathbf{v}_\alpha, t)$.

Of course the dotted line does not represent a particle but expresses a kind of relation between components of a diagram. This restriction to semi-connected diagrams has the important consequence that vertex (b) of Table 7.7.1 can never involve two "new" roman particles. Indeed one of them has to be semi-connected to a particle at its left. Therefore the order in N and L of this vertex is [see (7.7.4)]

$$\lambda N L^{-3}$$

instead of

$$\lambda N^2 L^{-3}$$

The order of the other vertices is not changed. We therefore obtain Table 7.8.1.

Table 7.8.1 Order in C of the basic diagrams.

	Vertex	*Order in C*		*Vertex*	*Order in C*
(a)		1	(d)	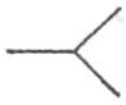	1
(b)		1	(e)	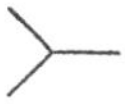	0
(c)		0	(f)		0

The results of Table 7.8.1 are very simple to understand: the diagrams (a), (b), and (d) introduce a single new particle at their "right", while (c), (e), and (f) introduce no new particles.

The important point is that the diagrams have now become completely independent of N, *they introduce only powers of the concentration.* Again this is a physically natural result: the change in reduced distributions should indeed become independent of the size of the system, once it is large enough.

The results of this paragraph show that if the reduced distribution functions exist and are finite at the initial $t = 0$ (see § 2) they will remain so for all subsequent times. In other words the singularity conditions (7.3.2) in the Fourier expansions of the phase distribution functions is preserved in time. This is of course a necessary requirement for its acceptance as one of the two basic assumptions of our theory.

It may be interesting to notice that in our theory the semi-connected diagrams play somewhat the role of the "linked" clusters in the quantum theory of the N-body problem (see Peierls, 1959 for an excellent discussion). There too the elimination of the unlinked clusters allows one to avoid spurious N factors.

CHAPTER 8

The Time Dependence of Diagrams

1. Effect of H_0 — Wave Packets

We shall now use the general theory developed in Chapter 7, especially our diagram representation, to obtain a deeper insight into the time evolution of the distribution functions (I. Prigogine and F. Henin, 1960; F. Henin, P. Résibois, and F. Andrews (1961).[1] This is, of course, a necessary step for the understanding of the mechanism of irreversibility.

Until now we have always considered the time evolution from an asymptotic point of view (see specially Chapter 2, § 5, and Chapter 6, § 1) and the interaction process was treated as an instantaneous event. This is insufficient for the general theory we want to develop now (see especially Chapter 11 where we shall have to consider the time evolution over arbitrary time intervals, long and short).

Let us first consider the case of freely propagating lines in our diagrams. We must therefore consider for the moment only the unperturbed Hamiltonian H_0. The contribution of a single line of wave vector $\boldsymbol{k}$ to the distribution function is (see Chapter 7, § 2)

$$\rho_k \exp\,[i\boldsymbol{k}\cdot(\boldsymbol{x}_j-\boldsymbol{v}_j t)] \tag{1}$$

As long as the effect of the perturbation V is neglected, ρ_k is time-independent.

Similarly, to

$$\rho_{\{k\}} \exp\,[i\sum \boldsymbol{k}_j\cdot(\boldsymbol{x}_j-\boldsymbol{v}_j t)] \tag{2}$$

where $\{\boldsymbol{k}\}$ represents a set of m non-vanishing $\boldsymbol{k}$ vectors, we associate a set of m lines.

[1] This chapter owes much to the work of Dr. P. Résibois (1960).

To a single line or to a set of lines corresponds a plane wave in phase space. Out of such plane waves we may form wave packets. Their properties are very similar to the properties of wave packets studied in quantum mechanics.

Let us consider two examples. A one-particle distribution which for $t = 0$ reduces to a δ function in coordinate space (for simplicity, we consider a single coordinate) may be represented by the formula

$$\begin{aligned}\rho(x, v, 0) &= 1/(2\pi\sigma^2)^{1/2}\, \delta(x) e^{-v^2/\sigma^2} \\ &= 1/(2\pi\sigma^2)^{1/2}\, 1/2\pi \int_{-\infty}^{+\infty} dk\, e^{ikx}\, e^{-v^2/\sigma^2}\end{aligned} \tag{3}$$

The parameter σ measures the dispersion of the velocity distribution. We then have at time t

$$\begin{aligned}\rho(x, v, t) &= 1/(2\pi\sigma^2)^{1/2}\, \delta(x-vt)\, e^{-v^2/\sigma^2} \\ &= 1/(2\pi\sigma^2)^{1/2}\, 1/2\pi \int_{-\infty}^{+\infty} dk\, e^{ik(x-vt)}\, e^{-v^2/\sigma^2}\end{aligned} \tag{4}$$

As t increases the corresponding density in coordinate space becomes more and more uniform. Indeed, if we integrate (8.1.4) over v we obtain

$$\rho(x, t) = 1/(2\pi\sigma^2 t^2)^{1/2}\, e^{-x^2/\sigma^2 t^2} \tag{5}$$

A characteristic feature of this case is that there exists nothing like a relaxation time independent of the initial conditions. The time necessary to achieve uniformity depends on the dispersion of the velocity σ. The distribution (8.1.4) is obtained by integrating over all values of k. It corresponds to a wave packet of "infinite extension" in wave vector space.

We may also consider wave packets of limited extension in wave vector space. Let us take for example in (8.1.1)

$$\rho_k \sim \exp\,(-k^2/\kappa^2) \tag{6}$$

The parameter κ measures the width of the wave packet in k space. We find now instead of (8.1.4)

$$\rho(x, v, t) \sim \exp\,[-\kappa^2(x-vt)^2]\, \exp\,[-v^2/\sigma^2] \tag{7}$$

The disappearance of the wave packet may be characterized by the time

$$t \sim 1/\kappa v \tag{8}$$

This is the time necessary for the wave packet to flow through a given point (or plane) in coordinate space.

Formula (8.1.8) is a special case of the relation between the time Δt that is required for a wave packet to pass a given point and $\Delta\nu$ the range of frequencies of the packet (see for example Bohm, 1951, p. 63)

$$\Delta\nu\, \Delta t \approx 1 \tag{9}$$

In our phase space description the frequency ν is related to k by $\nu = kv$.

Later we shall have to consider two limiting cases: wave packets with $1/\kappa$ of the order of the range of molecular interactions, and wave packets with $1/\kappa$ of macroscopic extension. In the first case (8.1.8) becomes of the order of the duration of an interaction.

2. Duration of a Collision

Let us consider the effect of a collision on ρ_0 in the weakly coupled approximation. [See Fig. 4.2.3]. The corresponding equation is [see (4.2.3), (6.5.1) and Chapter 2, § 3]

$$\int_0^t dt_1 \int_0^{t_1} dt_2 \int d\boldsymbol{l} |V_l|^2\, \boldsymbol{l} \cdot \boldsymbol{D}\, e^{-i\boldsymbol{l}\cdot\boldsymbol{g}(t_1-t_2)}\, \boldsymbol{l} \cdot \boldsymbol{D}\rho_0$$
$$= \int d\boldsymbol{l} |V_l|^2\, \boldsymbol{l} \cdot \boldsymbol{D}[t/i\boldsymbol{l}\cdot\boldsymbol{g} + (e^{-i\boldsymbol{l}\cdot\boldsymbol{g}t} - 1)/(i\boldsymbol{l}\cdot\boldsymbol{g})^2]\, \boldsymbol{l} \cdot \boldsymbol{D}\rho_0 \tag{1}$$

We have studied such expressions in Chapter 2, § 5 for times long with respect to the duration of a collision. We now want to understand their behavior for arbitrary values of t. For this purpose we may replace (8.2.1) by the integral

$$\int d\boldsymbol{l} |V_l|^2\, \boldsymbol{l} \cdot \boldsymbol{l}\, [t/i\boldsymbol{l}\cdot\boldsymbol{g} + (e^{-i\boldsymbol{l}\cdot\boldsymbol{g}t} - 1)/(i\boldsymbol{l}\cdot\boldsymbol{g})^2] \tag{2}$$

Let us for example study the x–x component

$$\int d\boldsymbol{l}\, |V_l|^2 l_x l_x [t/i\boldsymbol{l}\cdot\boldsymbol{g} + (e^{-i\boldsymbol{l}\cdot\boldsymbol{g}t} - 1)/(i\boldsymbol{l}\cdot\boldsymbol{g})^2] \tag{3}$$

We choose the simple exponential potential

$$V(r) = V_0 \exp(-\kappa r) \tag{4}$$

$1/\kappa$ gives the range of this potential. The Fourier components of this potential are given by

$$V_l = V_0 \int d\mathbf{r}\, e^{-i\mathbf{l}\cdot\mathbf{r}}\, e^{-\kappa r} = 8\pi\kappa V_0/(l^2+\kappa^2)^2 \tag{5}$$

We introduce into (8.2.3) cylindrical coordinates with a polar axis along $\mathbf{g}$,

$$l_x = b\cos\theta \qquad l_y = b\sin\theta, \qquad l_z \tag{6}$$

It is easy to integrate over θ and b, and we are left with an integral of the form

$$I = \int_{-\infty}^{+\infty} dl_z \{1/(\kappa^2+l_z^2)^2\} \left[t/il_z g + (e^{-il_z g t}-1)/(il_z g)^2\right] \tag{7}$$

The usual way to perform the integration is to consider the complex l_z plane and to close the contour of integration by a large semicircle in the lower half plane (see Fig. 8.2.1). One then uses the residue theorem. The only singularity in the lower half plane is the second order pole $l_z = -i\kappa$.

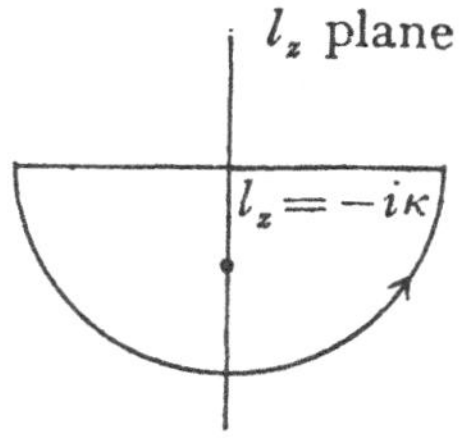

Fig. 8.2.1 Contour of integration for (8.2.7).

We shall therefore obtain three kinds of contributions.

a. A contribution $\sim t$; this is the only one retained in the theory of weakly coupled systems. The integrals (8.2.7) or (8.2.3) are inversely proportional to the relaxation time t_r, this contribution is therefore of the form

$$t/t_r \tag{8}$$

b. A time-independent contribution which corresponds to the

term δ'_+ already derived in Chapter 2, § 5. This term differs from (8.2.8) by a factor of the order of

$$1/t\kappa g \tag{9}$$

Now $1/\kappa g$ is clearly the duration t_{int} of the collision process; indeed $1/\kappa$ is the range of the potential. Therefore using (8.2.8) and (8.2.9) this contribution is of the form

$$(t_{\text{int}}/t) \times (t/t_r) = t_{\text{int}}/t_r \tag{10}$$

c. Exponentially decaying contributions containing the factor

$$\exp(-\kappa g t) = \exp(-t/t_{\text{int}}) \tag{11}$$

We have *two* characteristic times t_r and t_{int}. While t_r is related to the strength of the interaction [see (4.5.21)], t_{int} depends primarily on the range of the interaction. The situation is qualitatively the same for strong, short-range forces. Here also we have two characteristic times t_r and t_{int}. For dilute hard spheres of diameter a we have

$$t_r \sim 1/Ca^2 v \qquad t_{\text{int}} \sim a/v \tag{12}$$

Note that the ratio of these two times is

$$t_{\text{int}}/t_r = a^3 C \tag{13}$$

It is therefore clear that whenever we are interested in the effect of higher concentrations we shall not be allowed to treat collisions as instantaneous events.

All three terms are necessary to obtain a correct result for small t in (8.2.7). Then we may expand the exponential and we obtain $I \sim t^2$. Therefore I vanishes for $t \to 0$. This had to be expected. Indeed the value of this integral is determined by the collisions which start at t_2 and are terminated at $t_1 \leqq t$ [see (8.2.1)]. Precisely because of the finite duration of a collision the number of such collisions must decrease when t becomes of the order of or smaller than t_{int} and must go to zero for $t \to 0$.

One may say that the time-independent contribution (8.2.10) gives a correction to the asymptotic value of the cross section, while the time-dependent contribution (8.2.11) takes account of the fact that for small t not all collisions started at time $t = 0$ are yet completed.

The meaning of the duration of a collision may also be understood on the basis of the wave packet relation (8.1.9)

$$\Delta t \simeq 1/v\Delta k \tag{14}$$

where Δk is the range of the wave vectors represented in the intermolecular potential V_k. The whole collision process may be viewed as the formation of a wave packet in Fourier space, with a lifetime t_{int}, which is then destroyed by interference.

The simplicity of the theory of irreversible processes for weakly coupled or dilute systems comes from the fact that in those cases the collision processes may be treated as instantaneous. However, in the general theory we have to retain all three types of contribution (8.2.8), (8.2.10), and (8.2.11). It should be noticed that the time dependence of the contribution (8.2.11), which is essential for the short time behavior, depends critically on the nature of the singularity in (8.2.7) and its location in the complex plane. If for example we had a pole which was not on the imaginary axis, we would find a damped oscillatory behavior instead of (8.2.11). The relation between the analytic form of the interaction law and the singularities in the complex plane deserves a much more detailed study (see for another example Balescu, 1962).

Now we need a method for performing the time integrations in a more compact way than we had until now. It appears that the simplest method is to perform the integration over the wave vector $\boldsymbol{l}$ *first* and then to replace the successive time integrations by an integration over a variable z which is related to t by a Fourier (or more precisely by a Laplace) transform. In this way we shall be able to establish a direct link between our diagram technique and the time dependence of the distribution functions.

3. Resolvent Method

We introduce the Laplace transform $\tilde{\rho}(s)$ of the distribution function $\rho(t)$ (see for example Doetsch, 1943) which is defined by the relation

$$\tilde{\rho}(s) = \int_0^\infty dt\, e^{-st} \rho(t) \tag{1}$$

We multiply the Liouville equation (1.3.1) by e^{-st} and integrate over t from $t = 0$ to $t = \infty$. We then obtain the following equation for the Laplace transform $\tilde{\rho}(s)$:

$$-i\rho(0)+is\tilde{\rho} = L\tilde{\rho} \tag{2}$$

where $\rho(0)$ is the value of $\rho(t)$ at time $t = 0$. The formal solution of this equation is

$$\tilde{\rho}(s) = -i\, 1/(L-is)\rho(0) \tag{3}$$

It is shown in all books dealing with Laplace transforms that formula (8.3.1) may be inverted to express $\rho(t)$ in terms of its Laplace transform $\tilde{\rho}(s)$:

$$\rho(t) = (1/2\pi i)\int_{\gamma-i\infty}^{\gamma+i\infty} ds\, e^{ts}\,\tilde{\rho}(s) = -(i/2\pi i)\int_{\gamma-i\infty}^{\gamma+i\infty} ds\, e^{ts}[1/(L-is)]\,\rho(0) \tag{4}$$

The contour of integration is a straight line parallel to the imaginary axis with all singularities of $1/(L-is)$ to its left.[1] It is convenient to write

$$z = is \tag{5}$$

We then obtain

$$\rho(t) = -(1/2\pi i)\int_{-\infty+i\gamma}^{+\infty+i\gamma} dz\, e^{-izt}\,[1/(L-z)]\,\rho(0) \tag{6}$$

where the integration is now along a line parallel to the real axis, above all singularities of $1/(L-z)$.

In (8.3.6) there appears the operator

$$R(z) = 1/(L-z) \tag{7}$$

It is called the resolvent corresponding to the Liouville operator L. As L is an Hermitian operator and has only real eigenvalues, R exists and is bounded for all non-real z (see for example Stone, 1932). The only singularities of R are therefore on the real axis.

Because of the exponential factor $\exp(-izt)$, with $t \geqq 0$, in (8.3.6) we may replace the integration over a straight line in (8.3.6) by one over a large semicircle in the lower half plane

[1] The contour is chosen in such a way that $\rho(t)$ vanishes for $t < 0$.

[see Fig. 8.3.1(a)]. Moreover as the singularities of R are on the real axis we may also use any other closed contour C encircling the real axis.

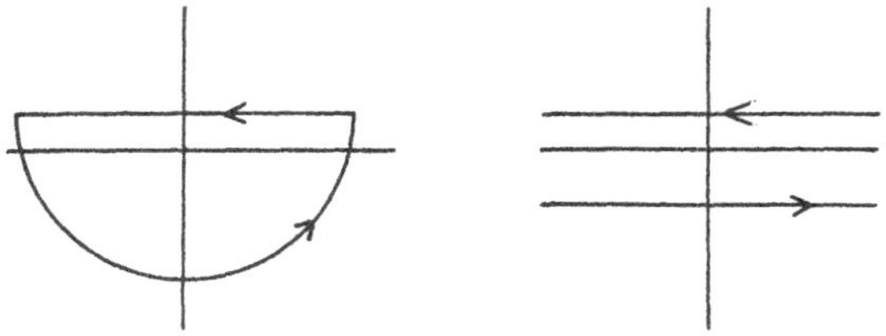

Fig. 8.3.1 Integration contours in (8.3.8).

We shall therefore write

$$\rho(t) = -(1/2\pi i)\oint_C dz\, e^{-izt}\, R(z)\rho(0) \tag{8}$$

We may also define a resolvent $(L_0-z)^{-1}$ for the unperturbed Liouville operator L_0; using the operator identity

$$A^{-1}-B^{-1} = A^{-1}(B-A)B^{-1} \tag{9}$$

we may write

$$(L-z)^{-1}-(L_0-z)^{-1} = -\lambda(L_0-z)^{-1}\delta L(L-z)^{-1} \tag{10}$$

which we solve formally by iteration

$$(L-z)^{-1} = \sum_{n=0}^{\infty}(-\lambda)^n(L_0-z)^{-1}[\delta L(L_0-z)^{-1}]^n \tag{11}$$

We now express our formal solution, given by (8.3.8) and (8.3.11), in the Fourier representation; with obvious notations we get [see (4.1.9)] [1]

$$\rho_{\{k\}}(t) = -(1/2\pi i)\oint_C dz\, e^{-izt}\sum_{\{k\}}\sum_{n=0}^{\infty}(-\lambda)^n\langle\{k\}|(L_0-z)^{-1} \times [\delta L(L_0-z)^{-1}]^n|\{k'\}\rangle\rho_{\{k'\}}(0) \tag{12}$$

We notice that the resolvent of the unperturbed system is diagonal in the Fourier representation [see (1.5.18) and (1.5.19)]

[1] In order to avoid too cumbersome a notation we include the complete time dependence in $\rho_{\{k\}}(t)$ and do not explicitly write the volume factors [see (7.2.1)].

$$\langle\{k\}\,|1/(L_0-z)|\{k'\}\rangle = (1/\sum k_j \cdot v_j - z)\,\delta_{\{k\}\{k'\}} \tag{13}$$

The main difference between this method and the one used in the preceding chapters is that now to each set of wave vectors $\boldsymbol{k}$ corresponds the time-independent propagator $1/(\sum k_j \cdot v_j - z)$ instead of the oscillating exponentials

$$\exp\,[i \sum_j k_j \cdot v_j(t_n - t_{n-1})]$$

We shall now study the analytic behavior of the resolvent in the z plane more closely.

4. Analytic Behavior of the Resolvent

To study the analytic behavior of the resolvent, let us consider the transition from $\rho_{\{0\}}$ to $\rho_{\{0\}}$ and first retain only the contributions up to order λ^2. In accordance with (8.3.12) and (8.3.13) we have

$$\langle 0|R(z)|0\rangle = -1/z + \lambda^2 \langle 0|(1/z)\,\delta L[1/(L_0-z)]\,\delta L\,(1/z)|0\rangle \tag{1}$$

The first term corresponds to the unperturbed motion, the second to the effect of cycles as studied for example in Chapter 2. This second term is of the form

$$1/z^2\,\Psi(z) \tag{2}$$

where Ψ is a sum of terms corresponding to all possible couples of particles. For example, the contribution of the collision between particles α and l is

$$\begin{aligned}\Psi_{\alpha l}(z) &= \langle 0|\delta L_{\alpha l}\,[1/(L_0-z)]\delta L_{\alpha l}|0\rangle \\ &= \sum_{\{k\}} \langle 0|\delta L_{\alpha l}|\{k\}\rangle 1/(k \cdot g_{\alpha l} - z)\langle\{k\}|\delta L_{\alpha l}|0\rangle\end{aligned} \tag{3}$$

Using (4.1.7) as well as the notation (6.5.2) and replacing the sum over $\boldsymbol{k}$ by an integral, we get

$$\Psi(z) = 1/8\pi^3 \int dk\, V_k\, k \cdot D\; 1/(k \cdot g - z) V_{-k}\, k \cdot D \tag{4}$$

where we have dropped the indices α and l.

This expression is (apart from the operators acting on the

velocities, which are irrelevant here) of the form of a so-called *Cauchy integral*

$$\varphi(z) = (1/2\pi i) \int_{-\infty}^{+\infty} dw\, f(w)/(w-z) \tag{5}$$

The denominator $1/(w-z)$ can only vanish for real values of z because the integration over w is performed along the real axis. Therefore this integral has a well-defined meaning for all non-real values of z. Let us call $\varphi^+(z)$, $\varphi^-(z)$ the values of φ corresponding to $\operatorname{Im} z > 0$ or $\operatorname{Im} z < 0$, respectively. The function φ equal to φ^+ in the upper half plane and to φ^- in the lower half plane is an analytic function everywhere except possibly on the real axis. There are some conditions that the function $f(w)$ in (8.4.5) must satisfy in order to obtain these results (see for example Muskelishwili 1953; Mikhlin, 1957). We shall however not discuss them here, (see also Balescu, 1962).

Let us study the way in which φ^+ and φ^- change when z approaches the real axis. We have seen that (cf. Chapter II, § 5)

$$1/(\alpha-i\varepsilon) = i\pi\delta_-(\alpha) = i\pi\delta(\alpha)+\mathscr{P}(1/\alpha) \tag{6}$$

$$1/(\alpha+i\varepsilon) = -i\pi\delta_+(\alpha) = -i\pi\delta(\alpha)+\mathscr{P}(1/\alpha) \tag{7}$$

Therefore $(z = x \pm i\varepsilon,\ \varepsilon > 0)$

$$\lim_{\varepsilon\to 0} \varphi^+(z) = \tfrac{1}{2}f(x)+(1/2\pi i) \int_{-\infty}^{+\infty} dw f(w)\, \mathscr{P}[1/(w-x)] \tag{8}$$

$$\lim_{\varepsilon\to 0} \varphi^-(z) = -\tfrac{1}{2}f(x)+(1/2\pi i) \int_{-\infty}^{+\infty} dw f(w)\, \mathscr{P}[1/(w-x)] \tag{9}$$

We see that φ has a finite discontinuity along the real axis. The behavior of the resolvent is completely similar. It is an analytic function of z except on the real axis where it has a discontinuity.

Note that

$$\lim_{\varepsilon\to 0} [\varphi^+(z)-\varphi^-(z)] = f(x) \tag{10}$$

We shall now extend the definition of $\varphi^+(z)$ to values of z in the lower half plane in such a way that it reduces to the function $\varphi^+(z)$ as defined above on the real axis and in the upper half plane. For this purpose it is sufficient to take

$$\varphi^+(z) = f(z)+\varphi^-(z) \qquad (\text{Im}\, z < 0) \tag{11}$$

Indeed when z approaches the real axis (from below) the right-hand side of (8.4.11) tends to

$$f(x)-\tfrac{1}{2}f(x)+(1/2\pi i)\int_{-\infty}^{+\infty} dw\, f(w)P(1/w-x)$$

which is precisely the limit of $\varphi^+(z)$ as given by (8.4.8). One says that (8.4.11) is the *analytic continuation* of $\varphi^+(z)$ into the lower half plane.

The situation is similar to that which occurs with the more familiar integral

$$\varphi(z) = (1/2\pi i)\oint_C dz'\, w(z')/(z'-z) \tag{12}$$

taken along a *closed* contour; $w(z')$ is some continuous function defined along the contour. According to whether z is inside or outside the contour we have two functions, say

$$\varphi^+(z), \qquad \varphi^-(z) \tag{13}$$

the first of which is analytic inside the contour and the other outside. An especially simple case occurs when $w(z')$ is equal to the value taken at z' by some function $\varphi(z)$ which is analytic within a domain bounded by the contour. Then the classical Cauchy formula gives

$$\begin{aligned} \varphi(z) &= (1/2\pi i)\oint_C dz'\, \varphi(z')/(z'-z) \qquad z \text{ inside } C \\ 0 &= (1/2\pi i)\oint_C dz'\, \varphi(z')/(z'-z) \qquad z \text{ outside } C \end{aligned} \tag{14}$$

In this case we have

$$\varphi^+(z) = \varphi(z) \qquad \varphi^-(z) = 0 \tag{15}$$

Now the function $\varphi^+(z)$ can generally be defined for values of z outside C as well. For example, if in (8.4.14) we take as our contour a circle of unit radius around the origin and let $\varphi(z) = 1/(z+2)$ we have by (8.4.15)

$$\varphi^+(z) = 1/(z+2) \tag{16}$$

While this function is analytic inside the contour, it has a pole for $z = -2$ *outside* the unit circle. It gives the analytic continuation of $\varphi^+(z)$ outside the contour C.

Let us now go back to the resolvent operator. As we did with the Cauchy integral (8.4.5), we may define two functions: $R^+(z)$ and $R^-(z)$. The first of these functions is analytic in the upper half plane, the second in the lower. We may also extend the definition of $R^+(z)$ to the lower half plane; the function $R^+(z)$, for $\text{Im}\, z < 0$, is the analytic continuation of $R^+(z)$ for $\text{Im}\, z \geqq 0$.

Using the path of Fig. 8.3.1 (b) we may write (8.3.8) as:

$$\rho(t) = (1/2\pi i) \lim_{\varepsilon \to 0} \int_{-\infty}^{+\infty} dx [e^{-i(x+i\varepsilon)t} R^+(z) - e^{-i(x-i\varepsilon)t} R^-(z)] \rho(0) \tag{17}$$

This expresses $\rho(t)$ in terms of the discontinuity of the resolvent along the real axis.

It is however more convenient to use instead of *both* $R^+(z)$ and $R^-(z)$, only $R^+(z)$, including its analytic continuation on the lower half plane. We then have using contour (a) of Fig. 8.3.1:

$$\rho(t) = -(1/2\pi i) \oint_C dz\, e^{-izt} R^+(z) \rho(0) \tag{18}$$

This integral may be evaluated by studying the singularities of $R^+(z)$ on the real axis and in the lower half plane. We shall consider an example in § 5.

Of course, $R(z)$ is an operator, not a function. When we speak about its analytic behavior in the z plane we really mean the analytic behavior of $R\rho(0)$ or, what is equivalent, that of its matrix elements.

Let us stress that the analytic behavior of the resolvent depends critically on the existence of a continuous spectrum for the Liouville operator L. It is only then that we may replace the sum (8.4.3) by the integral (8.4.4) and obtain a Cauchy integral of the form (8.4.5). In the case of a discrete spectrum corresponding to a finite volume, the resolvent would have only isolated singular points on the real axis. There would be no discontinuity when we cross the real axis and no question of analytic continuation, because the resolvent would be represented by a single function in the whole z plane.

5. Diagonal Fragments

We again first consider a cycle. As we have seen [cf. (8.4.1) and (8.4.2)] its contribution to the resolvent is given by

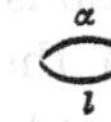

$$= -(1/2\pi i) \oint_C dz\, e^{-izt} (1/z) \Psi^+(z) (1/z)\, \rho_0(0) \qquad (1)$$

The contour of integration is, in accordance with (8.4.18), a straight line parallel to the real axis and a large semicircle in the lower z plane. For $\operatorname{Im} z < 0$ we use the analytic continuation Ψ^+ of the function Ψ defined by (8.4.4).

The evaluation of (8.5.1) is performed by the residue theorem. Inside the contour C we have a second-order pole at $z = 0$ and poles [1] arising from the function $\Psi^+(z)$ at points z_j in the lower half plane.

We have therefore

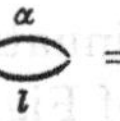

$$= it\Psi^+(0) - (d\Psi^+/dz)_0 + \sum_j (e^{-iz_j t}/z_j^2) \text{ res } \Psi^+(z_j) \qquad (2)$$

where the sum $\sum_j$ extends over all the poles of $\Psi^+(z_j)$. The three types of terms in (8.5.2) correspond exactly to the contributions (8.2.8), (8.2.10), and (8.2.11) that we studied in § 2. There we used an exponential intermolecular potential. $\Psi^+(z)$ then has a single pole of second order at

$$z_j = -i\kappa\bar{v} \qquad (3)$$

The important point is that the location of the poles in the lower half plane is related to the duration of the collision t_{int}. Therefore for times such that

$$|\operatorname{Im} z_j| t \sim t/t_{\text{int}} \gg 1 \qquad (4)$$

all the exponentials in (8.5.2) become negligible. If we consider the collision as if it were instantaneous and retain only the first term in (8.5.2) we come back to the expression (see Chapter 2, § 5) [2]

[1] Instead of poles other singularities like branch points or even essential singularities could arise, depending on the nature of the intermolecular potential. We shall limit our discussion to the case of poles.

[2] In formula (8.5.5), $\psi^+(0)$ means of course $\psi^+(z)$ for $z = 0$ and $\rho_0(0)$, the value of $\rho_0(t)$ for $t = 0$.

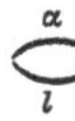

$$= it\Psi^{+}(0)\rho_0(0) = (1/8\pi^3 t)\int d\boldsymbol{k}V_k\boldsymbol{k}\cdot\boldsymbol{D}\delta_{-}(\boldsymbol{k}\cdot\boldsymbol{g})V_{-k}\boldsymbol{k}\cdot\boldsymbol{D}\rho_0(0) \quad (5)$$

We may say that the relaxation time is determined by the value of $\Psi^{+}(z)$ for $z = 0$, while the duration of the collision is determined by the position of the poles in the lower half plane.

The remainder of this chapter will be devoted to a qualitative study of the time dependence of arbitrary diagrams.

Let us call a "diagonal fragment" a set of at least three vertices such that the state preceding the fragment is identical to the state following it (see Fig. 8.5.1) [1].

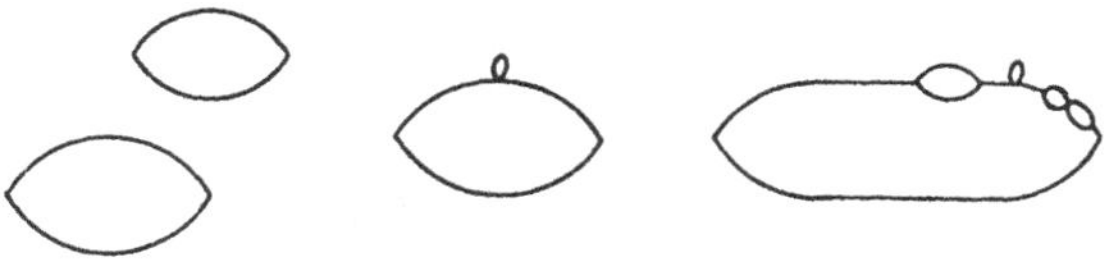

(a) (b) (c)

Fig. 8.5.1 Examples of diagonal fragments.

Whatever the precise form of the diagonal fragment may be it will give to the resolvent a contribution of the form [see (8.4.1) and (8.4.2)],

$$\langle 0|(1/z)\delta L\,[1/(L_0-z)]\,\delta L\,[1/(L_0-z)]\cdots\delta L(1/z)|0\rangle = 1/z^2\,\Psi(z) \quad (6)$$

The analytic form of $\Psi(z)$ depends on the precise structure of the diagonal fragment. On the other hand the factor z^{-2} indicates that we are considering a transition from a homogeneous state ρ_0 back to a homogeneous state. We shall make the following assumptions about the function $\Psi(z)$:

a. It is an analytic function of z for $\operatorname{Im} z \geqq 0$.

b. It has an analytic continuation for $\operatorname{Im} z < 0$. This analytic continuation has poles in the lower half plane at finite distances from the real axis.

Let us again call $\Psi^{+}(z)$ the function defined in this way for all values of z. The assumptions we have introduced are sufficient

[1] It is assumed in this definition that there is no other identical state inside the diagonal fragments.

but not necessary. Singularities other than poles may be treated as well.

The resolvent method immediately gives us the following theorems:

a. A single diagonal fragment gives an asymptotic time contribution of order t.

b. A succession of n diagonal fragments gives an asymptotic time contribution of order t^n (as well as terms of order $t^{n-1}, t^{n-2}, \ldots$).

Indeed a single diagonal fragment will give to the resolvent a contribution identical to (8.5.1) with a second-order pole at $z = 0$. We may then apply formula (8.5.2). Moreover a succession of n diagonal fragments gives to the resolvent a contribution of the form

$$-(1/2\pi i)\oint_C dz\, e^{-izt}(-1/z)\Psi^+(z)(-1/z)\Psi^+(z)\times\cdots\times\Psi^+(z)(-1/z)\rho_0(0)$$
$$= -(1/2\pi i)\oint_C dz\, e^{-izt}(-1/z)^{n+1}[\Psi^+(z)]^n\rho_0(0) \qquad (7)$$

This expression has a pole of order $n+1$ at $z = 0$ (we have two factors of z^{-1}, for the initial and the final state, and a factor of z^{-1} for each intermediate state between two diagonal fragments). To the pole of order $n+1$ at $z = 0$ corresponds, by the residue method, a contribution of t^n (as well as terms of order t^{n-1}, $t^{n-2}, \ldots$). For special cases this result has already been used in Chapters 2 and 6.

6. Free Propagation and Scattering

Let us first consider a freely propagating line (see § 1). Its contribution to (8.3.12) is [see also (8.3.13)]

$$\overset{\alpha}{\overline{\qquad\quad}} = -(1/2\pi i)\oint_C dz\, e^{-izt}\langle \boldsymbol{k}|1/(L_0-z)|\boldsymbol{k}\rangle\rho(\alpha_k, t=0)$$
$$= -(1/2\pi i)\oint_C dz\, e^{-izt}\,1/(\boldsymbol{k}\cdot\boldsymbol{v}_\alpha - z)\rho(\alpha_k, 0) \qquad (1)$$
$$= e^{-i\boldsymbol{k}\cdot\boldsymbol{v}_\alpha t}\rho(\alpha_k, 0)$$

in agreement with (8.1.1). As long as no integration over $\boldsymbol{k}$ is involved we have a plane wave. No Cauchy integral of the form (8.4.5) can be defined and no asymptotic time integration has a meaning. The situation changes, as explained in § 1, when we consider a wave packet of width κ obtained by integrating over $\boldsymbol{k}$. Then the lifetime of this packet is given by (8.1.8). Correspondingly in the resolvent formalism we obtain by integration over $\boldsymbol{k}$ the Cauchy integral:

$$-(1/2\pi i)\oint_C dz\, e^{-izt}\int d\boldsymbol{k}\; 1/(\boldsymbol{k}\cdot\boldsymbol{v}_\alpha - z)\rho(\alpha_k, 0) \tag{2}$$

The integral

$$I^+(z) = \int d\boldsymbol{k}\; 1/(\boldsymbol{k}\cdot\boldsymbol{v}_\alpha - z)\,\rho(\alpha_k, 0) \tag{3}$$

may be studied in exactly the same way as the integral $\Psi^+(z)$ that we considered in §§ 4–5. We may introduce its analytic continuation in the lower half plane. Again the position of the poles (or of other possible singularities) will be determined by the width of the wave packet. In this way the results obtained in § 1 may be recovered in the resolvent formalism.

Let us now consider a diagonal fragment inserted on a line. This corresponds to the scattering of particle α by a particle of the medium. We have already considered such a diagram in Chapter 5. We shall call it briefly a "*scattering diagram*".

The resolvent method permits us to separate the contribution of the diagonal fragment in a very direct way (see Fig. 8.6.1). Let us consider a cycle inserted on a line. Its contribution to (8.3.12) is

$$\begin{aligned}\rho(\alpha_k, t) &= -(1/2\pi i)\oint_C dz\, e^{-izt}[1/(\boldsymbol{k}\cdot\boldsymbol{v}_\alpha - z)]\\ &\times\int d\boldsymbol{l} V_l \boldsymbol{l}\cdot\boldsymbol{D}[1/(\boldsymbol{l}\cdot\boldsymbol{g}_{\alpha j} + \boldsymbol{k}\cdot\boldsymbol{v}_\alpha - z)]V_{-l}\boldsymbol{l}\cdot\boldsymbol{D}[1/(\boldsymbol{k}\cdot\boldsymbol{v}_\alpha - z)]\rho(\alpha_k, 0)\end{aligned} \tag{4}$$

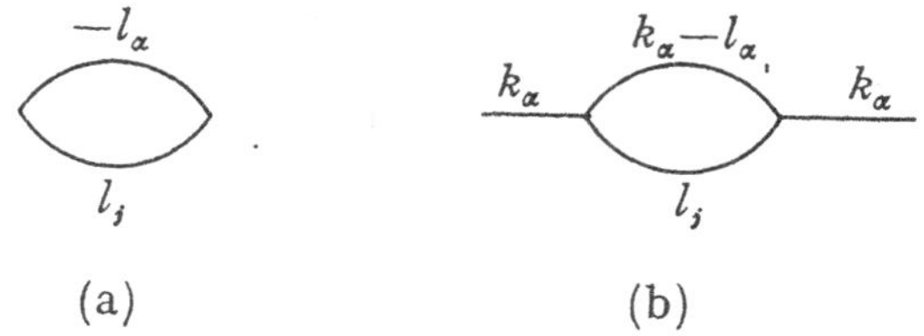

Fig. 8.6.1 (a) Cycle. (b) Cycle inserted on a line.

We introduce the following change of variables:

$$\begin{aligned} z - \boldsymbol{k}\cdot\boldsymbol{v}_\alpha &\to y\\ \partial/\partial\boldsymbol{v}_\alpha &\to (\partial/\partial\boldsymbol{v}_\alpha) - \boldsymbol{k}(\partial/\partial y). \end{aligned} \tag{5}$$

We then have

$$\rho(\alpha_k, t) = -(1/2\pi i)e^{-i\mathbf{k}\cdot\mathbf{v}_\alpha t}\oint_C dy\, e^{-ivt}(1/y)\int d\boldsymbol{l} V_l \boldsymbol{l}\cdot(\mathbf{D}-\mathbf{k}\,\partial/\partial y)$$
$$\times [1/(\boldsymbol{l}\cdot\mathbf{g}_{\alpha j}-y)]\, V_{-l}\boldsymbol{l}\cdot(\mathbf{D}-\mathbf{k}\,\partial/\partial y)(1/y)\rho(\alpha_k, 0) \quad (6)$$

The integration over $\boldsymbol{l}$ leads us again to a Cauchy integral (see 8.4.5). More precisely, performing the differentiation with respect to y we obtain three terms which have, respectively, poles of second, third, and fourth order at $y = 0$

$$\rho(\alpha_k, t) = -(1/2\pi i)\, e^{-i\mathbf{k}\cdot\mathbf{v}_\alpha t}\oint_C dy\, e^{-ivt}(1/y^2)$$
$$\times [\Psi_2^+(y)+(1/y)\Psi_3^+(y)+(1/y^2)\Psi_4^+(y)]\,\rho(\alpha_k, 0) \quad (7)$$

The poles of each of the Cauchy integrals Ψ_2^+, Ψ_3^+, Ψ_4^+ are determined by the range of the intermolecular forces.

If we consider the scattering process as instantaneous, we may again neglect the effect of all poles, except the one at $y = 0$, as well as corrections to the scattering cross section expressed by derivatives of the functions Ψ^+ at $y = 0$ [see (8.5.2)].

The contribution of the pole at $y = 0$ is

$$\begin{aligned}&-t\boldsymbol{l}\cdot\mathbf{D}\delta_+(\boldsymbol{l}\cdot\mathbf{g}_{\alpha j})\boldsymbol{l}\cdot\mathbf{D}\\ &\tfrac{1}{2}it^2[\boldsymbol{l}\cdot\mathbf{k}\,\delta_+(\boldsymbol{l}\cdot\mathbf{g}_{\alpha j})\boldsymbol{l}\cdot\mathbf{D}+\boldsymbol{l}\cdot\mathbf{D}\,\delta_+(\boldsymbol{l}\cdot\mathbf{g}_{\alpha j})\boldsymbol{l}\cdot\mathbf{k}]\\ &+\tfrac{1}{3}t^3\,\boldsymbol{l}\cdot\mathbf{k}\,\delta_+(\boldsymbol{l}\cdot\mathbf{g}_{\alpha j})\boldsymbol{l}\cdot\mathbf{k}\end{aligned} \quad (8)$$

We therefore obtain the asymptotic expression

$$\rho(\alpha_k, t) = e^{-i\mathbf{k}\cdot\mathbf{v}_\alpha t}\int d\boldsymbol{l}|V_l|^2\{t\boldsymbol{l}\cdot\mathbf{D}\,\delta_+(\boldsymbol{l}\cdot\mathbf{g}_{\alpha j})\boldsymbol{l}\cdot\mathbf{D}$$
$$-(t/2)\,(i\mathbf{k}t)\cdot\boldsymbol{l}\delta_+(\boldsymbol{l}\cdot\mathbf{g}_{\alpha j})\boldsymbol{l}\cdot\mathbf{D}-(t/2)\,\boldsymbol{l}\cdot\mathbf{D}\,\delta_+(\boldsymbol{l}\cdot\mathbf{g}_{\alpha j})\,\boldsymbol{l}\cdot(i\mathbf{k}t)$$
$$+(t/3)\,\boldsymbol{l}\cdot(i\mathbf{k}t)\delta_+(\boldsymbol{l}\cdot\mathbf{g}_{\alpha j})\boldsymbol{l}\cdot(i\mathbf{k}t)\}\rho(\alpha_k, 0) \quad (9)$$

This may also be written in the following way:

$$\rho(\alpha_k, t) = e^{-i\mathbf{k}\cdot\mathbf{v}_\alpha t}\int_0^t dt_1\int d\boldsymbol{l}|V_l|^2\boldsymbol{l}\cdot(\mathbf{D}-i\mathbf{k}t_1)\,\delta_+(\boldsymbol{l}\cdot\mathbf{g}_{\alpha j})$$
$$\times\boldsymbol{l}\cdot(\mathbf{D}-i\mathbf{k}t_1)\,\rho(\alpha_k, 0) \quad (10)$$

or

$$\rho(\alpha_k, t) = e^{-i\mathbf{k}\cdot\mathbf{v}_\alpha t}\int_0^t dt_1\, e^{i\mathbf{k}\cdot\mathbf{v}_\alpha t_1}\int d\boldsymbol{l}|V_l|^2\boldsymbol{l}\cdot\mathbf{D}\,\delta_+(\boldsymbol{l}\cdot\mathbf{g}_{\alpha j})\boldsymbol{l}\cdot\mathbf{D}\,e^{-i\mathbf{k}\cdot\mathbf{v}_\alpha t_1}$$
$$\times\rho(\alpha_k, 0) \quad (11)$$

The only difference between (8.6.11) and (8.5.5) is that in (8.6.11) a supplementary time dependence remains in the exponentials corresponding to free propagation from time $t = 0$ till t_1, when the interaction occurs, and then again from time t_1 to time t. This formula agrees with (5.3.14) which we obtained by a straightforward time integration.

7. Destruction of Correlations

Let us discuss in detail the asymptotic contiibution due to the simplest diagram corresponding to a destruction of correlations

$$= (1/2\pi i)\oint_C dz\, e^{-izt}(1/z)\int d\boldsymbol{k} V_k \boldsymbol{k}\cdot\boldsymbol{D}[1/(\boldsymbol{k}\cdot\boldsymbol{g}_{\alpha n} - z)]\,\rho(n_k, \alpha_{-k}, 0) \tag{1}$$

We again have to study the Cauchy integral

$$F^+(z) = \int d\boldsymbol{k}\, V_k\, \boldsymbol{k}\cdot\boldsymbol{D}\,[1/(\boldsymbol{k}\cdot\boldsymbol{g} - z)]\,\rho(k) \tag{2}$$

where we have written $\rho(k)$ for $\rho(n_k, \alpha_{-k}, t = 0)$. We are using the letter F because we shall keep Ψ for contributions due to diagonal fragments.

The important point is that this time dependence will now be determined by two characteristic lengths: the range of interaction $1/\kappa$, which appears here as well as in the case of diagonal fragments, and the *range of correlations at time* $t = 0$.

Indeed let us suppose that the spectral range corresponding to the correlation is of order

$$\Delta k \simeq \kappa_{\text{corr}} \tag{3}$$

We may for example represent $\rho(k)$ by a formula of the type $\exp(-\kappa_{\text{corr}}\, r)$. Following the procedure of § 2 we may show that now $F^+(z)$ will also have a pole at

$$|\text{Im } z| \simeq \kappa_{\text{corr}}\, v \tag{4}$$

The asymptotic time integration therefore requires

$$t\kappa_{\text{corr}}\, v \gg 1 \tag{5}$$

We may then neglect the effect of the decaying exponentials related to the poles (8.7.4) (see § 5).

In most situations we may expect that the correlations which exist initially are due to intermolecular interactions. Then $\kappa_{\text{corr}} \simeq \kappa_{\text{int}}$ and (8.7.5) reduces to the usual requirement that the time has to be long with respect to the duration of an interaction. But this is not always so. For example, in the problem of homogeneous turbulence one deals with long-range correlations. Then

$$\kappa_{\text{corr}} \ll \kappa_{\text{int}} \tag{6}$$

and macroscopic times may be required to ensure the validity of the asymptotic time integration for (8.7.1).

Using (8.7.2), we may write (8.7.1) in the following way [see (8.5.2), but here $z = 0$ is a first-order pole]:

$$\overset{n}{\underset{\alpha}{\Big\langle}} = -(1/2\pi i)\oint_C dz\, e^{-izt}(1/z)\,F^+(z) = F^+(0) + \sum_j (e^{-iz_jt}/z_j)\,\text{res}\, F^+(z_j) \tag{7}$$

The asymptotic time contribution is therefore time independent.

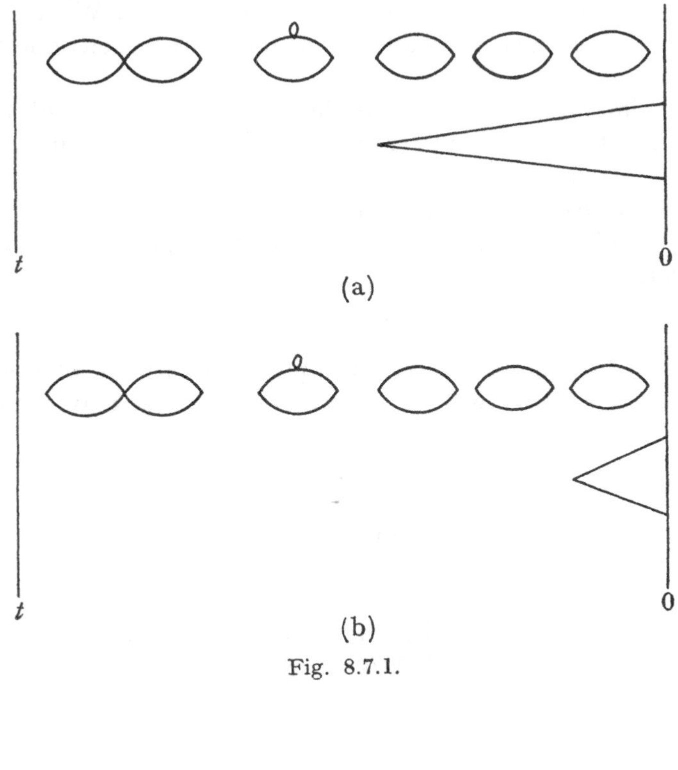

Fig. 8.7.1.

An important feature is that the asymptotic time dependence of diagrams involving both diagonal fragments and the destruction of correlations is critically determined by the relative position of the different parts of the diagram. Let us for example compare the two diagrams (a) and (b) of Fig. 8.7.1. We see immediately that the asymptotic time dependence of (a) is $\sim t^2$, while that of (b) is $\sim t^4$. Indeed in the first case $z = 0$ is a third-order pole, while in the second it is a fifth-order pole.

8. Creation of Correlations

Let us now consider diagrams corresponding to the creation of correlations. We shall again study in detail the simple case corresponding to

$$\overset{\alpha}{\underset{\beta}{>}} = (1/2\pi i) \oint_C dz\, e^{-izt} [1/(\boldsymbol{k}\cdot\boldsymbol{g}_{\alpha\beta} - z)]\, V_k \boldsymbol{k}\cdot\boldsymbol{D}(1/z)\, \rho_0(0) \qquad (1)$$

As in the case of free propagation (see § 6), as long as we remain in the Fourier representation we have no summation which would allow us to bring (8.8.1) into an integral of the Cauchy type. However if we are interested in distribution functions in phase space we still have to multiply (8.8.1) by exp $(i\boldsymbol{k}\cdot\boldsymbol{r})$ and then to sum over $\boldsymbol{k}$. In this way we obtain a Cauchy integral. The usual asymptotic contribution then arises as before from the first order pole at $z = 0$.

As in many other situations the physical meaning of a process becomes simpler when we go back to ordinary phase space. Let us therefore consider the expression[1]

$$\rho(\boldsymbol{r}, \boldsymbol{v}, t) = \rho_0(\boldsymbol{v}, 0) - \lambda \int_0^t dt\, \boldsymbol{F}(\boldsymbol{r} - \boldsymbol{v}t) \cdot \partial/\partial \boldsymbol{v}\, \rho_0(\boldsymbol{v}, 0) \qquad (2)$$

Its asymptotic time contribution corresponds to the limit $t \to \infty$ in the integral. This requires that the integral

$$\int_t^\infty \boldsymbol{F}(\boldsymbol{r} - \boldsymbol{v}t)\, dt = \int_{t-(r/v)}^\infty F(-\boldsymbol{v}t')\, dt' \qquad (3)$$

[1] Here $\boldsymbol{F}$ is the force between the two molecules α and β. This expression corresponds to the solution of the two-particle Liouville equation [see for example (6.1.3)] starting from the homogeneous state $\rho_0(\boldsymbol{v}, 0)$ and limited to first order in λ.

be negligible. This will be so if

$$t-(r/v) \gg t_{\text{int}} \qquad \text{or} \qquad t \gg t_{\text{int}}+(r/v) \tag{4}$$

where t_{int} is as usual the duration of an interaction.

We see therefore that as long as we are interested in the values of distribution functions for distances of the order of the range of intermolecular forces the effect of a creation diagram may be treated as time-independent, for times long with respect to t_{int}.

However if we are interested in much larger distances we have to take account of the effect of "retardation" r/v, in (8.8.4), which measures the time necessary for a correlation created at time $t = 0$ at the point $r = 0$ to travel to the point r. But then we can no longer neglect the scattering of particles α and β by the other particles of the medium. This corresponds to the diagrams represented in Fig. 8.8.1.

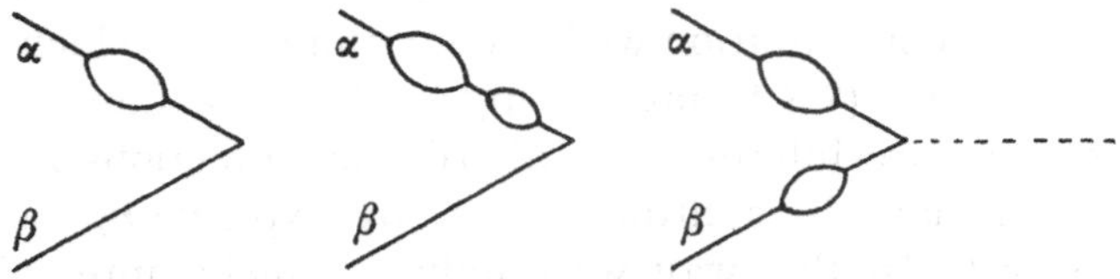

Fig. 8.8.1 Scattering of correlated particles α, β.

As could be expected, one can show (see Chapter 11, § 3) that the effect of this scattering on the correlation is to introduce a finite lifetime of the order of the relaxation time.

It is again important to understand the simple connection which exists between the topological structure of a diagram and its asymptotic time behavior. For example it is immediately verified that the asymptotic contribution of the diagram of Fig. 8.8.2 (a) vanishes while that of (b) is $\sim t^2$.

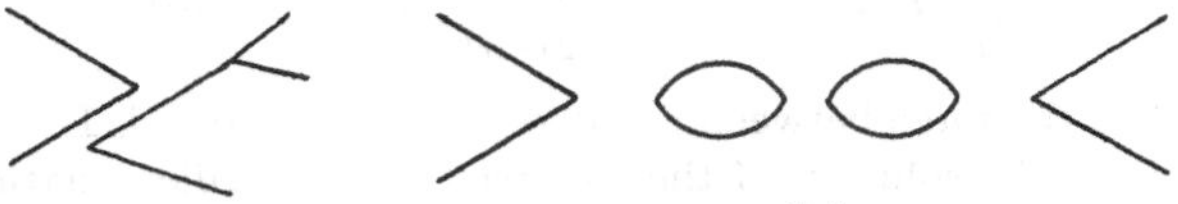

(a) (b)

Fig. 8.8.2.

Indeed in the first case there is no pole at $z = 0$. Therefore this diagram contains only decaying exponentials. In the second case $z = 0$ is a pole of third order.

9. Propagation of Correlations

Let us finally consider the asymptotic time integration for vertices involving the propagation of correlations [Fig. 7.6.1 (a) and (f)]. The asymptotic behavior of vertex (*f*) may be obtained by combining the results of § 7 and § 8. At the vertex (*f*) a correlation of wave vector $\boldsymbol{l}$ is destroyed. This destruction is followed by the creation of a correlation of wave vector $\boldsymbol{k}$ (see Fig. 8.9.1). When both the summations over $\boldsymbol{k}$ and $\boldsymbol{l}$ may be included in Cauchy's integrals, we obtain an asymptotically vanishing contribution. Indeed $z = 0$ is not a pole and we obtain only decaying exponentials, exactly as in the case of diagram (a) of Fig. 8.8.2.

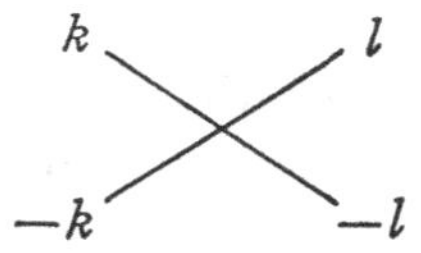

Fig. 8.9.1 Propagation of correlations.

On the contrary the vertex (a) (see also Fig. 8.9.2) introduces a new feature because a single wave vector $\boldsymbol{k}$ is involved. At the left and at the right of the vertex we have free propagating lines corresponding to a plane wave (see § 1). If however we want to obtain the reduced distribution function corresponding to particle α, we have to perform an *integration over the velocity of*

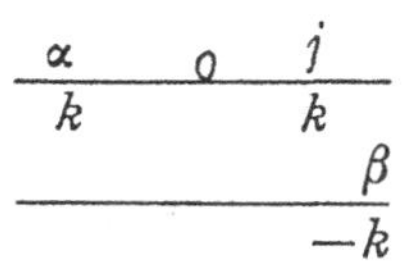

Fig. 8.9.2 Propagation of a correlation.

particle j. This integration provides us with the possibility of introducing an asymptotic time integration even if no summation over the wave vector $\boldsymbol{k}$ is involved. Indeed we then have the following contribution to the resolvent:

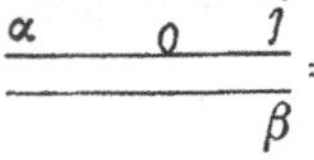

$$= -(1/2\pi i)\oint_C dz\, e^{-izt}\,[1/(\boldsymbol{k}\cdot\boldsymbol{g}_{\alpha\beta}-z)]\,V_k\,\boldsymbol{k}\cdot\boldsymbol{D}$$

$$\times \int d\boldsymbol{v}_j[1/(\boldsymbol{k}\cdot\boldsymbol{g}_{j\beta}-z)]\,\rho(j_k, \beta_{-k}, \boldsymbol{v}_j, \boldsymbol{v}_\alpha, \boldsymbol{v}_\beta, t=0) \quad (1)$$

The Cauchy integral which we have to study now is

$$I^+(z) = \int d\boldsymbol{v}_j\,[1/(\boldsymbol{k}\cdot\boldsymbol{g}_{j\beta}-z)]\,\rho(j_k, \beta_{-k}, \boldsymbol{v}_j, \boldsymbol{v}_\alpha, \boldsymbol{v}_\beta, t=0) \quad (2)$$

It differs from the Cauchy integrals we met before in that the integration extends over the velocity $\boldsymbol{v}_j$. This means that the time after which we may use an asymptotic time integration now depends on the way in which ρ varies with $\boldsymbol{v}_j$ (instead of on its variation with $\boldsymbol{k}$ as in § 7). A simple case in which all calculations may be performed explicitly corresponds to

$$\rho(j_k, \beta_{-k}, \boldsymbol{v}_j, \boldsymbol{v}_\alpha, \boldsymbol{v}_\beta, t=0) = f(j_k, \beta_{-k}, \boldsymbol{v}_\alpha, \boldsymbol{v}_\beta)\,(\gamma^{3/2}/\pi^2)\,[1/(1+\gamma v_j^2)^2] \quad (3)$$

The parameter γ determines the dispersion of the velocities; its physical meaning is therefore analogous to that of m/kT. One finds then that (8.9.2) has a pole in the lower half plane for $|\text{Im}\, z| \simeq k/\gamma^{1/2} = k\bar{v}$, where $\bar{v}$ is the average velocity of molecules β corresponding to (8.9.3). An asymptotic time integration is therefore possible for

$$tk\bar{v} \gg 1 \quad (4)$$

If $1/k$ is of the order of the range of the intermolecular forces this condition becomes identical with (8.5.4).

Using the change of variables (8.6.5) we may write (8.9.1) in the form

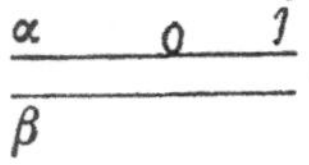

$$= -e^{-i\boldsymbol{k}\cdot\boldsymbol{g}_{\alpha\beta}t}(-1/2\pi i)\oint_C e^{-iyt}(1/y)\,V_k$$

$$\times\, \boldsymbol{k}\cdot(\boldsymbol{D}-\boldsymbol{k}\,\partial/\partial y)\,I^+(y+\boldsymbol{k}\cdot\boldsymbol{v}_\alpha) \quad (5)$$

For times for which (8.9.4) is satisfied this relation gives us the asymptotic time contribution due to the pole at $y = 0$,

$$= e^{-i\boldsymbol{k}\cdot\boldsymbol{g}_{\alpha\beta}t}\,V_k\,\boldsymbol{k}\cdot[\boldsymbol{D}I^+(\boldsymbol{k}\cdot\boldsymbol{v}_\alpha)-\boldsymbol{k}(\partial I^+/\partial y)_{y=\boldsymbol{k}\cdot\boldsymbol{v}_\alpha}] \quad (6)$$

It is interesting to compare more closely the diagrams of Figs. 8.9.1, 8.9.2, both of which correspond to the propagation of a correlation. While in Fig. 8.9.1 the particles which appear at the right and the left of the vertex are the same and the wave vector is changed, in Fig. 8.9.2. the particles are different and the wave vector is the same. The first diagram corresponds to a modification of the correlation between given particles and plays as we have seen in Chapter 6, § 6 a fundamental role in dilute gases. On the other hand Fig. 8.9.2 corresponds to the transmission of a given correlation from one molecule to another and plays as we shall see in Chapter 9 a fundamental role in collective processes involving long-range forces.

Let us now make a few remarks about the general relations which exist between the time dependence of the diagrams and their structure.

10. General Remarks About the Time Dependence of Diagrams

We have now identified the characteristic times to which we have to compare the time t in order to apply an asymptotic time integration.

It is remarkable that each type of diagram (diagonal fragments, creation or destruction fragments, propagation of correlations) introduces a new characteristic time. We therefore have to deal with four characteristic times. There are however two situations which are especially simple.

a. Weakly coupled systems or interacting particles at low concentrations; in these cases (which we have already discussed in Chapters 2 through 6) the evolution of the velocity distribution depends exclusively on diagonal fragments. Therefore, the only characteristic time that appears is the duration of an interaction.

b. The case in which all four characteristic times are of the order of the duration of an interaction t_{int}. As we shall see in § 12, there occurs then for times long with respect to t_{int}, an enormous reduction in the number of diagrams one has to consider.

Another essential point is that the resolvent method permits us to classify the diagrams according to their behavior for times long with respect to the characteristic times. We have two classes of diagrams: the first gives rise to effects that increase with

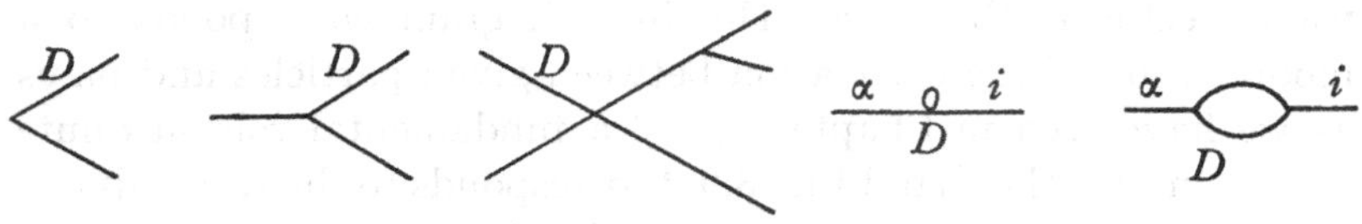

Fig. 8.10.1 Destruction vertex.

time, for example the diagonal fragment (8.5.2) or the scattering process (8.6.7). The second class gives asymptotic contributions which are at most time-independent[1] [for example the destruction diagram (8.7.1)], or even decay to zero, e.g., diagram (a) in Fig. 8.8.2. The first class has at least a second-order pole at $z = 0$ [eventually, after a change in variables, see (8.6.5)].

In a succession of diagrams it is very useful to set apart a region which gives at most a time-independent contribution. Now, from this point of view all the diagrams of Fig. 8.10.1 have the same behavior.

In every case the asymptotic contribution of the part of the diagram at the right of the point D is time-independent (note that diagram (e) is not a diagonal fragment but corresponds, as in diagram (d), to the "transfer" of the wave vector from particle i to particle α). We shall call such a point a *destruction vertex*. As may be verified in Fig. 8.10.1, it is either a vertex which has at its right two lines or the last vertex of a connected part of a diagram in which a wave vector is transferred from one particle to another.

Using this definition of a destruction vertex we now define what we call a *destruction region*; it is the part of the diagram

[1] Or involves a time-independent quantity multiplied by an oscillating exponential as in (8.9.6).

which is limited by the last destruction vertex (see Fig. 8.10.2).

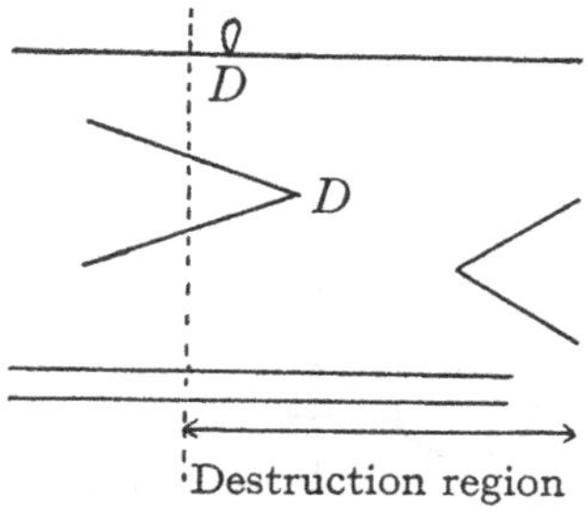

Fig. 8.10.2 Destruction region.

This whole region gives at most a time-independent asymptotic contribution.

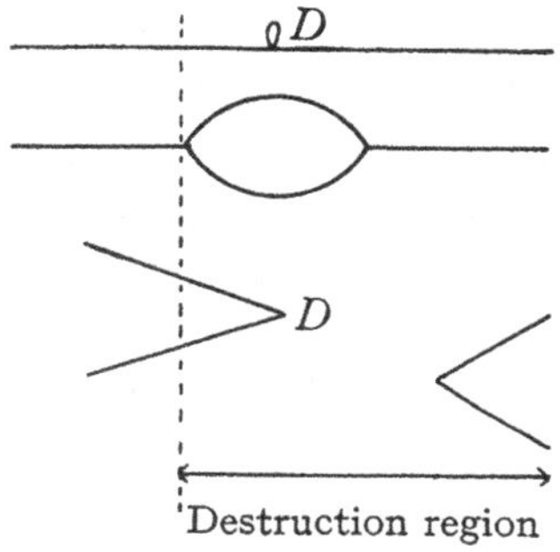

Fig. 8.10.3 Destruction region.

If a diagonal transition occurs at the same time as the last destruction vertex (see Fig. 8.10.3) it is convenient to absorb it in the destruction region.

After we have set apart the destruction region we are left with diagrams containing freely propagating lines (see §§ 1, 6), diagonal fragments (see § 5), scattering diagrams (see § 6), or vertices corresponding to a creation of correlations (see § 8). We obtain in this way for the general form of a diagram the Fig. 8.10.4.

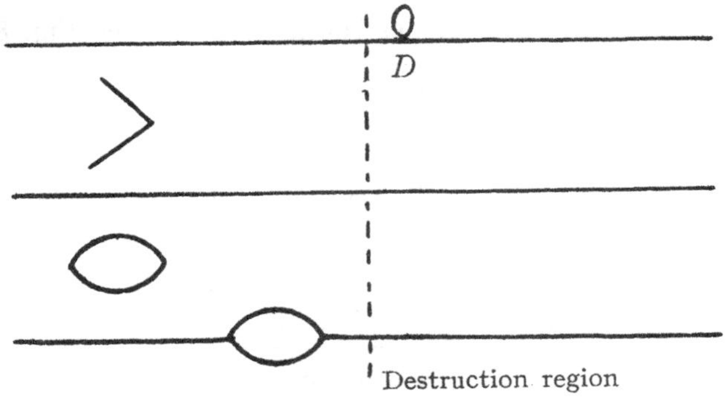

Fig. 8.10.4 General form of a diagram.

Let us consider the simplifications that occur in the structure of the diagrams when the only characteristic time we have to consider is t_{int}.

11. Velocity Distribution in Weakly Coupled or Dilute Systems

We may now consider this problem again, but from a much more general point of view than in Chapters 2 through 6. We shall first consider the evolution of the single-particle velocity distribution for weakly coupled systems.

The only contributions in $(\lambda^2 t)^n$ come from successive "semi-connected" cycles (see also Chapter 7, § 8). We may indeed verify that every more complicated diagonal fragment or every non-diagonal transition will introduce at least one factor λ uncompensated by a factor t. We are therefore back to our result (4.3.1) of (4.3.4). But we now see that it is valid under much more general conditions. There we assumed that only the velocity distribution ρ_0 was different from zero while all correlations vanished. We now see that this is not necessary, because in the weakly coupled case the other Fourier coefficients do not contribute to the evolution of the velocity distribution in the approximation we are considering (they give at most terms of order $\lambda^3 t$).

As we have already noticed, the evolution of the velocity distribution function proceeds entirely through "short living" binary correlations. This means that we never have more than two lines in any intermediate state. Therefore in order to construct the relevant contributions to ρ_0, we do not need the infinite set of Liouville equations for the Fourier coefficients as

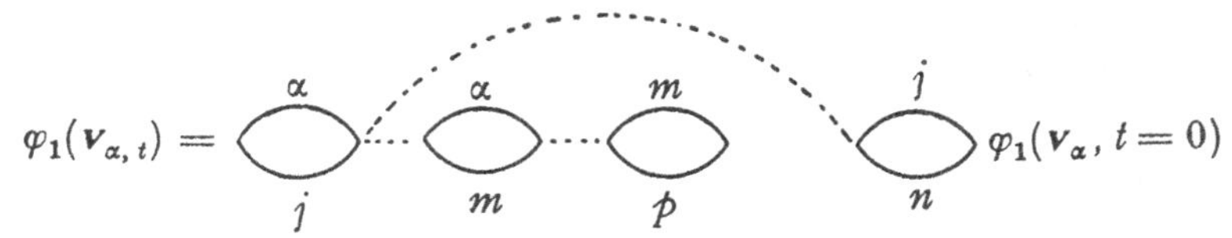

Fig. 8.11.1 Contributions to the velocity distribution.

represented in Fig. 7.6.2. The necessary contributions can be obtained from the closed set of the two equations which are represented by diagrams in Fig. 8.11.2. In them, the Fourier coefficient $\rho(l_k, j_{-k})$ plays a role somewhat like that of an intermediate virtual state in quantum mechanics.

While the total statistical "information" is carried by the Liouville equation for the complete phase distribution, the information corresponding to the velocity distribution is carried by a separate equation for ρ_0 (or φ_1).

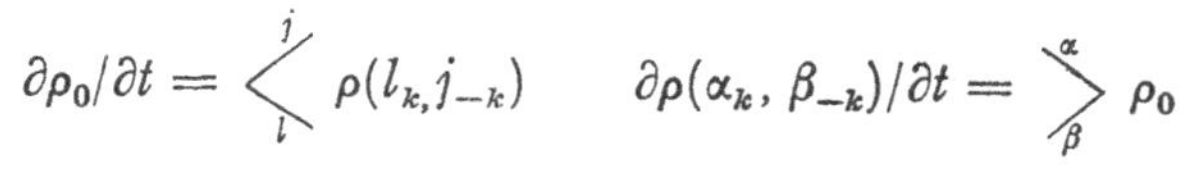

Fig. 8.11.2 Reduced set of equations for the evolution of the velocity distribution function of weakly coupled systems.

This fact is of course essential for the understanding of the mechanism of irreversibility and we shall discuss it again in Chapter 12, § 6.

A similar discussion can be presented for the velocity distribution for all particles. However, the master equation (4.2.2) still contains, as compared to the Boltzmann equation, an extra factor of N. This has to be so: the evolution of ρ_0 proceeds through collisions which may occur *anywhere* in the system, while that of $\varphi_1(v_\alpha)$ is influenced only by collisions in which particle α is involved (either directly or through the semiconnected diagrams).

In a time during which α undergoes one collision, there will be $O(N)$ collisions which will influence ρ_0.

The low concentration limit is similar to that of weak coupling. As we have seen in Chapter 6 we have only to replace the cycle by the sum of all successive Born approximations.

Let us finally consider the problem of the persistence of the molecular chaos condition (4.3.3). We already mentioned this problem in Chapter 4, § 3 but we have not yet given a proof of this persistence. Using the fact that only semiconnected diagrams contribute to the evolution of the velocity distribution (see Fig. 8.11.1) the proof becomes almost trivial.

We shall assume that at time $t = 0$ we have

$$\varphi_s(\mathbf{v}_1, \ldots, \mathbf{v}_s) = \varphi_1(\mathbf{v}_1)\varphi_1(\mathbf{v}_2)\cdots\varphi_1(\mathbf{v}_s) \tag{1}$$

where s is an arbitrary finite number independent of N. We want to show that this relation is maintained for all $t > 0$.

Let us give the proof for $s = 2$. The one-particle distribution function at time t will be expressed by the following series (see Fig. 8.11.1 and Chapter 6, § 6):

$\varphi_1(\mathbf{v}_\alpha, t) = \varphi_1(\mathbf{v}_\alpha, 0)$

$+t$ 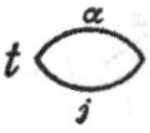 $\varphi_2(\mathbf{v}_\alpha, \mathbf{v}_j, 0)$

$+t^2/2$  $\varphi_3(\mathbf{v}_\alpha, \mathbf{v}_j, \mathbf{v}_n, 0) + \cdots$

$= \varphi_1(\mathbf{v}_\alpha, 0)$

$+t$ 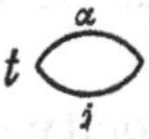 $\varphi_1(\mathbf{v}_\alpha, 0)\varphi_1(\mathbf{v}_j, 0)$

$+t^2/2$ 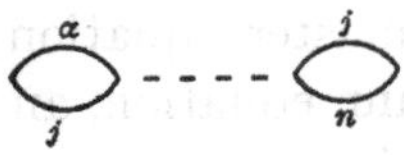 $\varphi_1(\mathbf{v}_\alpha, 0)\varphi_1(\mathbf{v}_j, 0)\varphi_1(\mathbf{v}_n, 0) + \cdots$ (2)

where we used the factorization condition (8.11.1) at time $t = 0$. We have a similar equation for $\varphi_1(\mathbf{v}_\beta, t)$ as well as for the two-particle velocity distribution $\varphi_2(\mathbf{v}_\alpha, \mathbf{v}_\beta, t)$. Indeed

$$
\begin{aligned}
\varphi_2(\mathbf{v}_\alpha, \mathbf{v}_\beta, t) = {} & \varphi_1(\mathbf{v}_\alpha, 0)\, \varphi_1(\mathbf{v}_\beta, 0) \\
& + t\Big[\overset{\alpha}{\underset{j}{\bigcirc}}\; \varphi_1(\mathbf{v}_\alpha, 0)\, \varphi_1(\mathbf{v}_j, 0)\, \varphi_1(\mathbf{v}_\beta, 0) \\
& + \overset{\beta}{\underset{n}{\bigcirc}}\; \varphi_1(\mathbf{v}_\beta, 0)\, \varphi_1(\mathbf{v}_n, 0)\, \varphi_1(\mathbf{v}_\alpha, 0)\Big] \\
& + t^2/2\Big[\overset{\alpha}{\underset{j}{\bigcirc}} \text{- - - -} \overset{j}{\underset{n}{\bigcirc}}\; \varphi_1(\mathbf{v}_\alpha, 0) \varphi_1(\mathbf{v}_j, 0) \varphi_1(\mathbf{v}_n, 0) \varphi_1(\mathbf{v}_\beta, 0) \\
& + \overset{\beta}{\underset{k}{\bigcirc}} \text{- - - -} \overset{k}{\underset{m}{\bigcirc}}\; \varphi_1(\mathbf{v}_\beta, 0)\, \varphi_1(\mathbf{v}_k, 0)\, \varphi_1(\mathbf{v}_m, 0)\, \varphi_1(\mathbf{v}_\alpha, 0) \\
& + \overset{\alpha}{\underset{j}{\bigcirc}}\; \overset{\beta}{\underset{k}{\bigcirc}}\; \varphi_1(\mathbf{v}_\alpha, 0)\, \varphi_1(\mathbf{v}_j, 0)\, \varphi_1(\mathbf{v}_\beta, 0)\, \varphi_1(\mathbf{v}_k, 0) \\
& + \overset{\beta}{\underset{k}{\bigcirc}}\; \overset{\alpha}{\underset{j}{\bigcirc}}\; \varphi_1(\mathbf{v}_\alpha, 0)\, \varphi_1(\mathbf{v}_j, 0)\, \varphi_1(\mathbf{v}_\beta, 0)\, \varphi_1(\mathbf{v}_k, 0)\Big] \\
& + \cdots
\end{aligned}
\tag{3}
$$

Comparing (8.11.3) with (8.11.2) we easily verify that

$$
\varphi_2(\mathbf{v}_\alpha, \mathbf{v}_\beta, t) = \varphi_1(\mathbf{v}_\alpha, t)\, \varphi_1(\mathbf{v}_\beta, t) \tag{4}
$$

This proves the persistence of (8.11.1). The essential step is of course the passage to the limit $L \to \infty$, $N \to \infty$, N/L^3 remaining finite. It is only then that we may neglect the direct interaction α, β and take into account only the interactions of α or β with the "dummy" particles j, n, k, and m. In agreement with the proof given by Kac (1954) the persistence of molecular chaos is an asymptotic property of large systems.

12. The Thermodynamic Case

We now consider the following situations:

a. The system is homogeneous [see (7.2.5)]. We therefore have only to study Fourier coefficients such that the sum of their wave vectors vanish. Lines corresponding to free propagation describe correlations between particles.

b. The characteristic times which appear in the asymptotic time integrations (see § 10) are all of the order of the duration of an interaction. This implies that the range of the correlations at the initial time t is of the order of the range of the interactions; also, we are only interested in the values of molecular correlations for distances of the order of the range of the forces. We may call this situation the "thermodynamic case." It is indeed the values of the distribution functions at distances of the order of the range of the intermolecular forces that determine the equation of state (see the virial theorem for the pressure).

In hydrodynamic problems the first restriction is of course not satisfied; in other problems, like homogeneous turbulence, it is the second restriction (b) which is not satisfied because of the macroscopic range of the initial correlations.

Let us now consider times long with respect to the duration of an interaction, t_{int}. In this case we may: (a) Suppress all lines corresponding to free propagation. Indeed as we are only interested in the values of molecular correlations, we have, as in § 8, to multiply the Fourier coefficient by $e^{i\mathbf{k}\cdot\mathbf{r}}$ (with $\mathbf{r}$ of the order of the range of the force $\sim 1/\kappa$) and to sum over $\mathbf{k}$. In this way we obtain wave packets of width κ which decay to zero for times long with respect to t_{int} (see the example treated in §1). (b) Suppress for the same reason all scattering diagrams (see §6). (c) Use the asymptotic expressions for both the destruction region and the vertices involving the creation of correlations.

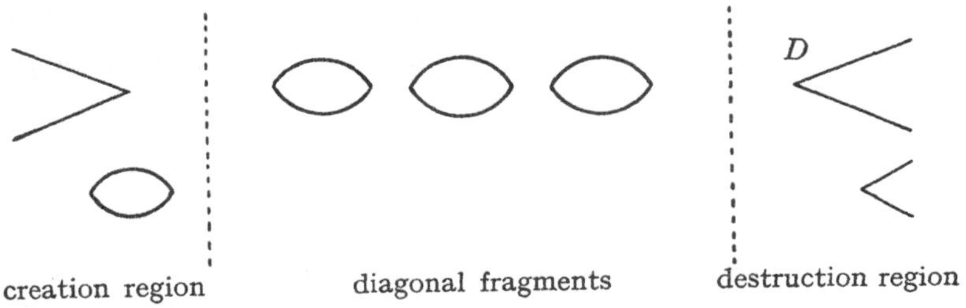

Fig. 8.12.1 General form of a diagram in the thermodynamic case for $t \gg t_{\text{int}}$.

The only diagrams we have to consider then are those which involve a succession of diagonal fragments embedded between the destruction region at the right and the creation of correlations at the left. The form of the diagrams obtained is represented in Fig. 8.12.1 which has to be compared to Fig. 8.10.4. This is a great simplification. We shall study in detail the corresponding kinetic equations in Chapter 11, § 5.

13. Dynamics of Correlations and Time Dependence

The results obtained in this chapter illustrate the close link which exists between the dynamics of correlations and their asymptotic time dependence. Indeed, the asymptotic time dependence is determined by the structure of the diagrams and therefore by the sequence of the states of correlation.

This may be considered as a rather wide generalization of a well-known result of scattering theory. A single scattering process gives an asymptotic contribution of order t whatever the details of the process involved (two- or many-body scattering, classical or quantum mechanics). A scattering process corresponds to the transition $\rho_0 \rightarrow \rho_0$, from the "vacuum of correlations" back to the vacuum of correlations (see Chapter 6, § 1).

It is therefore indeed a special case of the theorem, that the asymptotic time dependence is determined uniquely by the sequence of the states of correlation.

Once more we see that the laws of mechanics become surprisingly simple for large systems. For systems in which the boundaries play an important role no laws of such simplicity can be valid. The results derived in this chapter may therefore be considered as a step towards the formulation of a mechanics dealing specifically with large systems, a *statistical* mechanics in the true sense of the word.

CHAPTER 9

Approach to Equilibrium in Ionized Gases [1]

1. Choice of Diagrams

In Chapters 4, 6, and 8 we studied various situations in gases and we derived equations describing irreversible processes in these systems. We have studied weakly coupled gases and dilute gases of hard spheres, characterized respectively by the "Fokker-Planck" equation and by the Boltzmann equation. There are, however, systems which cannot be idealized by either of these two extreme models. We have already seen in Chapter 4, § 7, that *ionized gases* (or plasmas), in which the particles interact through Coulomb forces, give rise to difficulties when treated as weakly coupled gases, although the interactions are not strong on the average. The reason for this is the very slow decay of the interaction potential $1/r$ for large r. In all situations involving long-range forces we may expect the appearance of such difficulties, because in these cases configurations involving many correlated particles play a dominant role in the approach to equilibrium. The importance of these *collective effects* is the reason for the breakdown of the former methods which are essentially based on two- or three-body processes.

With our diagram technique these collective effects will appear through diagrams involving a large number of particles. We shall now consider the simplest situation, which is the case of a homogeneous one-component gas [2] of charged particles, in the lowest

[1] The statistical problems involved in the approach to equilibrium in ionized gases are studied in much greater detail in Balescu's monograph (Balescu, 1962).

[2] In fact, we have to assume the existence of a uniform background of opposite charge in this case in order to neutralize the overall electrostatic energy. But this background plays no role in the derivation of the evolution equation.

approximation [i.e., the approximation which in equilibrium corresponds to $\rho^{\text{eq}} = \rho(H_0)$]; we shall derive an equation for the evolution of the reduced one-particle momentum distribution function $\varphi(\boldsymbol{p}_\alpha, t) \equiv \varphi(\alpha)$ for this case. We shall essentially follow the treatment given by Balescu (1960). Our aim is to give a supplementary example of the use of the methods developed in the preceding chapters, in order to gain an increased familiarity with the physical meaning of diagrams. However the results we shall derive in this chapter are not used later and the reader interested in the general theory may proceed directly to Chapter 10.

Our system is defined by the Hamiltonian

$$H = \sum_j (p_j^2/2m) + \sum_{j<n} (e^2/|\boldsymbol{x}_j - \boldsymbol{x}_n|) = \sum_j (p_j^2/2m) + \sum_{j<n} (e^2/2\pi^2) \int d\boldsymbol{l}(1/l^2)\, e^{i\boldsymbol{l}\cdot(\boldsymbol{x}_j-\boldsymbol{x}_n)} \quad (1)$$

The first important point is that in the case of the plasma neither e^2 nor the concentration C are significant expansion parameters, as they are in the case of weakly coupled or hard-sphere gases. Indeed, as shown in Chapter 4, § 7 such expansions give rise to diagrams which diverge for small values of the wave vector $\boldsymbol{l}$. In order to find a key for the determination of the right parameter we shall be guided by the well-known results of the equilibrium theory. It has been shown rigorously [Mayer (1950)], that in that case the long-range interactions are in fact screened by a collective process which leads to an effective Debye-Hückel interaction potential (see Fig. 4.7.5)

$$V_{\text{eff}} = (e^2/r) \exp(-\kappa r) \quad (2)$$

where the Debye length κ^{-1} is defined by

$$\kappa^2 = 4\pi e^2 C/kT \quad (3)$$

We see that the combination e^2C appears in the exponential. Thus, in order to obtain a description that is to the same degree of approximation as the Debye equilibrium theory, we cannot limit ourselves to the first powers of e^2 or of C in any expansions. Rather, we have to retain *all* the terms proportional to any power of the combination e^2C in the expansions.

Now consider the case of a *homogeneous* plasma and examine the contributions to the Fourier component $\rho_0(\boldsymbol{p}_\alpha) \equiv \varphi(\alpha)$, i.e., the reduced distribution function of the momentum of one particle α. We then have to consider, as in Chapter 7, § 8 (see also Chapter 8, § 11) semi-connected diagrams, the components of which are diagonal fragments. Let us examine the general form of the possible diagonal fragments.

Among the six elementary diagrams (see Table 7.8.1), three are proportional to e^2C [(a), (b), (d)] and the three others are proportional to e^2 [(c), (e), (f)]. We want to construct diagrams containing an arbitrary number of e^2C-vertices, but a minimum number of e^2-vertices (we suppose that e^2 is small).

In order to have no external lines (homogeneous systems) the diagram has to be terminated at the left by a vertex (*b*) and at the right by a vertex (*c*). Because the latter introduces a factor e^2, the dominant diagrams will be obtained by inserting only e^2C-vertices between these two extremes. But the only vertex of that type that can be connected to the two extremes without any further adjunction of an e^2-vertex is the loop type (*a*). Therefore, the most general diagonal fragment in this problem is the *ring*, Fig. 9.1.1. In order that this diagram may give a non-vanishing

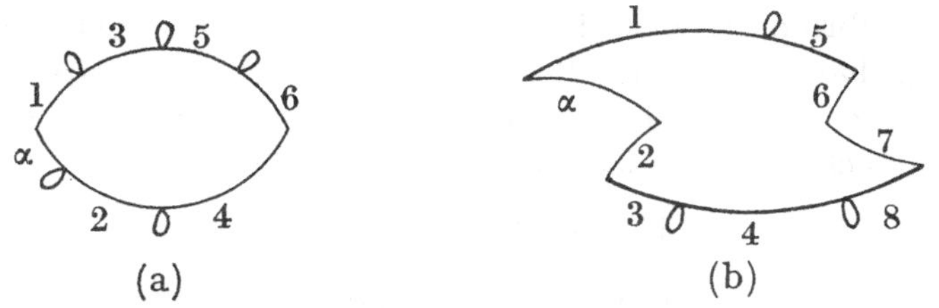

Fig. 9.1.1 Rings.

contribution after integration over all the particles except α, one of the two lines starting from the last vertex at the left has to be labeled α (see Chapter 7, § 8[1]). Moreover the asymptotic time dependence of any single diagonal fragment is always

[1] The "roman" indices of the preceding chapters are replaced in the present one by numerals 1, 2, ...

proportional to t (see Chapter 8, § 5). Therefore the contribution of a ring is of order $e^2(e^2/C)^n t$.

Let us remark that other diagrams of the same order are of type (b) Fig. 9.1.1 (concave rings); but these diagrams all give vanishing contributions after integration over all particles except α (see the second rule of Chapter 7, § 8).

Once we have obtained the significant diagrams, the general methods established in Chapter 7 allow us to derive the equation of evolution of $\varphi(\alpha)$ in the form:

$$\partial\varphi(\alpha)/\partial t = \int \prod{}' d\boldsymbol{v}\, R\varphi_N(\boldsymbol{v}_1 \cdots \boldsymbol{v}_N, t) \tag{4}$$

very easily, where φ_N is the distribution function of all the momenta and R is an operator corresponding to the sum of all possible rings; see Fig. 9.1.2.

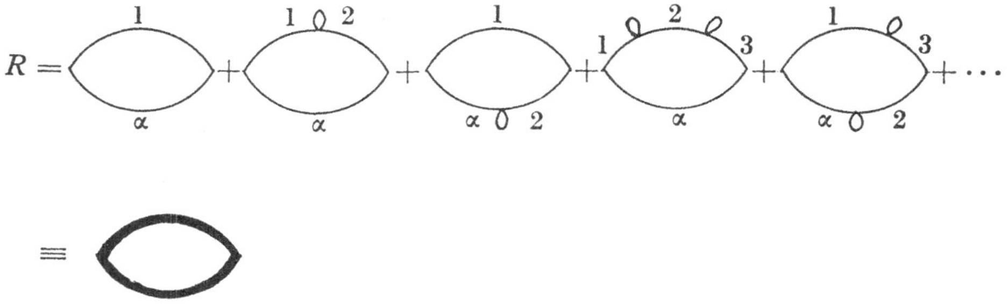

Fig. 9.1.2 The definition of the operator R.

It is interesting to note here that two other arguments could be used for the choice of the diagrams. First, one could classify all possible diagrams according to their dependence on C; then, choose in each of these classes the diagram proportional to the minimum power of the charge. This is the procedure first used by Mayer (1950) in the equilibrium theory. A second type of argument derives from the fact, already mentioned in Chapter 4, that each individual diagram diverges at small values of the wave vector $\boldsymbol{l}$. One may then retain, among all diagrams of the same order in e^2 the ones which are most divergent. This is the argument used by Gell-Mann and Brueckner (1957) in treating the quantum mechanical problem of the electron gas. We could use both of these arguments in our problem and find the same result as before, i e., the rings. This explains why there is such a strong topological analogy between our theory and the equilibrium theory, classical as well as quantum mechanical.[1]

Let us now construct the mathematical expression for the contribution to $\varphi(\alpha)$ corresponding to an arbitrary ring. First consider as an example the simple, typical ring of Fig. 9.1.3. The general rules of the previous chapters give the following expression for this contribution to the equation of evolution of the velocity distribution of α [see (4.1.9), (4.7.3), and Chapter 8, §§ 3–5):

$$e^8 C^3(-1/2\pi i) \oint_C dz\, e^{-izt} \int d\boldsymbol{v}_1 d\boldsymbol{v}_2 d\boldsymbol{v}_3 \int d\boldsymbol{l} (1/2\pi^2 l^2)^4 (1/z)^2$$
$$\times i\boldsymbol{l}\cdot \boldsymbol{D}_{\alpha 1}(\boldsymbol{l}\cdot\boldsymbol{g}_{\alpha 1}-z)^{-1} i\boldsymbol{l}\cdot\boldsymbol{D}_{12}(\boldsymbol{l}\cdot\boldsymbol{g}_{\alpha 2}-z)^{-1}(-i\boldsymbol{l}\cdot\boldsymbol{D}_{\alpha 3})(\boldsymbol{l}\cdot\boldsymbol{g}_{32}-z)^{-1}$$
$$\times i\boldsymbol{l}\cdot\boldsymbol{D}_{23}\,\varphi_4(\boldsymbol{v}_\alpha, \boldsymbol{v}_1, \boldsymbol{v}_2, \boldsymbol{v}_3, t) \tag{5}$$

The asymptotic time contribution comes as usual from the pole $z = 0$, and we use the molecular chaos condition (8.11.1) for the velocity distribution function.

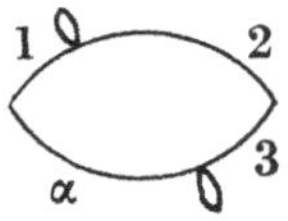

Fig. 9.1.3 A typical ring.

Observing that certain terms in the difference of derivatives, $\boldsymbol{D}_{ij}$, vanish by integration over the momenta, we rewrite (9.1.5) as follows:

$$e^8 C^3\, t \int d\boldsymbol{v}_1 d\boldsymbol{v}_2 d\boldsymbol{v}_3 \int d\boldsymbol{l} [(1/2\pi^2 l^2)^4] [i\boldsymbol{l}\cdot(\partial/\partial\boldsymbol{v}_\alpha)]\, \pi\delta_+(\boldsymbol{l}\cdot\boldsymbol{g}_{\alpha 1})$$
$$\times\ [i\boldsymbol{l}\cdot(\partial/\partial\boldsymbol{v}_1)]\,\pi\delta_+(\boldsymbol{l}\cdot\boldsymbol{g}_{\alpha 2})[-i\boldsymbol{l}\cdot(\partial/\partial\boldsymbol{v}_\alpha)]\,\pi\delta_+(\boldsymbol{l}\cdot\boldsymbol{g}_{32})[i\boldsymbol{l}\cdot\boldsymbol{D}_{23}]$$
$$\times\ \varphi(\alpha)\,\varphi(1)\,\varphi(2)\,\varphi(3) \tag{6}$$

[1] With minor modifications the method of this chapter applies to quantum mechanical plasmas as well (Balescu, 1961).

with, as usual [see (2.5.10)],

$$\pi\delta_+(x) = \pi\delta(x) + i\mathscr{P}(x^{-1}) \tag{7}$$

Let us introduce the following convenient abbreviations:

$$\begin{aligned} 4\pi i e^2 C l^{-2} \boldsymbol{l}\cdot(\partial/\partial \boldsymbol{v}_j) &\equiv d_j \\ (2\pi^2)^{-1} i e^2 l^{-2} \boldsymbol{l}\cdot \boldsymbol{D}_{jm} &\equiv d_{jm} \qquad j, m = \alpha, 1, 2, \ldots \\ \pi\delta_+(\boldsymbol{l}\cdot \boldsymbol{g}_{jm}) &\equiv \delta_+^{jm} \end{aligned} \tag{8}$$

The contribution (9.1.6) will then take the form:

$$t\int d\boldsymbol{v}_1 d\boldsymbol{v}_2 d\boldsymbol{v}_3 \int d\boldsymbol{l}\, d_\alpha \delta_+^{\alpha 1} d_1 \delta_+^{\alpha 2}(-d_\alpha)\delta_+^{32} d_{23}\varphi(\alpha)\varphi(1)\varphi(2)\varphi(3) \tag{9}$$

Studying other examples in a similar manner, one easily finds the following general rules of construction for the contribution corresponding to an arbitrary ring. (Let us use the convention of always drawing the rings in such a way that the line labeled α, with which all the diagrams end, is situated on the *lower* part of the ring.)

(a) To the extreme left vertex corresponds the expression $d_\alpha \delta_+^{\alpha 1}$.

(b) One writes the symbol d_j for each loop, successively, j being the label of the line situated at the *left* of the loop. These factors are given the sign + or − according to whether the loop is situated on the upper or on the lower side of the ring, respectively.

(c) One introduces factors δ_+^{is} between the factors d_j, where i is the label of the lower line and s the index of the upper line, in the state situated at the *right* of the loop corresponding to the preceding d_j factor.

(d) To the last vertex at the right corresponds a factor d_{si}, where s is the label of the upper line and i that of the lower line in the preceding state.

(e) The operator thus obtained acts on a product of momentum distribution functions $\varphi(\alpha)\varphi(1)\ldots\varphi(n)$; $\alpha, 1, \ldots, n$ are the indices of all the particles appearing in the ring.

(f) One integrates the expression thus obtained over the velocities $\boldsymbol{v}_1, \ldots, \boldsymbol{v}_N$, and over the wave vector $\boldsymbol{l}$.

2. Summation of Rings

In order to perform the summation of Fig. 9.1.2 explicitly we proceed in the following way.

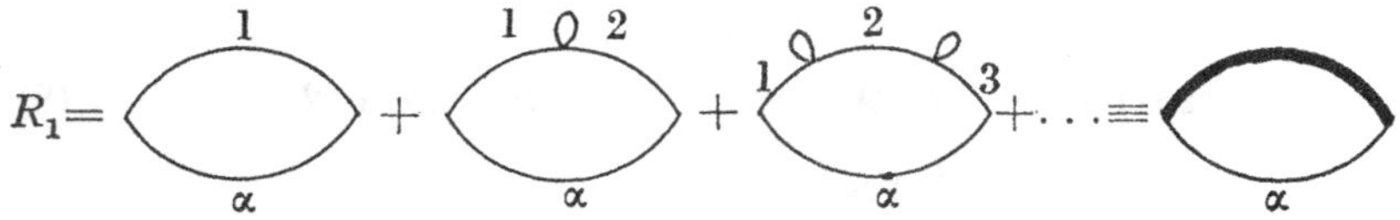

Fig. 9.2.1 The first step in the summation.

In the first step we sum all the rings having loops on the upper side alone (see Fig. 9.2.1). From the rules derived above, the following sum of contributions corresponds to this series:

$$\begin{aligned} R_1 = \int dl \{ & \int d\boldsymbol{v}_1 d_\alpha \delta_+^{\alpha 1} d_{1\alpha} \varphi(1) \varphi(\alpha) \\ & + \int d\boldsymbol{v}_2 d_\alpha \left[\int d\boldsymbol{v}_1 \delta_+^{\alpha 1} d_1 \varphi(1) \right] \delta_+^{\alpha 2} d_{2\alpha} \varphi(2) \varphi(\alpha) \\ & + \int d\boldsymbol{v}_3 d_\alpha \left[\int d\boldsymbol{v}_1 \delta_+^{\alpha 1} d_1 \varphi(1) \right] \left[\int d\boldsymbol{v}_2 \delta_+^{\alpha 2} d_2 \varphi(2) \right] \delta_+^{\alpha 3} d_{3\alpha} \varphi(3) \varphi(\alpha) \\ & + \cdots \} \end{aligned} \tag{1}$$

This is simply a geometrical progression, the ratio of which is

$$\mathcal{J}_\alpha = \int d\boldsymbol{v}_1 \delta_+^{\alpha 1} d_1 \varphi(1) \tag{2}$$

One can thus sum the series formally and obtain:

$$R_1 = \int dl \int d\boldsymbol{v}_1 d_\alpha (1 - \mathcal{J}_\alpha)^{-1} \delta_+^{\alpha 1} d_{1\alpha} \varphi(1) \varphi(\alpha) \tag{3}$$

To this class of rings we now add those which are obtained in the second step, in which we take the sum of all rings having an arbitrary number of upper loops and one lower loop at the right of all the others, plus the sum of all rings having an arbitrary number of upper loops and two lower ones at the right of all the others, etc. (see Fig. 9.2.2). These diagrams can be summed in the same way and give:

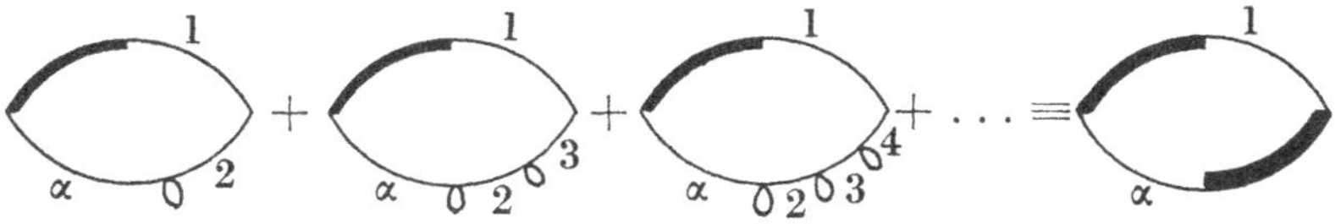

Fig. 9.2.2 The second step in the summation.

$$R_2 = \int d\boldsymbol{l} \int d\boldsymbol{v}_1\, d_\alpha (1-\mathscr{J}_\alpha)^{-1}\, \delta_+^{\alpha 1}(-d_\alpha)\varphi(\alpha)(1-\mathscr{J}_1^*)^{-1} \int d\boldsymbol{v}_2\, \delta_+^{21}\, d_{12}\,\varphi(1)\varphi(2) \tag{4}$$

where $\mathscr{J}_1^*$ is the complex conjugate of $\mathscr{J}_1$.

The way to continue is evident. One successively adds loops on the upper side, then lower loops to the right of the former, then again upper loops to the right of the preceding ones, etc. One sums over the number of loops added in each step. This grouping is shown graphically in Fig. **9.2.3**. We immediately see that we thus obtain the whole series of Fig. **9.1.2**. Proceeding as before, we can write the final result in the form:

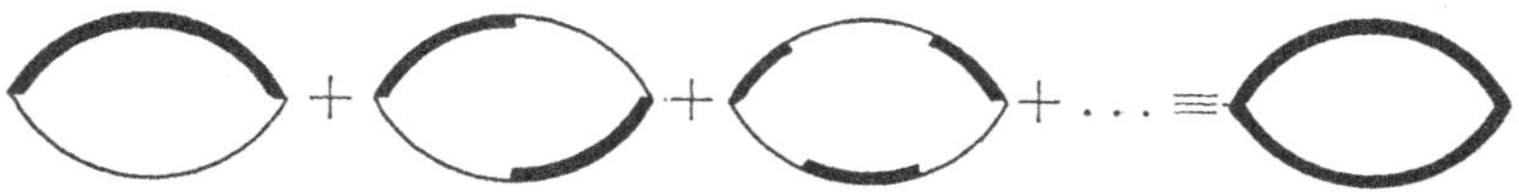

Fig. 9.2.3 The summation of the rings.

$$\partial\varphi(\alpha)/\partial t = \int \mathrm{d}\boldsymbol{l}\, d_\alpha\, F(\alpha) \tag{5}$$

where $F(\alpha)$ is the sum:

$$\begin{aligned} F(\alpha) &= \int d\boldsymbol{v}_1 (1-\mathscr{J}_\alpha)^{-1}\, \delta_+^{\alpha 1} d_{1\alpha}\, \varphi(1)\,\varphi(\alpha) \\ &+ \int d\boldsymbol{v}_1 \int d\boldsymbol{v}_2\, (1-\mathscr{J}_\alpha)^{-1}\, \delta_+^{\alpha 1}\, d_\alpha^*\, \varphi(\alpha)\, (1-\mathscr{J}_1^*)^{-1}\, \delta_+^{*12}\, d_{21}^*\, \varphi(2)\,\varphi(1) \\ &+ \int d\boldsymbol{v}_1 \int d\boldsymbol{v}_2 \int d\boldsymbol{v}_3 (1-\mathscr{J}_\alpha)^{-1}\, \delta_+^{\alpha 1}\, d_\alpha^*\, \varphi(\alpha)\, (1-\mathscr{J}_1^*)^{-1}\, \delta_+^{*12}\, d_1\, \varphi(1) \\ &\qquad \times\ (1-\mathscr{J}_2)^{-1}\, \delta_+^{23}\, d_{32}\,\varphi(3)\,\varphi(2) \\ &+ \cdots \end{aligned} \tag{6}$$

(remember that $d_\alpha^* = -d_\alpha$, $d_{ij}^* = d_{ji}$, $\delta_+^{*ij} = \delta_+^{ji}$).

The series (9.2.6) is simply the iteration solution of the following integral equation:

$$F(\alpha) = \mathscr{G}(\alpha) + \int dv_1 \mathscr{H}(\alpha, 1) F^*(1) \quad (7)$$

where:

$$\mathscr{G}(\alpha) = \int dv_1 (1-\mathscr{I}_\alpha)^{-1} \delta_+^{\alpha 1} d_{1\alpha} \varphi(1) \varphi(\alpha) \quad (8)$$

$$\mathscr{H}(\alpha, 1) = (1-\mathscr{I}_\alpha)^{-1} \delta_+^{\alpha 1} d_\alpha^* \varphi(\alpha) \quad (9)$$

Let us rewrite these formulae in full, replacing the abbreviations by their definitions (9.1.8) and (9.2.2):

$$\partial\varphi(\alpha)/\partial t = 4\pi e^2 C \int d\boldsymbol{l}\, l^{-2} i\boldsymbol{l} \cdot (\partial/\partial \boldsymbol{v}_\alpha) F(\alpha) \quad (10)$$

$$\begin{aligned} F(\alpha) = {} & [2\pi^2 \eta(\alpha) l^2]^{-1} e^2 \int d\boldsymbol{v}_1\, \delta_+[\boldsymbol{l}\cdot(\boldsymbol{v}_\alpha - \boldsymbol{v}_1)]\, i\boldsymbol{l}\cdot \boldsymbol{D}_{\alpha 1} \varphi(\alpha)\varphi(1) \\ & - 4\pi^2 e^2 C [l^2 \eta(\alpha)]^{-1} i\boldsymbol{l}\cdot(\partial\varphi(\alpha)/\partial \boldsymbol{v}_\alpha) \int d\boldsymbol{v}_1\, \delta_+[\boldsymbol{l}\cdot(\boldsymbol{v}_\alpha - \boldsymbol{v}_1)]\, F^*(1) \end{aligned} \quad (11)$$

where the function $\eta(\alpha)$ is defined by:

$$\eta(\alpha) = 1 - 4\pi^2 i e^2 C l^{-2} \int d\boldsymbol{v}_2\, \delta_+(\boldsymbol{l}\cdot \boldsymbol{g}_{\alpha 2})\, \boldsymbol{l}\cdot[\partial\varphi(2)/\partial \boldsymbol{v}_2] \quad (12)$$

We have thus reduced our problem to the system of two equations, (9.2.10) and (9.2.11). In order to have an explicit equation of evolution we now have to eliminate the function $F(\alpha)$, i.e., to solve the integral equation (9.2.11) in a closed form. This equation would be of a standard type (Muskelishwili, 1953) if we had $F(1)$ instead of $F^*(1)$ on the right-hand side. The appearance of the complex conjugate in that term means that (9.2.11) is in fact a system of *two* integral equations, the real part of F depending on its imaginary part and vice-versa. Such a system of equations cannot, in general, be solved in a closed form. However, we shall show in the following paragraph that in our particular case the difficulty is only apparent, and we shall solve (9.2.11) in a very elementary way.

3. Solution of the Integral Equation

Let us separate $F(\alpha)$ into real and imaginary parts:

$$F(\alpha) = \Phi(\alpha) + i\Psi(\alpha) \quad (1)$$

A basic remark is the following one; *only the imaginary part of F gives a non-vanishing contribution to the equation of evolution* (9.2.10), so that the latter can be written

$$\partial\varphi(\alpha)/\partial t = -4\pi e^2 C \int d\boldsymbol{l}\, l^{-2}\boldsymbol{l}\cdot(\partial/\partial \boldsymbol{v}_\alpha)\, \Psi(\alpha) \tag{2}$$

This fact is easily understandable; the function $\varphi(\alpha)$, because it is the momentum distribution function, has to be a *real* function; the *real* part Φ of F, substituted into (9.2.10) would give an imaginary contribution which would be unphysical. Mathematically, one can verify that the imaginary part of each ring vanishes for reasons of symmetry by integration over $\boldsymbol{l}$, or else, one can solve the integral equation for the real part [see (9.3.16)] and verify the statement a posteriori.

Let us now study the dependence of the various functions on their argument. The function $F(\alpha) \equiv F(\boldsymbol{v}_\alpha)$ depends (explicitly and through φ) on the *vector* $\boldsymbol{v}_\alpha$, i.e., on *three* scalar variables. On the other hand, the kernel of the integral equation, $\delta_+[\boldsymbol{l}\cdot(\boldsymbol{v}_\alpha-\boldsymbol{v}_1)]$ depends on $\boldsymbol{v}_\alpha$ only through the scalar product $\boldsymbol{l}\cdot\boldsymbol{v}_\alpha$, i.e., it depends only on one scalar variable. In a coordinate system the z-axis of which coincides with $\boldsymbol{l}$ the integral of the second term in the right-hand side of (9.2.11) would be:

$$\int d\boldsymbol{v}_1 \delta_+[\boldsymbol{l}\cdot(\boldsymbol{v}_\alpha-\boldsymbol{v}_1)]\, F^*(\boldsymbol{v}_1) = \int dv_{1z}\,\delta_+[l\cdot(v_{\alpha z}-v_{1z})] \\ \times\left\{\iint dv_{1x}\,dv_{1y}\, F^*(v_{1x}v_{1y}v_{1z})\right\} \tag{3}$$

One could say, loosely, that the dependence of F on v_z is determined by an integral equation, while its dependence on v_y and on v_x is determined "algebraically," by the coefficients of equation (9.2.11). In order to solve equation (9.2.11) we first have to bring it into a form in which all quantities depend on a single variable. This can be done essentially by integrating the equation over the components of $\boldsymbol{v}$ normal to $\boldsymbol{l}$. In order to keep the symmetry of the expressions we shall introduce the following definition of the "reduced" functions $\bar{g}(\nu_\alpha)$ corresponding to any function $g(\boldsymbol{v}_\alpha)$:

$$\bar{g}(\nu_\alpha) = l \iiint d\boldsymbol{v}_1\, \delta(\nu_\alpha - \boldsymbol{l}\cdot\boldsymbol{v}_1/l)\, g(\boldsymbol{v}_1) \tag{4}$$

where

$$\nu_\alpha = l^{-1}\, \boldsymbol{l}\cdot\boldsymbol{v}_\alpha \tag{5}$$

Let us now introduce the following abbreviations:

$$\begin{aligned} q(\boldsymbol{v}_\alpha) &= q_1 + iq_2 \\ q_1 &= (e^2/2\pi^2 l^2) \int d\boldsymbol{v}_1 [\boldsymbol{l}\cdot(\boldsymbol{v}_\alpha - \boldsymbol{v}_1)]^{-1}\, \boldsymbol{l}\cdot\boldsymbol{D}_{\alpha 1}\, \varphi(\alpha)\, \varphi(1) \\ q_2 &= (-e^2/2\pi l^2) \int d\boldsymbol{v}_1\, \delta[\boldsymbol{l}\cdot(\boldsymbol{v}_\alpha - \boldsymbol{v}_1)]\, \boldsymbol{l}\cdot\boldsymbol{D}_{\alpha 1}\, \varphi(\alpha)\, \varphi(1) \end{aligned} \tag{6}$$

$$\begin{aligned} \eta(\nu_\alpha) &= \eta_1 + i\eta_2 \\ \eta_1 &= 1 + (4\pi e^2 C/l^2) \int d\boldsymbol{v}_1 [\boldsymbol{l}\cdot(\boldsymbol{v}_\alpha - \boldsymbol{v}_1)]^{-1}\, \boldsymbol{l}\cdot[\partial\varphi(1)/\partial\boldsymbol{v}_1] \\ \eta_2 &= (-4\pi^2 e^2 C/l^2) \int d\boldsymbol{v}_1\, \delta[\boldsymbol{l}\cdot(\boldsymbol{v}_\alpha - \boldsymbol{v}_1)]\, \boldsymbol{l}\cdot[\partial\varphi(1)/\partial\boldsymbol{v}_1] \end{aligned} \tag{7}$$

$$D(\boldsymbol{v}_\alpha) = (-4\pi e^2 C/l^2)\, \boldsymbol{l}\cdot[\partial\varphi(\alpha)/\partial\boldsymbol{v}_\alpha] \tag{8}$$

In all these expressions, and also later, all integrals involving $(\boldsymbol{l}\cdot\boldsymbol{g}_{ij})^{-1}$ are understood as principal parts. Remembering (9.1.7), equation (9.2.11) can be written:

$$F(\boldsymbol{v}_\alpha) = q(\boldsymbol{v}_\alpha)/\eta(\nu_\alpha) + i[\pi D(\boldsymbol{v}_\alpha)/\eta(\nu_\alpha)] \int d\boldsymbol{v}_1\, \delta_+(\boldsymbol{l}\cdot\boldsymbol{g}_{\alpha 1})\, F^*(\boldsymbol{v}_1) \tag{9}$$

Multiply both sides by $\delta(l\nu_\beta - \boldsymbol{l}\cdot\boldsymbol{v}_\alpha)$ and integrate over $\boldsymbol{v}_\alpha$. Using (9.3.4), the result can be written (replacing the label β by α):

$$\bar{F}(\nu_\alpha) = \bar{q}(\nu_\alpha)/\eta(\nu_\alpha) + i[\eta_2(\nu_\alpha)/\eta(\nu_\alpha)] \int d\nu_1\, \delta_+(\nu_\alpha - \nu_1)\, \bar{F}^*(\nu_1) \tag{10}$$

(we have used the property: $\pi\bar{D}/l \equiv \eta_2$). This is now an ordinary integral equation in the unknown function $\bar{F}(\nu)$. If we know the solution of (9.3.10), we shall automatically have the solution of (9.3.9); indeed, by our preceding remark, and by (9.3.3), we have:

$$\int d\boldsymbol{v}_1\, \delta_+(\boldsymbol{l}\cdot\boldsymbol{g}_{\alpha 1}) F^*(\boldsymbol{v}_1) = (1/l) \int d\nu_1\, \delta_+(\nu_\alpha - \nu_1)\, \bar{F}^*(\nu_1) \tag{11}$$

We can therefore eliminate this integral between (9.3.9) and

(9.3.10) and thus get the following algebraic relation between $F(v_\alpha)$ and $\bar{F}(\nu_\alpha)$

$$F = (\pi D/l\eta_2)\,\bar{F} + \eta^{-1}[q - (\pi D/l\eta_2)\,\bar{q}] \tag{12}$$

The problem is thus reduced to the solution of equation (9.3.10). Using (9.3.1), (9.3.6), and (9.3.7), we separate (9.3.10) into real and imaginary parts, thus obtaining the following two equations for $\bar{\Phi}$ and $\bar{\Psi}$:

$$\begin{aligned} |\eta|^{-2}\{\eta_1^2\bar{\Phi} - \eta_1\eta_2\bar{\Psi} - \eta_1\eta_2 P\bar{\Phi} + \eta_2^2 P\bar{\Psi} - q_1\eta_1\} &= 0 \\ |\eta|^{-2}\{-\eta_1\eta_2\bar{\Phi} + (|\eta|^2 + \eta_2^2)\,\bar{\Psi} + \eta_2^2 P\bar{\Phi} + \eta_1\eta_2 P\bar{\Psi} + q_1\eta_2\} &= 0 \end{aligned} \tag{13}$$

where we have used the notation

$$Pf \equiv \pi^{-1}\int d\nu_1\, f(\nu_1)/(\nu_\alpha - \nu_1) \tag{14}$$

Moreover, we have used the property $\bar{q}_2 = 0$, which is easily demonstrated, using (9.3.4) and (9.3.6).

At this point the great simplicity of our problem appears; the solution of equation (9.3.10) is *real*. In other words:

$$\bar{\Psi} = 0 \tag{15}$$

In order to see this, put $\bar{\Psi} = 0$ in (9.3.13) and divide the first equation by $\eta_1/|\eta|^2$ and the second one by $-\eta_2/|\eta|^2$. The two equations (9.3.13) then reduce to the unique equation:

$$\eta_1\bar{\Phi} - \eta_2 P\bar{\Phi} = \bar{q}_1 \tag{16}$$

Since the equations are linear, this solution ($\bar{\Phi} \neq 0$, $\bar{\Psi} = 0$) is unique. Equation (9.3.16) could be solved immediately by the standard methods of the theory of singular integral equations (Muskelishwili, 1953) but we shall not need $\bar{\Phi}$ here.

Indeed, in going from $\bar{F}$ to F by formula (9.3.12) we see that the imaginary part, Ψ, of F does not depend on $\bar{\Phi}$. We are thus left with:

$$\Psi = |\eta|^{-2}\{-\eta_2 q_1 + (\pi D/l)\,q_1 + \eta_1 q_2\} \tag{17}$$

Applying formulae (9.3.4) through (9.3.8) one can very

easily show that Ψ can be written more simply as:

$$\Psi = q_2|\eta|^{-2} \tag{18}$$

Introducing this expression into (9.3.2), we finally obtain the equation of evolution for the homogeneous plasma:

$$\partial\varphi(\alpha)/\partial t = 2e^4 C \int d\boldsymbol{l} \int d\boldsymbol{v}_1 \boldsymbol{l} \cdot (\partial/\partial \boldsymbol{v}_\alpha)$$

$$\times [\delta(\boldsymbol{l} \cdot \boldsymbol{g}_{\alpha 1})/|l^2 - 4\pi i e^2 C \int d\boldsymbol{v}_2 \delta_+(\boldsymbol{l} \cdot \boldsymbol{g}_{\alpha 2}) \boldsymbol{l} \partial\varphi(2)/\partial \boldsymbol{v}_2|^2] \boldsymbol{l} \cdot \boldsymbol{D}_{\alpha 1} \varphi(\alpha) \varphi(1) \tag{19}$$

This is the basic evolution equation for particles interacting through electrostatic forces first derived by Balescu (1960). As has been shown by Lenard (1960), Balescu's equation (9.3.19) can also be derived by Bogolioubov's method, which we shall discuss briefly in Chapter 11, § 5.

4. Discussion of the Transport Equation

Equation (9.3.19) has a structure which is radically different from any of the classical transport equations. All the equations that we have studied until now are characterized by the occurrence in the collision operator of a product of two distribution functions $\varphi(\alpha)\varphi(1)$ [see, e.g., (4.3.4), (6.6.9)]. The new feature of equation (9.3.19) is the occurrence of the distribution function in the *denominator* of the right-hand side. This is the characteristic structure introduced by the collective effects. If we expand the *denominator* in a power series in e^2C, we would find a sum of terms, each involving a product of a larger and larger number of distribution functions (see Fig. 9.2.1). This shows very clearly that (9.3.19) explicitly describes an N-body process, with N arbitrarily large.

The fact that an N-body problem can be handled explicitly follows from the very simple structure of the diagrams. The ring structure already shows that at each instant during the collision process only two particles are correlated. However, this correlation is continually transferred during the collision from one couple of particles to another (see also Chapter 8, § 9).

Equation (9.3.19) also has, however, a formal similarity to the "Fokker-Planck" equation (4.3.4). This is due to the fact that in deriving it, we have made the assumption of weak coupling (we neglected terms of order $e^4 (e^2 C)^n t$); but the treatment of weakly coupled systems has to be modified because of the long range of the interactions. Comparing (9.3.19) to (4.3.4) we easily see that the former can be obtained by inserting into the latter the expression:

$$8\pi^3 \lambda V_l = 4\pi e^2/[l^2 - 4\pi^2 i e^2 C \int d\mathbf{v}_2 \delta_+(\boldsymbol{l}\cdot\mathbf{g}_{\alpha 2})\, \boldsymbol{l}\cdot(\partial/\partial\mathbf{v}_2)\varphi(\mathbf{v}_2, t)] \quad (1)$$

This result shows that the summation over the rings is equivalent to a "renormalization" of the interaction potential. The effective potential (9.4.1) means that a particle, α, of the gas feels not only the exact Coulomb interaction due to another particle, but also an average field due to the medium. The latter effect is, of course, characteristic of long-range forces; in ordinary gases the interactions fall off so rapidly that particles other than the collision partner of α cannot influence its behavior. The physical effect of (9.4.1) is the formation of a polarization cloud around each particle. It is important to note that this cloud does not have a permanent character as does the Debye cloud of equilibrium theory. It is in fact a functional of $\varphi(\mathbf{v}_\alpha, t)$, so that its shape changes while the system approaches equilibrium. Moreover, even if the medium in which particle α moves were in equilibrium (as in a Brownian motion problem), the effective potential (9.4.1) would still be a function of the velocity of particle α. A detailed analysis, which we shall not give here (see Balescu, 1962), would show that for velocities small with respect to the thermal velocity of the medium, the polarization cloud is localized around the particle, is slightly flattened in the direction of motion, and accompanies it in its motion. For very large velocities, on the other hand, the cloud degenerates into a wake of plasma oscillation behind the particle. There will be an additional frictional force, due to the exchange of energy between the particle and the *collective* degrees of freedom of the gas.

It is interesting to study the relation of (9.3.19) to the "Fokker

Planck'' equation (4.3.4) more closely. In order to do this we again consider the Brownian motion of a charged particle α in a plasma in equilibrium, described by

$$\varphi(\boldsymbol{v}_1) = A^3 \exp(-mv^2/2kT)$$

with

$$A = (m/2\pi kT)^{1/2}$$

We introduce this expression into (9.4.1); this becomes

$$8\pi^3 \lambda V_l = 4\pi e^2/\{l^2 - 4\pi^2 i e^2 C A^3 \int d\boldsymbol{v}_2 \delta_+(\boldsymbol{l}\cdot\boldsymbol{g}_{\alpha 2}) \times(-\boldsymbol{l}\cdot\boldsymbol{v}_2/kT)\exp(-mv_2^2/2kT)\} \quad (2)$$

Writing $\boldsymbol{l}\cdot\boldsymbol{v}_2 = \boldsymbol{l}\cdot\boldsymbol{v}_\alpha - \boldsymbol{l}\cdot\boldsymbol{g}_{\alpha 2}$, we obtain

$$8\pi^3 \lambda V_l = 4\pi e^2/\{l^2 + (4\pi e^2 C/kT) + [\boldsymbol{l}\cdot\boldsymbol{v}_\alpha/l(kT/m)^{1/2}] \times(4\pi e^2 C/kT) F(\boldsymbol{l}\cdot\boldsymbol{v}_\alpha/l(kT/m)^{1/2})\} \quad (3)$$

The function $F(x)$ which appears in (9.4.3) is rather complicated but well-known (it is related to the error function of a complex argument, see for example Balescu, 1962). It is sufficient for our purpose to know that it approaches a constant when x tends to zero. Therefore, the third term of the denominator is very small compared to the first two terms if $v_\alpha/(kT/m)^{1/2} \ll 1$, and we can then write

$$8\pi^3 \lambda V_l = 4\pi e^2/(l^2 + \kappa^2) \quad (4)$$

where $1/\kappa$ is the Debye length defined by (9.1.3). Now, (9.4.3) is precisely the Fourier transform of the Debye potential (9.1.2). This calculation shows, therefore, that equation (9.3.19) reduces exactly to the Fokker-Planck equation with a Debye potential in the situation considered here (*slow* Brownian motion). This calculation then indicates the domain of validity of the approximation we used in Chapter 4, § 7.

The great interest of Balescu's equation (9.3.9) is, however, to provide a tool for the calculation of transport properties in situations for which the use of the effective Debye potential is not permitted.

CHAPTER 10

Statistical Hydrodynamics

1. Introduction

In Chapters 4 and 6 we considered the approach to equilibrium in homogeneous, weakly coupled, or dilute, systems. We now want to extend this treatment to deal specifically with non-homogeneous systems. A clear understanding of the domain of validity of the usual transport equations, like the Boltzmann equation, for such situations is of special importance because they form the statistical basis for the whole of hydrodynamics (see for example Chapman and Cowling, 1939; Hirschfelder, Curtiss, and Bird, 1954).

A statistical theory is of course necessary to calculate the transport coefficients (like the viscosity coefficient) which appear in the macroscopic equations of motion. However an even much more basic problem is the determination of the domain of validity of the hydrodynamic equations, that are derived from a continuum theory.

The special feature of non-homogeneous situations is that a new characteristic time appears. Indeed, until now we had to consider the following characteristic times:

(a) the duration of an interaction [see (6.1.5) and Chapter 8, § 2]

$$t_{\text{int}} \sim 1/\kappa v \tag{1}$$

where $1/\kappa$ is of the order of the range of the potential.

(b) the relaxation time, t_r, which for example for weakly coupled systems is of the form [see (2.3.9) and Chapter 8, § 2]

$$t_r \sim 1/\lambda^2 \tag{2}$$

We have seen in Chapter 8 that there exist other characteristic times which appear in the evaluation of the time dependence of diagrams. We shall however suppose in this chapter that they are all of the order of the duration of an interaction (10.1.1) (see Chapter 8, §§ 10, 12). We have now to introduce a characteristic time related to the *scale or range of the inhomogeneities*. The Fourier coefficient $\rho(\alpha_k)$ of the phase distribution function corresponding to a single non-vanishing wave vector determines the space variation of the density [we shall write the briefer ρ_k instead of $\rho(\alpha_k)$]. Indeed (7.2.4) shows that ρ_k is the Fourier transform of the deviation of the density from homogeneity. The function ρ_k therefore describes the shape of the inhomogeneity in space (see Fig. 10.1.1). If the inhomogeneity is a slowly varying function

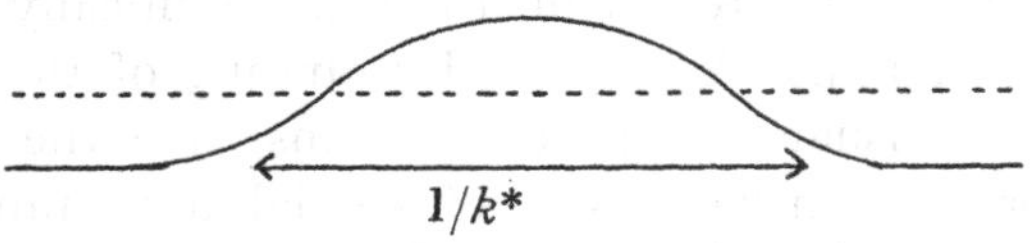

Fig. 10.1.1 Spatial inhomogeneity.

in space, its Fourier transform will be peaked sharply around $k = 0$ and fall off rapidly to zero with increasing values of k. If k^* is a characteristic value for which ρ_k is still appreciably different from zero [1] then $1/k^*$ is of the order of the range of the inhomogeneity and we have the characteristic time

$$t_h \sim 1/k^* v \tag{3}$$

Instead of the three times (10.1.1) through (10.1.3) we may as well define the three characteristic lengths:

(1) the range of interactions L_{int}

(2) the mean free path L_r

(3) the range of inhomogeneity L_h.

It is clear that for weakly coupled or dilute systems

$$L_r \gg L_{\text{int}} \quad \text{or} \quad t_r \gg t_{\text{int}} \tag{4}$$

as well as

$$L_h \gg L_{\text{int}} \quad \text{or} \quad t_h \gg t_{\text{int}} \tag{5}$$

But we have to be more careful in comparing L_r to L_h. For weakly

[1] See the treatment of wave packets in quantum mechanics (for example in Bohm, 1951).

coupled systems we have always considered the limit

$$\lambda \to 0 \tag{6}$$

This means that the free path becomes very long

$$L_r \to \infty \tag{7}$$

If at the same time we consider the extension $1/k^*$ of the inhomogeneity as *finite*, we would have

$$L_r \gg L_h \qquad \text{or} \quad t_r \gg t_h \tag{8}$$

We then have situations in which the free path is much longer than the scale of the inhomogeneities. Such situations may really exist, especially in rarified gases, however, the usual case treated in hydrodynamics corresponds not to (10.1.8) but to

$$L_h \gg L_r \qquad \text{or} \quad t_h \gg t_r \tag{9}$$

The scale of the inhomogeneity is then the *largest* characteristic length, larger than any characteristic molecular length. It is only in this case that hydrodynamic equations (like the Stokes-Navier equation) deduced from a continuum theory may be expected to be valid.

In § 2 we shall treat the evolution of the distribution function in the case (10.1.8). In the rest of this chapter we shall study the extension of our theory to (10.1.9).

We shall thus find a generalization of the diagram technique which can be expressed briefly as follows: If the conditions (10.1.9) are satisfied, the diagonal fragments (see Chapter 8, § 5, 6, 10) are no longer the only contributions giving a systematic increase with time. There are also certain classes of non-diagonal diagrams called *pseudo-diagonal diagrams* which give asymptotic contributions of the same order in time as the diagonal ones (Prigogine and Balescu, 1960).

The idea of three different and widely separated time scales was first put forward by Bogolioubov (1946). As we shall see they appear in a very explicit way in our diagram technique.

For convenience, in this chapter we shall always consider the case of weakly coupled gases. There would be no difficulty in discussing the case of dilute gases interacting through strong forces in a similar way (see also Andrews, 1960). The general situation corresponding to an arbitrary strength of interaction, or to arbitrary concentrations, will be considered briefly in Chapter 11. As may be expected the problem is much more complicated then, precisely because the different time scales can no longer be so clearly separated.

2. Transport Equation in the Limit of Large Free Paths

Let us discuss the evolution of the Fourier coefficient $\rho(\alpha_k)$ in the weakly coupled case. The only contributions then come from the diagram of the type represented in Fig. 10.2.1 (see Chapter 8, §§ 5, 6, 10, 11)

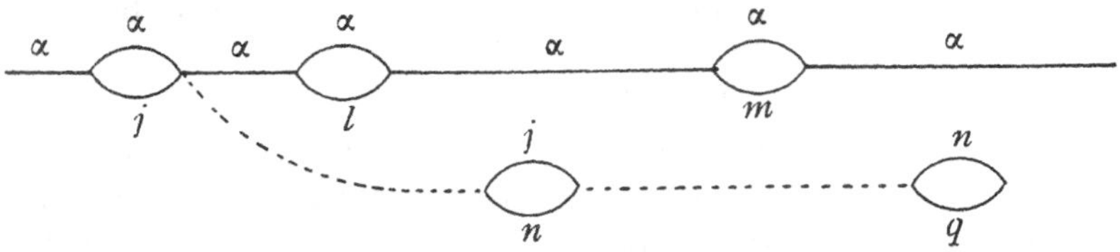

Fig. 10.2.1 Contribution to $\rho(\alpha_k)$ in the weakly coupled case.

Indeed we need diagrams which end (at their left) with a single line α. No "destruction region" as defined in Chapter 8, § 10 can appear in these diagrams. Indeed such a region gives, as we have seen, an asymptotic time-independent contribution and would therefore leave powers of λ uncompensated by powers of t.

The only possibility is therefore a succession of scattering diagrams (see Chapter 8, § 6) which are semi-connected to diagonal fragments. This is precisely the structure of the diagrams represented in Fig. 10.2.1. We therefore obtain the equations

$\partial\rho(\alpha_k, \mathbf{v}_\alpha, t)/\partial t =$ 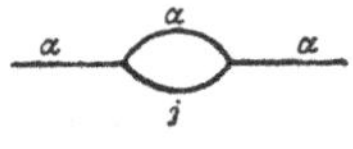 $\rho(\alpha_k, \mathbf{v}_\alpha, \mathbf{v}_j, t)$ (1)

which describes the evolution of $\rho(\alpha_k)$ due to the scattering of α with an arbitrary particle j of the medium.

The explicit form of the diagram in (10.2.1) has been discussed in detail in Chapter 8, § 6 as well as in Chapter 5, § 3.

The final result [see (5.3.14) or (8.6.11)] may be obtained rapidly in the following way: we use, as in Chapter 5, a time iteration formalism. Diagram (a) of Fig. 10.4.1 then gives as usual (see Chapter 5, § 3)

$$\lambda^2 \int d\boldsymbol{l}\, V_l V_{-l} \int_0^t dt_2 \int_0^{t_2} dt_1\, e^{i\boldsymbol{k}\cdot\boldsymbol{v}_\alpha t_2}\, e^{-i\boldsymbol{k}\cdot\boldsymbol{v}_\alpha (t_2-t_1)}\, e^{-i\boldsymbol{l}\cdot\boldsymbol{g}_{\alpha j}(t_2-t_1)} \times e^{-i\boldsymbol{k}\cdot\boldsymbol{v}_\alpha t_1}\, \rho(\alpha_k, \boldsymbol{v}_\alpha, \boldsymbol{v}_j, t=0) \qquad (1')$$

where we have written only the factors of interest. The exponential $\exp[-i\boldsymbol{k}\cdot\boldsymbol{v}_\alpha(t_2-t_1)]$ may be replaced by one. Indeed t_2-t_1 is of the order of the duration of the collision and $\boldsymbol{k}\cdot\boldsymbol{v}_\alpha$ of the order of $1/t_h$ [(see (10.1.3)]. For the same reasons we may replace t_1 by t_2 in $\exp[-i\boldsymbol{k}\cdot\boldsymbol{v}_\alpha t_1]$.

Integrating over t_1 we obtain

$$\lambda^2 \int_0^t dt_2\, e^{i\boldsymbol{k}\cdot\boldsymbol{v}_\alpha t_2} \int d\boldsymbol{l} |V_l|^2 \delta(\boldsymbol{l}\cdot\boldsymbol{g}_{\alpha j})\, e^{-i\boldsymbol{k}\cdot\boldsymbol{v}_\alpha t_2} \qquad (1'')$$

The characteristic feature of this contribution is that the homogeneous operator is "sandwiched" between the exponentials corresponding to free propagation. We therefore obtain

$$\partial\rho(\alpha_k, \boldsymbol{v}_\alpha, t)/\partial t = 8\pi^4\lambda^2 C \int d\boldsymbol{v}_j \int d\boldsymbol{l}\, e^{i\boldsymbol{k}\cdot\boldsymbol{v}_\alpha t} |V_l|^2\, \boldsymbol{l}\cdot\boldsymbol{D}_{j\alpha} \times \delta(\boldsymbol{l}\cdot\boldsymbol{g}_{j\alpha})\, \boldsymbol{l}\cdot\boldsymbol{D}_{j\alpha}\, e^{-i\boldsymbol{k}\cdot\boldsymbol{v}_\alpha t} \rho(\alpha_k, \boldsymbol{v}_\alpha, \boldsymbol{v}_j, t) \qquad (2)$$

We then go back to phase space through [see (7.2.3) and (7.1.7)]

$$f_1(\boldsymbol{x}_\alpha, \boldsymbol{v}_\alpha, t) = C[\varphi_1(\boldsymbol{v}_\alpha) + \int d\boldsymbol{k}\, \rho(\alpha_k, \boldsymbol{v}_\alpha, t)\, e^{i\boldsymbol{k}\cdot(\boldsymbol{x}_\alpha - \boldsymbol{v}_\alpha t)}]$$

$$f_{1,2}(\boldsymbol{x}_\alpha, \boldsymbol{v}_\alpha, \boldsymbol{v}_j, t) = C[\varphi_2(\boldsymbol{v}_\alpha, \boldsymbol{v}_j, t) + \int d\boldsymbol{k}\, \rho(\alpha_k, \boldsymbol{v}_\alpha, \boldsymbol{v}_j, t)\, e^{i\boldsymbol{k}\cdot(\boldsymbol{x}_\alpha - \boldsymbol{v}_\alpha t)}] \qquad (3)$$

This gives us the following equation for the evolution of the distribution function $f_1(\boldsymbol{x}_\alpha, \boldsymbol{v}_\alpha)$,

$$(\partial f_1/\partial t) + \boldsymbol{v}_\alpha(\partial f_1/\partial \boldsymbol{x}_\alpha) = 8\pi^4\lambda^2 C \int d\boldsymbol{v}_j \int d\boldsymbol{l} |V_l|^2\, \boldsymbol{l}\cdot(\partial/\partial\boldsymbol{v}_\alpha)\, \delta(\boldsymbol{l}\cdot\boldsymbol{g}_{\alpha j})\, \boldsymbol{l}\cdot\boldsymbol{D}_{\alpha j} \times f_{1,2}(\boldsymbol{x}_\alpha, \boldsymbol{v}_\alpha, \boldsymbol{v}_j, t) \qquad (4)$$

Using the factorization condition (7.4.2) we get

$$(\partial f_1/\partial t)+\boldsymbol{v}_\alpha\cdot(\partial f_1/\partial \boldsymbol{x}_\alpha) = 8\pi^4\lambda^2 C\int d\boldsymbol{v}_j\int d\boldsymbol{l}|V_l|^2\boldsymbol{l}\cdot(\partial/\partial \boldsymbol{v}_\alpha)\delta(\boldsymbol{l}\cdot\boldsymbol{g}_{\alpha j}) \times\boldsymbol{l}\cdot\boldsymbol{D}_{\alpha j}f_1(\boldsymbol{x}_\alpha, \boldsymbol{v}_\alpha, t)\varphi_1(\boldsymbol{v}_j, t) \quad (5)$$

In some aspects, this equation is indeed similar to the usual Boltzmann equation. The evolution of f_1 is described by the superposition of a flow term (on the left) and a collision term (on the right). There is no interference between these two processes in the approximation considered. However in the right-hand side of this equation we have the factor [see (7.1.7)]

$$f_1(\boldsymbol{x}_\alpha, \boldsymbol{v}_\alpha, t)\varphi_1(\boldsymbol{v}_j, t) = (1/N)f_1(\boldsymbol{x}_\alpha, \boldsymbol{v}_\alpha, t)\int d\boldsymbol{x}_j f_1(\boldsymbol{x}_j, \boldsymbol{v}_j, t) \quad (6)$$

while in the Boltzmann equation we have (see also § 5)

$$f_1(\boldsymbol{x}_\alpha, \boldsymbol{v}_\alpha, t)f_1(\boldsymbol{x}_\alpha, \boldsymbol{v}_j, t) = f_1(\boldsymbol{x}_\alpha, \boldsymbol{v}_\alpha, t)\int d\boldsymbol{x}_j\delta(\boldsymbol{x}_j-\boldsymbol{x}_\alpha) \times f_1(\boldsymbol{x}_j, \boldsymbol{v}_j, t) \quad (7)$$

One may say that equation (10.2.5) corresponds to a *non-local* theory. This is in complete agreement with condition (10.1.8) which implies that the mean free path is very long with respect to the scale of the inhomogeneities. Molecule α interacts, so to speak, with an "average molecule" coming from an arbitrary distance.

On the contrary, Boltzmann's equation corresponds to a *localized* theory. Molecule α interacts with a molecule j which *on the scale of the inhomogeneities* is situated at the same point as α.

As we shall see in § 6 this introduces a fundamental difference into the structure of the transport equations. Equation (10.2.5) does *not* lead to a hydrodynamic equation of the Stokes-Navier form.

We may also notice that equation (10.2.5) occupies, in a sense, an intermediate position between the usual Boltzmann equation and the theory of Brownian motion (see Chapter 3 and Chapter 5). If in (10.2.5) we replace φ by the equilibrium velocity distribution function we are back to the Brownian motion problem for molecule α.

Our present more general method also gives us, therefore, a justification for the theory of Brownian motion for weakly coupled systems which we developed in Chapters 3 and 5.

3. Factorization Theorems for Fourier Coefficients

In Chapter 7, § 4 we expressed the Fourier coefficients in terms of the u functions. We have seen that the u functions factorize for independent particles [see (7.4.6), (7.4.7)]. Let us show that this property combined with (10.1.9) gives rise to an important theorem on the factorization properties of the Fourier coefficients.

Indeed (7.4.11) shows that for $\boldsymbol{k}+\boldsymbol{k}' \neq 0$

$$\begin{aligned} \rho(1_k, 2_{k'}) &= \int d\boldsymbol{x}_1 d\boldsymbol{x}_2 e^{-i(\boldsymbol{k}\cdot\boldsymbol{x}_1+\boldsymbol{k}'\cdot\boldsymbol{x}_2)} u(1, 2) \\ &= 2\int d\boldsymbol{R}\, d\boldsymbol{r}\, e^{-i(\boldsymbol{k}+\boldsymbol{k}')\cdot\boldsymbol{R}} e^{-i(\boldsymbol{k}-\boldsymbol{k}')\cdot\boldsymbol{r}} u(1, 2) \end{aligned} \tag{1}$$

with $\boldsymbol{R} = \frac{1}{2}(\boldsymbol{x}_1+\boldsymbol{x}_2)$ and $\boldsymbol{r} = \frac{1}{2}(\boldsymbol{x}_1-\boldsymbol{x}_2)$.

Let us consider a system in which (10.1.9) is satisfied. The hydrodynamical length L_h is of a macroscopic order of magnitude, and much larger than the correlation length we shall call L_{co}.

Now we want to study $\rho(1_k, 2_{k'})$ for values of the wave vectors $\boldsymbol{k}$ and $\boldsymbol{k}'$ in the macroscopic range, that is for values of order L_h^{-1}. The range of integration for $\boldsymbol{R}$ and $\boldsymbol{r}$ in (10.3.1) will be of order $1/k \sim L_h$, because for much larger values of the integration variables the exponentials will oscillate rapidly and the integral will become very small.

We may therefore divide the domain of integration over $\boldsymbol{r}$ into a sphere of radius L_{co} around the origin and a spherical shell of a radius of order L_h outside the first sphere. Now outside the sphere of radius L_{co} the function $u(1, 2)$ is factorized into $u(1)u(2)$. Therefore, we get

$$\rho(1_k, 2_{k'}) = \rho(1_k)\rho(2_{k'})[1+O(L_{\text{co}}^3/L_h^3)]$$

for
$$\boldsymbol{k}, \boldsymbol{k}' = O(L_h^{-1}) \tag{2}$$

This property is easily generalized and we obtain the important *factorization theorem*:

$$\rho(1_{k_1}, 2_{k_2} \cdots s_{k_s}) = \rho(1_{k_1}) \cdots \rho(s_{k_s})$$

for *all*
$$\boldsymbol{k}_j \approx O(L_h^{-1}) \qquad \text{and} \qquad \sum_1^s \boldsymbol{k}_j \neq 0 \tag{3}$$

This factorization theorem extends to other types of Fourier coefficients involving wave vectors in the macroscopic range. Take, for example, [see (7.5.2)]

$$\rho(1_k, 2_{-k}, 3_{k'}) = \int d(\mathbf{x}_1-\mathbf{x}_2) \int d\mathbf{x}_3 e^{-i\mathbf{k}\cdot(\mathbf{x}_1-\mathbf{x}_2)} e^{-i\mathbf{k}'\cdot\mathbf{x}_3} u(1, 2, 3) \quad (4)$$

Again if k' is of order L_h^{-1}, the range of integration for $\mathbf{x}_3$ covers a volume of macroscopic extension and we have $u(1, 2, 3) = u(1, 2)u(3)$ except for a region of order $(L_{co}/L_h)^3$.

Therefore,

$$\rho(1_k, 2_{-k}, 3_{k'}) = \rho(1_k, 2_{-k})\, \rho(3_{k'})$$

for $$k' = O(L_h^{-1}) \quad (5)$$

These factorization theorems state that, for the large distances involved, characteristic lengths of molecular origin play no role.

Let us now consider the modification of the diagram technique introduced by the existence of a macroscopic hydrodynamic length.

4. Time Scales and Diagrams

In § 2 we have taken account only of the diagram (a) of Fig. 10.4.1. We shall now show that when condition (10.1.9) is satisfied we have to take account of all three diagrams of Fig. 10.4.1.

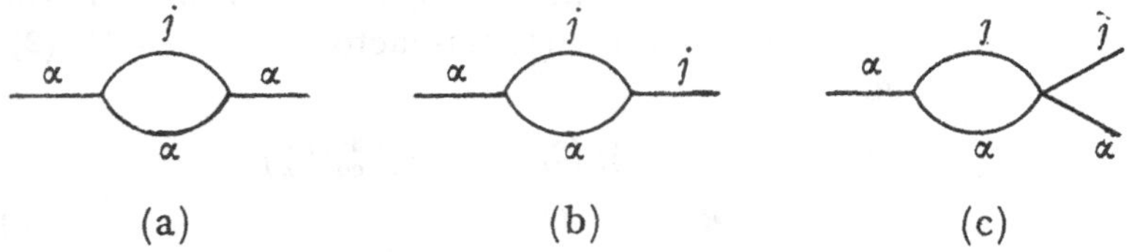

Fig. 10.4.1 Diagrams giving contributions of order $\lambda^2 t$ in the hydrodynamic case (10.1.9).

Diagram (a) has been discussed in detail in § 2. Let us discuss diagrams (b) and (c) in a similar manner.

We have for diagram (b), instead of (10.2.1′), again neglecting contributions of order t_{int}/t_h:

$$\lambda^2 \int d\boldsymbol{l} V_l V_{k-l} \int_0^t dt_2 \int_0^{t_2} dt_1\, e^{i\boldsymbol{k}\cdot\boldsymbol{v}_\alpha t_2} e^{-i\boldsymbol{k}\cdot\boldsymbol{v}_j(t_2-t_1)}$$
$$\times e^{-i\boldsymbol{l}\cdot\boldsymbol{g}_{\alpha j}(t_2-t_1)}\, e^{-i\boldsymbol{k}\cdot\boldsymbol{v}_j t_1} \rho(j_k,\, \boldsymbol{v}_\alpha,\, \boldsymbol{v}_j,\, t=0)$$
$$= \lambda^2 \int d\boldsymbol{l}\, V_l V_{k-l} \int_0^t dt_2 \int_0^{t_2} dt_1\, e^{i\boldsymbol{k}\cdot\boldsymbol{v}_\alpha t_2} e^{-i\boldsymbol{l}\cdot\boldsymbol{g}_{\alpha j}(t_2-t_1)}$$
$$\times e^{-i\boldsymbol{k}\cdot\boldsymbol{v}_j t_2} \rho(j_k,\, \boldsymbol{v}_\alpha,\, \boldsymbol{v}_j,\, t=0)$$
$$= \lambda^2 \int_0^t dt_2\, e^{i\boldsymbol{k}\cdot\boldsymbol{v}_\alpha t_2} \int d\boldsymbol{l}\, |V_l|^2 \delta(\boldsymbol{l}\cdot\boldsymbol{g}_{\alpha j})\, e^{-i\boldsymbol{k}\cdot\boldsymbol{v}_j t_2}$$
$$\times \rho(j_k,\, \boldsymbol{v}_\alpha,\, \boldsymbol{v}_j,\, t=0) \tag{1}$$

To obtain the last equality, we used the separation between the molecular scale and the hydrodynamic scale

$$V_{k-l} \approx V_{-l} + O(t_{\text{int}}/t_h) \tag{2}$$

The only difference from (10.2.1″) is in the exponential at the right, corresponding to particle j instead of α.

In the same way we obtain for diagram (c)

$$\lambda^2 \int d\boldsymbol{l} V_l V_{k-k'-l} \int_0^t dt_2 \int_0^{t_1} dt_1\, e^{i\boldsymbol{k}\cdot\boldsymbol{v}_\alpha t_2} e^{i\boldsymbol{k}\cdot\boldsymbol{v}_j(t_1-t_2)}$$
$$\times e^{i\boldsymbol{k}'\cdot\boldsymbol{g}_{\alpha j}(t_1-t_2)}\, e^{i\boldsymbol{l}\cdot\boldsymbol{g}_{\alpha j}(t_1-t_2)}\, e^{-i[\boldsymbol{k}'\cdot\boldsymbol{v}_\alpha + (\boldsymbol{k}-\boldsymbol{k}')\boldsymbol{v}_j]t_1}$$
$$\times \rho(\alpha_{k'},\, j_{k-k'},\, \boldsymbol{v}_\alpha,\, \boldsymbol{v}_j,\, t=0)$$
$$= \lambda^2 \int_0^t dt_2\, e^{i\boldsymbol{k}\cdot\boldsymbol{v}_\alpha t_2} \int d\boldsymbol{l} |V_l|^2 \delta(\boldsymbol{l}\cdot\boldsymbol{g}_{\alpha j})\, e^{-i[\boldsymbol{k}'\cdot\boldsymbol{v}_\alpha + (\boldsymbol{k}-\boldsymbol{k}')\boldsymbol{v}_j]t_2}$$
$$\times \rho(\alpha_{k'},\, j_{k-k'},\, \boldsymbol{v}_\alpha,\, \boldsymbol{v}_j,\, t=0) \tag{3}$$

All the diagrams of Fig. 10.4.1 differ therefore only by the incoming lines, as long as terms of order t_{int}/t_h are neglected. Now we have seen that diagram (a) gives a contribution of order [see (8.6.8)]

$$\lambda^2 t[1 + t/t_h + (t/t_h)^2] \tag{4}$$

We are interested in times of order t_r. Therefore in the hydrodynamic case studied here [(see 10.1.9)], (10.4.4) reduces essentially to a $\lambda^2 t$ contribution.

This is however not so in the situations studied in § 2 where we took $t_r/t_h \gg 1$. Diagram (b) contains, compared to diagram (a), an extra factor: $\exp i\boldsymbol{k}\cdot\boldsymbol{g}_{\alpha j}t \approx \exp it/t_h$, which becomes rapidly oscillating for $t = t_r \gg t_h$. This is why we have to neglect diagrams (b) and (c) when t_r/t_h is large (and retain them when t_r/t_h is small).

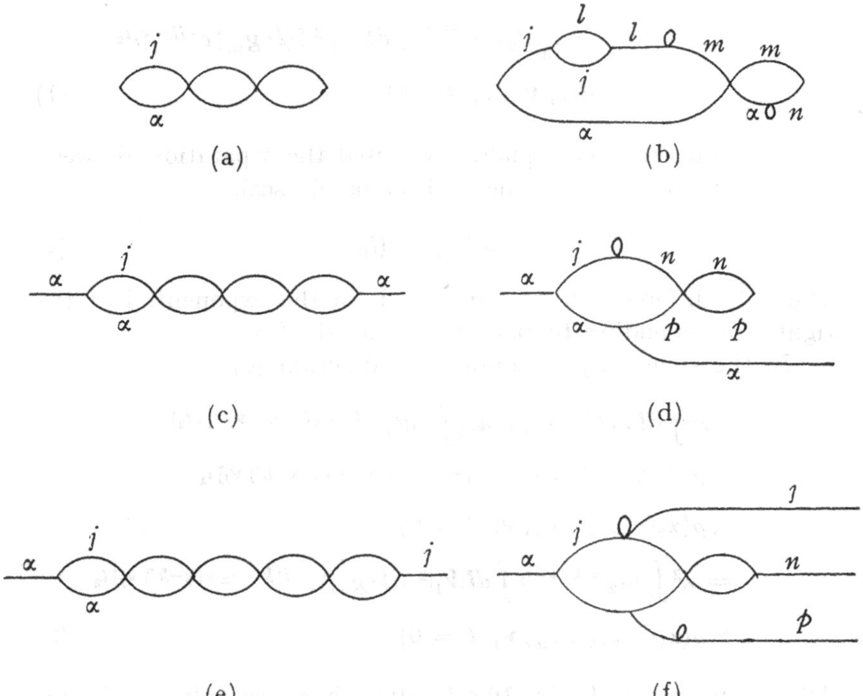

Fig. 10.4.2 (a), (b) Homogeneous diagonal fragments. (c), (d) Inhomogeneous diagonal fragments. (e), (f) Pseudo-diagonal fragments.

The detailed calculations we have gone through for the diagrams of Fig. 10.4.1 will enable us to understand the general properties to be stated below.

We shall distinguish between "homogeneous" and "inhomogeneous" diagonal fragments (or scattering diagrams) according to whether there are no outgoing lines [for example, Fig. 10.4.2

(a), (b)] or there is one outgoing line bearing the same label at each side of the diagram [Fig. 10.4.2 (c), (d)].

We now define *pseudo-diagonal fragments*: They are diagrams which can be obtained from a diagonal fragment by permutation of a particle in the external line to the right [in Fig. 10.4.1 (b), α is replaced by j] or by the addition of new external lines [as in Fig. 10.4.1(c)]. These outgoing lines may start at any vertex of the diagonal fragment which is not of type (f) in Fig. 7.6.1. *All wave vectors which correspond to lines that have been modified or added in this way have to be in the macroscopic range.*

In Fig. 10.4.2 (e) and (f) we give two examples of pseudo-diagonal fragments.

We thus generate a class of pseudo-diagonal fragments all of which have the same asymptotic time dependence, apart from terms of order t/t_h.

The existence of the hydrodynamic scale has to be taken account of in the asymptotic time evolution of other types of diagrams as well. An example may suffice. In Fig. 10.4.3 we have

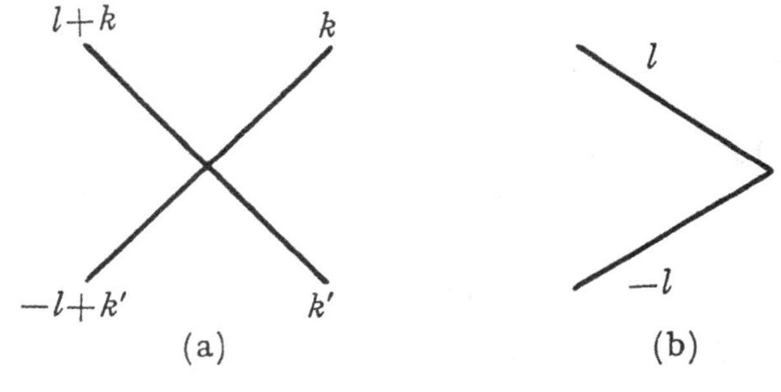

Fig. 10.4.3 Comparison between diagram (a), corresponding to a propagation of correlations, and (b), to a creation of correlations; wave vectors $\boldsymbol{k}$, $\boldsymbol{k}'$ in the hydrodynamic range.

indicated the single vertex diagram (a) which corresponds to a propagation of correlations (see Fig. 8.9.1), as well as diagram (b) corresponding to the creation of a correlation. If the wave vectors $\boldsymbol{k}$ and $\boldsymbol{k}'$ are in the hydrodynamic range, the time dependence of these two diagrams can differ only by terms of order t/t_h.

These terms are small for $t \approx t_r$ in the case we are studying here $(t_r \ll t_h)$.

5. Transport Equation in the Hydrodynamic Case — the Boltzmann Equation

In accordance with the discussion of § 4 we shall replace the single diagram in equation (10.2.1) by the sum of the three diagrams of Fig. 10.4.1. We then obtain, instead of (10.2.2), [see also (10.4.1), (10.4.3)]

$$\begin{aligned}\partial\rho(\alpha_k, \mathbf{v}_\alpha, t)/\partial t = {} & 8\pi^4\lambda^2 C\int d\mathbf{v}_j \int d\mathbf{l}\, e^{i\mathbf{k}\cdot\mathbf{v}_\alpha t}\,|V_l|^2\,\mathbf{l}\cdot(\partial/\partial\mathbf{v}_\alpha)\delta(\mathbf{l}\cdot\mathbf{g}_{\alpha j})\mathbf{l}\cdot\mathbf{D}_{\alpha j}\\ & \times e^{-i\mathbf{k}\cdot\mathbf{v}_\alpha t}\rho(\alpha_k, \mathbf{v}_\alpha, \mathbf{v}_j, t)\\ & + 8\pi^4\lambda^2 C\int d\mathbf{v}_j \int d\mathbf{l}\, e^{i\mathbf{k}\cdot\mathbf{v}_\alpha t}\,|V_l|^2\,\mathbf{l}\cdot(\partial/\partial\mathbf{v}_\alpha)\,\delta(\mathbf{l}\cdot\mathbf{g}_{\alpha j})\,\mathbf{l}\cdot\mathbf{D}_{\alpha j}\\ & \times e^{-i\mathbf{k}\cdot\mathbf{v}_j t}\rho(j_k, \mathbf{v}_\alpha, \mathbf{v}_j, t)\\ & + 8\pi^4\lambda^2 C\int d\mathbf{v}_j \int d\mathbf{l}\int d\mathbf{k}'\, e^{i\mathbf{k}\cdot\mathbf{v}_\alpha t}\,|V_l|^2\,\mathbf{l}\cdot(\partial/\partial\mathbf{v}_\alpha)\,\delta(\mathbf{l}\cdot\mathbf{g}_{\alpha j})\\ & \times \mathbf{l}\cdot\mathbf{D}_{\alpha j}\, e^{-i[\mathbf{k}'\cdot\mathbf{v}_\alpha + (\mathbf{k}-\mathbf{k}')\cdot\mathbf{v}_j]t}\rho(\alpha_{k'}, j_{k-k'}, \mathbf{v}_\alpha, \mathbf{v}_j, t) \qquad (1)\end{aligned}$$

The transition to phase space is performed as in § 2 and gives

$$\begin{aligned}(\partial f_1/\partial t) + \mathbf{v}_\alpha\cdot(\partial f_1/\partial\mathbf{x}_\alpha) & \\ = 8\pi^4\lambda^2 C^2 & \int d\mathbf{v}_j \int d\mathbf{l}\,|V_l|^2\,\mathbf{l}\cdot(\partial/\partial\mathbf{v}_\alpha)\,\delta(\mathbf{l}\cdot\mathbf{g}_{\alpha j})\,\mathbf{l}\cdot\mathbf{D}_{\alpha j}\\ \times \{[\varphi_2(\mathbf{v}_\alpha, \mathbf{v}_j, t) & + \int d\mathbf{k}\,\rho(\alpha_k, \mathbf{v}_j, \mathbf{v}_\alpha, t)\, e^{i\mathbf{k}\cdot(\mathbf{x}_\alpha - \mathbf{v}_\alpha t)}\\ + \int d\mathbf{k}\,\rho(j_k, & \mathbf{v}_j, \mathbf{v}_\alpha, t)\, e^{i\mathbf{k}\cdot(\mathbf{x}_\alpha - \mathbf{v}_j t)}\\ + \iint d\mathbf{k}\, d\mathbf{k}'\,\rho(\alpha_{k'}, & j_{k-k'}, \mathbf{v}_j, \mathbf{v}_\alpha, t)\, e^{i[\mathbf{k}\cdot\mathbf{x}_\alpha - \mathbf{k}'\cdot\mathbf{v}_\alpha t - (\mathbf{k}-\mathbf{k}')\cdot\mathbf{v}_j t]}\} \qquad (2)\end{aligned}$$

Seemingly, this equation relates the single-particle distribution function $f_1(\alpha)$ to the two-particle Fourier coefficients $\rho(\alpha_k, j_{k'})$ so that the equation would not in general be closed. However these Fourier coefficients originate only in the pseudo-diagonal fragment (c) of Fig. 10.4.1 in which both vectors $\mathbf{k}$ and $\mathbf{k}'$ belong to the macroscopic range. Therefore the conditions for the application of the fundamental factorization theorem of § 3 are satisfied and properties (7.4.1), (7.4.2), and (10.3.2) imply that the bracketed expression, on the right-hand side is simply (see also 7.2.3)

$$\begin{aligned} &[\varphi_1(\mathbf{v}_\alpha, t) + \int d\mathbf{k}\, \rho(\alpha_k, \mathbf{v}_\alpha, t)\, e^{i\mathbf{k}\cdot(\mathbf{x}_\alpha - \mathbf{v}_\alpha t)}] \\ &\times [\varphi_1(\mathbf{v}_j, t) + \int d\mathbf{k}\, \rho(j_k, \mathbf{v}_j, t)\, e^{i\mathbf{k}\cdot(\mathbf{x}_\alpha - \mathbf{v}_j t)}] \\ &= (1/C^2)\, f_1(\mathbf{v}_\alpha, \mathbf{x}_\alpha)\, f_1(\mathbf{v}_j, \mathbf{x}_\alpha) \end{aligned} \tag{3}$$

We may therefore write our evolution equation in the final form

$$\begin{aligned} (\partial f_1/\partial t) + \mathbf{v}_\alpha \cdot (\partial f/\partial \mathbf{x}_\alpha) &= 8\pi^4 \lambda^2 \int d\mathbf{x}_j\, d\mathbf{v}_j \int d\boldsymbol{l}\, |V_l|^2\, \boldsymbol{l} \cdot (\partial/\partial \mathbf{v}_\alpha) \\ &\times \delta(\boldsymbol{l}\cdot\mathbf{g}_{\alpha j})\, \boldsymbol{l}\cdot\mathbf{D}_{\alpha j}\, \delta(\mathbf{x}_\alpha - \mathbf{x}_j)\, f_1(\mathbf{x}_\alpha, \mathbf{v}_\alpha, t)\, f_1(\mathbf{x}_j, \mathbf{v}_j, t) \end{aligned} \tag{4}$$

This is indeed the Boltzmann equation for weakly coupled systems.

The case of dilute gases with strong interactions could be treated in the same way by everywhere replacing the cycles by the sum of all possible chains.

We have thus completed the proof of the Boltzmann equation for inhomogeneous systems. As we have seen, the existence of the macroscopic hydrodynamic length introduces deep changes into the structure of the transport equations. This deserves a more detailed discussion which we shall present in the next paragraph of this chapter.

6. Discussion of the Approach to Equilibrium in Inhomogeneous Systems

The influence of the scale of the inhomogeneity on the approach to equilibrium is clearly reflected in the structure of the transport equations (10.2.5) and (10.5.4). When the scale of the inhomogeneity is small compared to the free path the transport equations contain the "non-local" expression (10.2.6). On the contrary, the usual "local" form (10.2.7) appears in the limit of a large hydrodynamic scale.

This difference is also clearly reflected in the structure of the diagrams. Taking account of the diagrams of Fig. 10.4.1 we see that a typical contribution to the Boltzmann equation will come from diagrams which have more and more lines to the right (see Fig. 10.6.1).

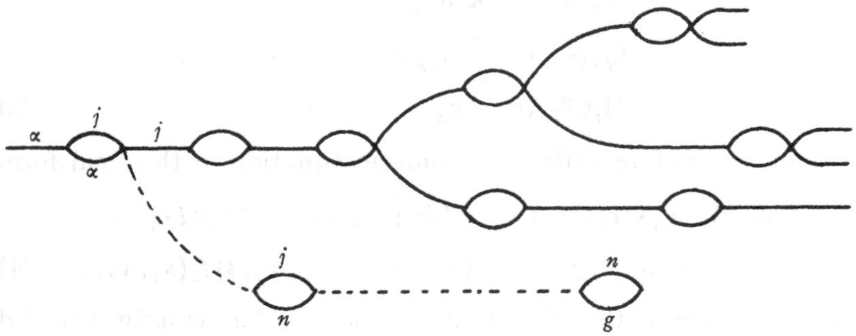

Fig. 10.6.1 A typical contribution to $\rho(\alpha_k)$ under hydrodynamic conditions

This has to be contrasted with Fig. 10.2.1. Contrary to what happens in the limit of large free paths which was studied in § 2, under hydrodynamic conditions $\rho(\alpha_k)$ originates from Fourier coefficients $\rho(\alpha_k, \beta_{k'}, \gamma_{k''}, \ldots)$ with an arbitrary large number of indices. As a consequence, the distribution function $f_1(\boldsymbol{x}_\alpha, \boldsymbol{v}_\alpha)$ is determined by distribution functions involving an arbitrarily large number of points in the space.

At first this seems to prevent us from obtaining a closed equation for f_1. But this difficulty appears only for macroscopic values of wave vectors, just those, in fact, for which the basic factorization theorem (10.3.3) is valid. The multiple Fourier coefficients $\rho(\alpha_k, \beta_{k'}, \gamma_{k''}, \ldots)$ are then products of the $\rho(\alpha_k)$'s and we obtain a *closed non-linear equation* without invoking any new principle or any consideration alien to the strict formulation of our problem.

The difference in the behavior of Fourier coefficients in the molecular and in the macroscopic range simply reflects the difference in their physical meaning. Molecular correlations arise from intermolecular forces, while macroscopic correlations or inhomogeneities express externally imposed constraints or boundary conditions.

From the Boltzmann equation (10.5.4) we may deduce the usual macroscopic equations expressing the conservation of mass, of momentum (Stokes-Navier equation), and of energy. This is done in detail in all textbooks on the kinetic theory of gases and will not be repeated here (see for example Chapman and Cowling, 1939; Hirschfelder, Curtiss, and Bird, 1954; Balescu, 1962). For example the time change of the local momentum is determined by the divergence of the pressure tensor.

CHAPTER 11

General Kinetic Equations

1. Evolution of the Velocity Distribution

In this chapter we shall derive the general form of the kinetic equations that describe the evolution of the system (Prigogine and Résibois, 1961). No assumption will be introduced about the strength of the interaction or the concentration. The only restrictions are, as in the rest of this book:

a. the existence of reduced distribution functions in the limit of a large system, as discussed in Chapter 7;

b. the absence of correlations involving particles at infinite distances.

These assumptions are "self-perpetuating:" If valid at $t = 0$, they remain valid for all $t > 0$ (see Chapter 7, § 8, Chapter 8, § 11).

In Chapter 12, we shall study the long-time behavior of the solution of the kinetic equations and shall show that they drive the system to equilibrium (generalized H-theorem). In this way the link with equilibrium statistical mechanics is established.

We shall use the following notation:

represents the sum of all possible diagonal fragments containing any number of vertices and particles. Similarly

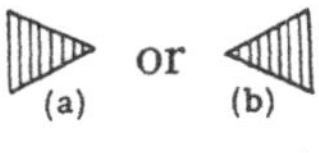

represents the sum of all possible creation or destruction diagrams (see Chapter 8, §§ 7, 8) having a given number of lines at the left (or at the right). We shall also use the symbol (b) to indicate an arbitrary destruction region as defined in Chapter 8, § 10.

Let us first consider the evolution of the velocity distribution function $\rho_0(t)$. It can be expressed as the sum of an arbitrary number of diagonal fragments preceded, or not, by a destruction region [1]:

$\rho_0(t) = \sum_{n=0}^{\infty}$  $^{n}\,[\rho_0(0) + \sum_{k}$  $\rho_k(0)]$ (1)

where  ^{n} represents a sequence of n arbitrary diagonal fragments. Such a sequence gives rise, in the resolvent formalism developed in Chapter 8, to the contribution [see Chapter 8, §§ 5, 7, especially (8.5.7) and (8.7.7)]

$$\rho_0(t) = -(1/2\pi i)\oint_C dz\, e^{-izt}(1/z)\sum_{n=0}^{\infty}(-1)^n[\Psi^+(z)/z]^n[\rho_0(0)+F^+(z)] \quad (2)$$

where $F^+(z)$ represents the effect of the destruction region. After differentiating with respect to time, we obtain

$$\begin{aligned}\partial\rho_0/\partial t = &-(1/2\pi)\oint_C dz\, e^{-izt}F^+(z)\\ &+(1/2\pi)\oint_C dz\, e^{-izt}(\Psi^+(z)/z)\sum_{n=0}^{\infty}(-1)^n[\Psi^+(z)/z]^n[\rho_0(0)+F^+(z)] \quad (3)\end{aligned}$$

We now use the convolution theorem for Laplace transforms. This theorem states that the Laplace transform of the integral

$$\Phi(t) = \int_0^t dt_1 f_1(t-t_1)\, f_2(t_1) \quad (4)$$

is the product of the Laplace transforms of the functions f_1 and f_2 [see (8.3.1)]. Indeed we have

$$\begin{aligned}\tilde{\Phi}(s) &= \int_0^{\infty} dt\, e^{-st}\Phi(t) = \int_0^{\infty} dt\int_0^t dt_1\, e^{-st} f_1(t-t_1)\, f_2(t_1)\\ &= \tilde{f}_1(s)\,\tilde{f}_2(s) \quad (5)\end{aligned}$$

Let us write

$$F^+(t) = -(1/2\pi)\oint_C dz\, e^{-izt}F^+(z) \quad (6)$$

$$G_0^+(t) = (1/2\pi i)\oint_C dz\, e^{-izt}\Psi^+(z) \quad (7)$$

[1] For the calculation of reduced distribution functions we have to retain only semi-connected diagrams in (11.1.1). This has already been discussed (see Chapter 7, § 8; Chapter 8, § 11).

and apply the convolution theorem to the product

$$-\Psi^+(z)\{(1/z)\sum_{n=0}^{\infty} -(1)^n[\Psi^+(z)/z]^n[\rho_0(0)+F^+(z)]\} \tag{8}$$

which appears in (11.1.3). $\Psi^+(z)$ is the Laplace transform of $G_0^+(t)$ according to (11.1.7) and the second factor is according to (11.1.2) the Laplace transform of $\rho_0(t)$. Therefore the convolution theorem states that (11.1.8) is the Laplace transform of the integral $\int_0^t dt_1\, G_0^+(t-t_1)\,\rho_0(t_1)$ and (11.1.3) becomes

$$\partial\rho_0/\partial t = F^+(t) + \int_0^t dt_1\, G_0^+(t-t_1)\,\rho_0(t_1) \tag{9}$$

This is the general equation for the evolution of the velocity distribution function. A remarkable feature of this equation is its non-Markowian character. The change of ρ_0 at time t depends on the previous values of ρ_0. This non-Markowian character is a direct consequence of the finite duration of a collision process, measured by the characteristic time t_{int} (see Chapter 8, § 2). The change in ρ_0 at time t depends on collisions which started at some previous time, of order $t-t_{\text{int}}$, and are terminated at time t. For this reason an integration "over the past" is involved. As we shall see for times long with respect to the duration of an interaction t_{int}, (11.1.9) reduces to a Markowian equation. The velocity distribution has a "memory" which extends over a time of order t_{int}.

The physical meaning of the inhomogeneous term $F^+(t)$ in (11.1.9) is clear: it expresses the effect of the "destruction region" and shows the influence of the correlations that existed at the initial time $t = 0$, on the evolution of the velocity distribution function.

In order to study this equation we shall, as we did in Chapter 8 (see especially § 5), make the following assumptions about $F^+(z)$ and $\Psi^+(z)$. Both functions are analytic for $\operatorname{Im} z > 0$. For $\operatorname{Im} z < 0$ they represent the analytic continuation of $F^+(z)$ and $\Psi^+(z)$ into the lower half of the complex z plane. They have poles in the lower half plane at points z_j ($|\operatorname{Im} z_j| \neq 0$). The position of these poles is, in the case of $\Psi^+(z)$, related to the duration of an interaction, and in the case of $F^+(z)$ to the range of the initial correlations too.[1]

[1] The behavior of the resolvent may become more complicated when collective interactions are involved (see Balescu, 1962).

As we already noted in Chapter 8, § 5, other singularities (branch points, . . .) could also arise, depending on the form of the intermolecular potential. The conditions we have enumerated are sufficient but not necessary for the applicability of our theory. As the extension to other types of singularities seems not to present any problem, we shall not discuss this matter further.

If we suppose, in order to avoid too cumbersome a notation, that all poles of $\Psi^+(z)$ and $F^+(z)$ are of first order we get [see (8.5.2), and (8.7.7); $z = 0$ is not a pole in (11.1.6), (11.1.7)]

$$G_0^+(t) = \sum_j e^{-iz_j t} \operatorname{res} \Psi^+(z_j) \tag{10}$$

$$F^+(t) = -i \sum_k e^{-iz_k t} \operatorname{res} F^+(z_k) \tag{11}$$

Therefore (11.1.9) can also be written

$$\partial \rho_0/\partial t = -i \sum_k e^{-iz_k t} \operatorname{res} F^+(z_k) + \sum_j \operatorname{res} \Psi^+(z_j) \int_0^t dt_1 e^{-iz_j(t-t_1)} \rho_0(t_1) \tag{12}$$

As $|\operatorname{Im} z_j|$ is of order t_{int}^{-1} this expression shows in an especially clear way that the "memory" of the system extends over a time of the order of the duration of an interaction.

Let us now investigate how the non-Markowian equation (11.1.9) reduces to a Markowian one for times $t \gg t_{\text{int}}$. Consider the effect of a single diagonal fragment on the value of the velocity distribution ρ_0 at time t. We have seen that this effect vanishes for $t \to 0$ because of the finite duration of any interaction (see Chapter 8, §§ 2, 5). Moreover for short times it becomes proportional to t^2. This implies that its first derivative with respect to t also vanishes for $t = 0$. The effect of an arbitrary diagonal fragment is expressed by formula (8.5.2) in which the specific structure of the diagonal fragment is reflected only by the form of the function $\Psi^+(z)$. Therefore our two conditions give us

$$(d\Psi^+/dz)_{z=0} + \sum_j (1/z_j^2) \operatorname{res} \Psi^+(z_j) = 0 \tag{13}$$

$$-i\Psi^+(0) + \sum_j (1/iz_j) \operatorname{res} \Psi^+(z_j) = 0 \tag{14}$$

These relations express the sums over the residues in terms of the behavior of Ψ^+ at $z = 0$. Using (11.1.10) we immediately obtain the identities

$$\int_0^\infty d\tau\, G_0^+(\tau) = \sum_j \int_0^\infty d\tau\, e^{-iz_j\tau} \operatorname{res} \Psi^+(z_j) = i\Psi^+(0) \tag{15}$$

and similarly

$$\int_0^\infty d\tau\, \tau\, G_0^+(\tau) = -(d\Psi^+/dz)_{z=0} \tag{16}$$

Now for sufficiently long times we may neglect the inhomogeneous term $F^+(t)$ in (11.1.9). Moreover $\rho_0(t_1)$ will then vary slowly over times of the order of t_{int} and we may write

$$\rho_0(t_1) = \rho_0(t) - (t - t_1)\, \partial\rho_0/\partial t + \ldots \tag{17}$$

Therefore (11.1.9) becomes, for $t \gg t_{\text{int}}$,

$$\begin{aligned} \partial\rho_0/\partial t &= \int_0^t dt_1\, G_0^+(t-t_1)\, \rho_0(t) - \int_0^t dt_1 (t-t_1)\, G_0^+(t-t_1)\, \partial\rho_0/\partial t + \ldots \\ &= \int_0^\infty d\tau\, G_0^+(\tau)\, \rho_0(t) - \int_0^\infty d\tau\, \tau\, G_0^+(\tau)\, \partial\rho_0/\partial t + \ldots \\ &= i\Psi^+(0)\, \rho_0(t) + (d\Psi^+/dz)_{z=0}\, i\Psi^+(0)\, \rho_0(t) + \ldots \end{aligned}$$

$= [1 + \Psi^{+(1)}(0)]$  $\rho_0(t) + \ldots$ (18)

where we use for the first derivative of Ψ^+ the notation

$$\Psi^{+(1)}(0) = (d\Psi^+/dz)_{z=0}$$

In the first term appear all diagonal fragments corresponding to all possible collision processes. In this term the effect of the collisions is taken asymptotically, that is, completely neglecting the duration of the collision process. If moreover we retain only a special class of collisions in $\Psi^+(0)$ we come back to the case of weakly coupled gases (Chapter 4), dilute gases (Chapter 6), or plasmas (Chapter 9). The next term appears as a correction, due to the finite duration of the collision. As we have seen [see (8.2.13)] such terms have to be retained in general if one is interested in the effect of higher concentrations. However as we shall show in § 6 these contributions are negligible in processes involving the hydrodynamic scale as defined in Chapter 10. For example the phenomenological transport coefficients involve only the asymptotic cross sections expressed by our diagonal fragment, and no terms related to the finite duration of collisions appear.

Equation (11.1.18) gives a Markowian approximation to the non-Markowian equation (11.1.9). We shall come back to it in § 2. The main physical effect due to its non-Markowian character is to produce through the time integration a kind of time smoothing over intervals of order t_{int}. The importance of such an effect of course depends critically on the variation of ρ_0 over such time intervals. It becomes negligible when ρ_0 may be treated as a constant over times of order t_{int}. This is precisely the reason why all terms related to the finite duration of collisions drop in the description of stationary (equilibrium or non-equilibrium) situations (see § 6).

It is interesting to note that Van Hove has also derived a kind of master equation of a non-Markowian character valid to all orders in the coupling constants (Van Hove, 1957). It is however not easy to compare Van Hove's equation with ours. Indeed Van Hove has derived his equation in the quantum mechanical case and its classical analogue is not obvious. Also, his equation does not apply to the velocity distribution (or to the diagonal elements of the Von Neumann density matrix, see Chapter 13), but to a function related in a rather complicated way to ρ_0. Another difference is that Van Hove uses the assumption of random phases, which is not valid except to the lowest order in the coupling constant or the concentration (see Philippot, 1961, and Appendix IV of this book).[1]

2. Markowian Form of the Evolution Equation for the Velocity Distribution

In § 1 we derived the general form of the evolution equation for the velocity distribution (11.1.9) and we showed that for sufficiently long times it reduces to the Markowian equation (11.1.18). The procedure we used based on the Taylor expansion (11.1.17) is however not very convenient and we shall now derive the Markowian approximation to (11.1.9) starting directly from (11.1.3). We shall assume that the initial correlations are

[1] In a recent paper Fujita (1962) discusses the relation between (11.1.9) and Van Hove's master equation.

of a molecular origin and take

$$t \gg t_{\text{int}} \tag{1}$$

We may then neglect the first term in the right-hand side of (11.1.3) and in the second term limit the contour to a small circle around the point $z = 0$. All other singularities give decaying exponentials which vanish for times which satisfy the inequality (11.2.1). We shall therefore write

$$\partial \rho_0/\partial t = (1/2\pi) \oint_O dz\, e^{-izt} (\Psi^+(z)/z) \sum_{n=0}^{\infty} (-1)^n [\Psi^+(z)/z]^n [\rho_0(0) + F^+(z)] \tag{2}$$

where O is a contour encircling the origin. Let us expand $\Psi^+(z)$ in a Taylor series near $z = 0$,

$$\Psi^+(z) = \Psi^+(0) + z\Psi^{(1)+}(0) + (z^2/2)\Psi^{(2)+}(0) + \cdots \tag{3}$$

Then (11.2.2) becomes

$$\begin{aligned} \partial \rho_0/\partial t = {} & (1/2\pi)\Psi^+(0) \oint_O dz\, e^{-izt} (1/z) \sum_{n=0}^{\infty} (-1)^n [\Psi^+(z)/z]^n [\rho_0(0) + F^+(z)] \\ & - (1/2\pi)\Psi^{(1)+}(0) \oint_O dz\, e^{-izt} (\Psi^+(z)/z) \sum_{n=0}^{\infty} (-1)^n [\Psi^+(z)/z]^n [\rho_0(0) + F^+(z)] \\ & + (1/2\pi)(1/2)\Psi^{(2)+}(0) \oint_O dz\, e^{-izt} ([\Psi^+(z)]^2/z) \sum_{n=0}^{\infty} (-1)^n [\Psi^+(z)/z]^n \\ & \times [\rho_0(0) + F^+(z)] + \cdots \end{aligned} \tag{4}$$

To write this formula we have systematically suppressed all terms which have no singularity at $z = 0$, and give therefore a vanishing contribution when we integrate over the contour O. In (11.2.4) we again expand $\Psi^+(z)$ and all powers of $\Psi^+(z)$ which are at the left of the sum $\sum_{n=0}^{\infty}$, in a Taylor series according to (11.2.3). Moreover using (11.1.2) we obtain in this way

$$\begin{aligned} \partial \rho_0/\partial t &= [i\Psi^+(0) + \Psi^{(1)+}(0)\, i\Psi^+(0) + [\Psi^{(1)+}(0)]^2 i\Psi^+(0) \\ &\qquad + \tfrac{1}{2}\Psi^{+(2)}(0)\, \Psi^+(0)\, i\Psi^+(0) + \ldots]\, \rho_0(t) \\ &= \Omega\, i\, \Psi^+(0)\, \rho_0(t) \end{aligned}$$

$= \Omega$ $\rho_0(t)$ (5)

with

$$\Omega = 1+\Psi^{(1)+}(0)+\Psi^{(1)+}(0)\Psi^{(1)+}(0)+\tfrac{1}{2}\Psi^{(2)+}(0)\,\Psi^{+}(0)+\cdots \qquad (6)$$

It is easy to give a general rule for the formation of the operator Ω. We may write

$$\Omega = \sum_{\alpha=0}^{\infty} A_\alpha \qquad (7)$$

with

$$A_0 = 1,\quad A_1 = \Psi^{(1)+}(0),\quad A_2 = \Psi^{(1)+}(0)\Psi^{(1)+}(0)+\tfrac{1}{2}\Psi^{(2)+}(0)\Psi^{+}(0) \qquad (7')$$

In general A_α contains the sum of all terms of the form

$$\Psi^{(i)+}(0)\,\Psi^{(j)+}(0)\cdots\Psi^{(l)+}(0)\,\Psi^{+}(0)$$

with

$$i+j+\cdots+l = \alpha$$

The operator A_α may be expressed in terms of $A_{\alpha-1}, A_{\alpha-2}, \cdots$ through the recurrence relation

$$A_\alpha = (1/\alpha)\{(A_{\alpha-1}\Psi^{+})^{(1)}A_0+(A_{\alpha-2}\Psi^{+})^{(1)}A_1+\cdots$$
$$+(A_0\Psi^{+})^{(1)}A_{\alpha-1}\} \qquad (8)$$

where $(A_{\alpha-1}\Psi^{+})^{(1)}\cdots$ represent the first derivative of $A_{\alpha-1}\Psi^{+}$, always taken at $z = 0$. This relation is valid for $\alpha = 0, 1, 2$, as shown by (11.2.7′). For an arbitrary α it is easy to prove it by induction, supposing its validity for $\alpha-1$.

The first two terms in (11.2.6) are identical to (11.1.18). Again the first term contains the effect of all collisions taken asymptotically while all other terms express corrections due to the finite duration of the collision process.

Let us consider the order of magnitude of the different terms that appear in (11.2.5) for the case of hard spheres, taking account of (8.2.12) and (8.2.13). We have for $\Psi^{+}(0)$ an expression of the form

$$\begin{aligned}\Psi^{+}(0) &\sim 1/t_r\\ &\sim a^2 Cv\,[1+a^3C+(a^3C)^2+\cdots]\end{aligned} \qquad (9)$$

The first term comes from binary collisions, the next from ternary, and so on.

On the other hand as we have seen Ω takes account of the finite duration of a collision [see (8.2.10)]. For example $\Psi^{(1)+}$ corresponds to the δ'_+ contribution we studied in Chapter 8, § 2. Its order of magnitude is

$$\Psi^{(1)+} \sim t_{\text{int}}/t_r \sim a^3 C \tag{10}$$

Similarly the other terms of (11.2.6) give [1]

$$\begin{aligned} \Omega &\sim 1 + t_{\text{int}}/t_r + (t_{\text{int}}/t_r)^2 + \cdots \\ &\sim 1 + a^3 C + (a^3 C)^2 + \cdots \end{aligned} \tag{11}$$

We see therefore that the Boltzmann equation correct to the first power in $a^3 C$ will be of the form

$\partial \rho_0/\partial t = [(1+$  $^{(1)}_{2}$ $)$  $_2 +$  $_3] \rho_0$ (12)

In this equation  $_n$ represents the sum of all possible diagonal fragments evaluated asymptotically and involving n particles (that is taking $\Omega = 1$). Similarly  $^{(1)}_{n}$ represents the contribution of n-body collisions to the operator $\Psi^{(1)+}(0)$ in (11.2.6). Equation (11.2.12) contains therefore "corrected" two-body collisions as well as three-body collisions taken asymptotically.

An essential point is that the velocity distribution function satisfies a separate equation. This introduces an enormous simplification.

In Chapter 12 we shall show that (11.2.5) indeed implies the validity of the equilibrium velocity distribution for sufficiently long times.

3. Evolution of Correlations

The method we used in § 1 for the velocity distribution may easily be extended to the time evolution of correlations. We shall denote by ρ_γ a correlation involving γ particles [for example, we shall write ρ_2 for $\rho(\alpha_k \beta_{-k})$, ρ_3 for $\rho(\alpha_k \beta_{k'} \gamma_{-k-k'}) \cdots$].

It is very useful to subdivide ρ_γ into two classes (Prigogine

[1] A differentiation with respect to z is equivalent to a multiplication by t_{int} [see (11.1.13), (11.1.14)].

and Henin, 1960)

$$\rho_\gamma = \rho'_\gamma + \rho''_\gamma \tag{1}$$

ρ'_γ includes all diagrams such that their last part corresponds to a diagonal fragment inserted on a line [see Fig. 11.3.1(a)], while ρ''_γ includes all diagrams that end with a creation fragment [see Fig. 11.3.1(b)].[1]

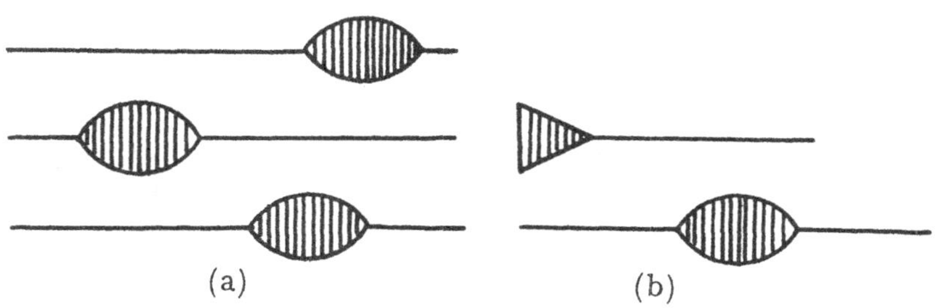

Fig. 11.3.1 (a) Contribution to ρ'_γ. (b) Contribution to ρ''_γ.

The physical meaning of these two classes is rather different. As we have seen in the first case we have a correlation which during its propagation through the medium is scattered by the other particles (see Chapter 8, § 6). In the second case the correlation is "created" out of a lower order correlation.

The types of equation satisfied by ρ'_γ and ρ''_γ are also different. For ρ'_γ we may expect a diagonal equation very similar to (11.1.9) in which the other Fourier coefficients enter only through the initial conditions. On the contrary, ρ''_γ will satisfy a relation which permits us to express its value at time t in terms of lower order correlations at earlier times.

Let us derive these equations briefly. With a slight change in notation (11.1.2) is still valid for ρ'_γ. We have only to replace the z in the denominator by $z-\boldsymbol{k}\cdot\boldsymbol{v}$ with $\boldsymbol{k}\cdot\boldsymbol{v} = \sum_j \boldsymbol{k}_j\cdot\boldsymbol{v}_j$. We also have to replace $\Psi^+(z)$ by $\Psi^+_\gamma(z)$, which corresponds to the sum of all diagonal transitions relating the state $\{k\}$ to $\{k\}$. We then obtain

[1] Both ρ'_γ and ρ''_γ may contain a destruction region. Diagrams which contain *only* a destruction region will be included in ρ'_γ.

$$\rho'_\gamma(t) = (1/2\pi i) \oint_C dz\, e^{-izt} [1/(z-\boldsymbol{k}\cdot\boldsymbol{v})] \sum_{n=0}^{\infty} (-1)^n [\Psi^+_\gamma(z)/(z-\boldsymbol{k}\cdot\boldsymbol{v})]^n \times [\rho_\gamma(0)+F^+_\gamma(z)] \quad (2)$$

Again F^+_γ corresponds to the "destruction region" in which initial correlations γ' are transferred into correlations γ (with $\gamma' > \gamma$). We then proceed as in § 1; we take the time derivative and apply the convolution theorem. In this way we obtain

$$\partial\rho'_\gamma/\partial t + i \sum_j \boldsymbol{k}_j \cdot \boldsymbol{v}_j \rho'_\gamma = F^+_\gamma(t) + \int_0^t dt_1\, G^+_\gamma(t-t_1)\, \rho'_\gamma(t_1) \quad (3)$$

The main difference from (11.1.9) is the presence of the flow term $i \sum \boldsymbol{k}_j \cdot \boldsymbol{v}_j \rho'_\gamma$. This equation generalizes equation (5.3.16) that we obtained for the Brownian motion problem in weakly coupled gases. (Here ρ'_γ includes the complete time dependence.) For times long with respect to the duration of the collision we may again neglect the non-Markowian character as well as the influence of the initial correlations, which we suppose to be of molecular range. We then obtain an equation of the form [see (11.2.5)]

$$\partial\rho'_\gamma/\partial t + i \sum \boldsymbol{k}_j \cdot \boldsymbol{v}_j \rho'_\gamma = \Omega_\gamma -$$  $$\rho'_\gamma \quad (3')$$

Here again Ω_γ takes account of the finite duration of the collisions. In the weakly coupled case we may replace Ω_γ by one and retain only the two-vertex diagonal fragment. We then come back to (5.3.16).

By a generalization of the argument used in Chapter 5 we shall show in Chapter 12, § 2, that

$$\rho'_\gamma \underset{t\to\infty}{\to} 0 \quad (4)$$

Thus initial correlations are destroyed by scattering.

Let us now consider ρ''_γ. We have in the resolvent formalism

$$\rho''_\gamma(t) = (1/2\pi i) \oint_C dz\, e^{-izt} \sum_{\gamma'<\gamma} C^+_{\gamma\gamma'}(z)\, [1/(z-\boldsymbol{k}'\cdot\boldsymbol{v})] \times \sum_{n=0}^{\infty} (-1)^n [\Psi^+_{\gamma'}(z)/(z-\boldsymbol{k}'\cdot\boldsymbol{v})]^n [\rho_{\gamma'}(0)+F^+_{\gamma'}(z)] \quad (5)$$

$C^{+}_{\gamma\gamma'}(z)$ is the contribution of the creation fragment transforming a correlation γ' to a correlation γ. We shall assume (as we have done for $\Psi^{+}_{\gamma}(z)$ and $F^{+}_{\gamma}(z)$) that $C^{+}_{\gamma\gamma'}(z)$ has an analytic continuation into the lower half of the complex plane and that its only singularities are poles situated at z_j with $|\text{Im}\, z_j| < 0$.

Let us write

$$\varphi^{+}_{\gamma\gamma'}(t) = -(1/2\pi) \oint_C dz\, e^{-izt} C^{+}_{\gamma\gamma'}(z) \tag{6}$$

and apply the convolution theorem (11.1.5) to the product

$$-C^{+}_{\gamma\gamma'}(z)\{[1/(z-\boldsymbol{k}'\cdot\boldsymbol{v})] \sum_{n=0}^{\infty} (-1)^n [\Psi^{+}_{\gamma'}/(z-\boldsymbol{k}'\cdot\boldsymbol{v})]^n [\rho_{\gamma'}(0)+F^{+}_{\gamma'}(z)]\} \tag{7}$$

Taking account of (11.3.2) we obtain

$$\rho''_{\gamma}(t) = \sum_{\gamma'<\gamma} \int_0^t dt_1\, \varphi^{+}_{\gamma\gamma'}(t-t_1)\, \rho'_{\gamma'}(t_1) \tag{8}$$

This relation indeed expresses ρ''_{γ} at time t in terms of the $\rho'_{\gamma'}$ (for $\gamma' < \gamma$) at earlier times. For times such that (11.3.4) is valid (11.3.8) reduces to [see also (11.3.1)]

$$\rho_{\gamma}(t) = \int_0^t dt_1\, \varphi^{+}_{\gamma 0}(t-t_1)\, \rho_0(t_1) \tag{9}$$

The correlations are then functionals of the velocity distribution function. If we may replace ρ_0 by its equilibrium value we obtain

$$\rho_{\gamma}(t)_{t\to\infty} = \int_0^{\infty} d\tau\, \varphi^{+}_{\gamma 0}(\tau)\, \rho_0^{\text{eq}} \tag{10}$$

The integral that appears in (11.3.10) can be expressed in terms of $C^{+}_{\gamma 0}$ $(z = 0)$. Indeed, exactly as in (8.7.7) we have

$$= (1/2\pi i) \oint_C dz\, e^{-izt} (1/z)\, C^{+}_{\gamma 0} = C^{+}_{\gamma 0}(0) + \sum (e^{-iz_j t}/z_j)\, \text{res}\, C^{+}_{\gamma 0}(z_j) \tag{11}$$

Now this expression has to vanish for $t \to 0$; obviously the interactions had no time to make themselves felt then. Therefore

$$C^{+}_{\gamma 0}(0) + \sum_j (1/z_j)\, \text{res}\, C^{+}_{\gamma 0}(z_j) = 0 \tag{12}$$

As in § 1 this permits us to write (11.3.10) in the form

$$\rho_{\gamma}(t)_{t\to\infty} = C^{+}_{\gamma 0}(0)\, \rho_0^{\text{eq}} \tag{13}$$

This expression will be used in Chapter 12 to show that the long time correlations are identical with those derived by equilibrium statistical mechanics.

A great simplification arises in the "thermodynamic" case of Chapter 8, § 12. As we have seen, we may in this case neglect all lines corresponding to free propagation and all scattering diagrams expressed by diagonal fragments inserted on lines. In other words we then have

$$\rho_\gamma = \rho_\gamma'' \tag{14}$$

and equation (11.3.9) is valid for all times $t \gg t_{\text{int}}$. This means that the correlations are functionals of the velocity distribution (see § 5). Moreover as we want to calculate the values of the distribution functions only for distances of the order of the range of the forces, we may use an asymptotic time evaluation for the creation fragment (see Chapter 8, § 8). Therefore in the "thermodynamic case" we have simply

$$\rho_\gamma(t) = C^+_{\gamma 0}(0)\, \rho_0(t) =$$  $$\rho_0(t) \tag{15}$$

This equation relates the values of the correlations to the velocity distribution function taken at the same time.

4. Hydrodynamic Situations

We now want to extend these results to the hydrodynamic case. We shall suppose, as we did at the end of § 3 that all the characteristic times which enter the asymptotic evolution of the diagrams are of the order of the duration of an interaction (the "thermodynamic case," see Chapter 8, § 12). But in addition we introduce, as in Chapter 10, § 1, a characteristic time (or length) related to the scale of the inhomogeneity. We shall immediately consider the case in which this time is much longer than any other characteristic time. We may therefore apply the factorization theorems and introduce the pseudo-diagonal diagrams, as we did in Chapter 10 (see §§ 3, 4).

The calculations are so similar to those developed in Chapter 10 that we shall be very brief. Let us consider the evolution of the one-particle distribution function $f_1(\boldsymbol{x}_\alpha\, \boldsymbol{v}_\alpha\, t)$. As in Chapter 10, § 5 we have only to add to each diagonal fragment all possible pseudo-diagonal fragments. We shall now consider times long with respect to t_{int} [see (11.2.1)] and study the hydrodynamic extension of equation (11.2.5).

Let us in equation (11.2.5) retain only two- and three-particle diagrams. To this approximation the equation for f_1 is, by a straightforward extension of (11.2.12),

$\partial f_1/\partial t + \boldsymbol{v}_\alpha \cdot \partial f_1/\partial \boldsymbol{x}_\alpha = 8\pi^4 \int d\boldsymbol{x}_j d\boldsymbol{v}_j$  $_{(2)}\, \delta(\boldsymbol{x}_\alpha - \boldsymbol{x}_j)\, f_1(\alpha)\, f_1(j)$

$+ 8\pi^4 C^{-1} \int d\boldsymbol{x}_j d\boldsymbol{v}_j d\boldsymbol{x}_n d\boldsymbol{v}_n [$  $_{(3)} +$  $^{(1)}_{(2)}$  $_{(2)}]$

$\times \delta(\boldsymbol{x}_\alpha - \boldsymbol{x}_j)\, \delta(\boldsymbol{x}_\alpha - \boldsymbol{x}_n)\, f_1(\alpha)\, f_1(j)\, f_1(n)$ (1)

This is clearly the extension of the Boltzmann equation (10.5.4) to include triple collisions and "corrected" binary collisions. A similar generalized Boltzmann equation can be written to any order in terms of either the strength of interaction or the concentration or both.

The essential point is that it is now the singlet distribution function instead of the velocity distribution which satisfies a separate equation.

Let us now consider the evolution of the correlations to obtain the generalization of (11.3.15). We have [see also (7.4.9) and (7.4.3)] for the two-particle distribution function[1]

$$\begin{aligned} f_2(\alpha\beta) = C^2[&\varphi_1(\alpha)\,\varphi_1(\beta) + \varphi_1(\beta)\,(1/\Omega) \sum_k{}' \rho(\alpha_k)\, e^{i\boldsymbol{k}\cdot\boldsymbol{x}_\alpha} \\ &+ \varphi_1(\alpha)\,(1/\Omega) \sum_k{}' \rho(\beta_k)\, e^{i\boldsymbol{k}\cdot\boldsymbol{x}_\beta} + (1/\Omega) \sum_k{}' \rho(\alpha_k \beta_{-k}) e^{i\boldsymbol{k}\cdot(\boldsymbol{x}_\alpha - \boldsymbol{x}_\beta)} \\ &+ (1/\Omega^2) \sum_{kk'}{}' \rho(\alpha_k \beta_{k'})\, e^{i(\boldsymbol{k}\cdot\boldsymbol{x}_\alpha + \boldsymbol{k}'\cdot\boldsymbol{x}_\beta)}] \end{aligned} \quad (2)$$

We now separate the contribution to the last term from values of k and k' that are in the macroscopic region; using the factorization theorem (10.3.2) we then obtain

[1] In (11.4.2) and (11.4.3), Ω means $L^3/(2\pi)^3$ and should not be confused with the operator Ω defined in (11.2.6).

$$f_2(\alpha\beta) = f_1(\alpha)\,f_1(\beta) + C^2[(1/\Omega)\sum_n{}' \rho(\alpha_n\beta_{-n})\, e^{i\mathbf{n}\cdot(\mathbf{x}_\alpha - \mathbf{x}_\beta)} + (1/\Omega^2)\sum_{nn'}{}' \rho(\alpha_n\beta_{n'})\, e^{i(\mathbf{n}\cdot\mathbf{x}_\alpha + \mathbf{n}'\cdot\mathbf{x}_\beta)} \tag{3}$$

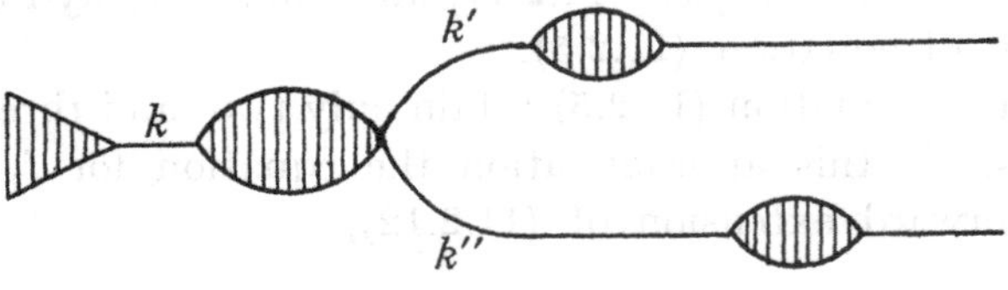

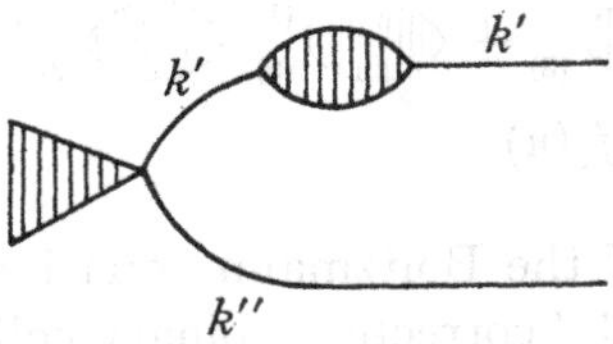

Fig. 11.4.1 Contributions to correlation functions in the hydrodynamic case; all wave vectors $\boldsymbol{k}$ are in the macroscopic range.

We have used the indices n and n' instead of k and k' to emphasize the fact that the contribution of the macroscopic spectral region to the last term has been included in the product $f_1 f_1$. Now for $\rho(\alpha_n\beta_{-n})$ expression (11.3.15) is valid. On the other hand in the sum over $\rho(\alpha_n\beta_{n'})$ we have to retain all diagrams in which all the lines connecting different fragments correspond to macroscopic wave vectors (see Fig. 11.4.1 and Fig. 10.6.1). Indeed lines that correspond to a molecular wave vector would form wave packets which give a vanishing contribution after times of the order of the duration of an interaction (see Chapter 8, § 12). This implies that the sum $n+n'$ in $\rho(\alpha_n\beta_{n'})$ is in the macroscopic range and also that the creation fragment that creates the final two-particle correlation acts at the left of all other vertices.

Again using the factorization theorem we may express (11.4.3) as

$$f_2(\alpha\beta) = f_1(\alpha)\,f_1(\beta) + \triangleright\, f_1(\alpha)\,f_1(\beta) \tag{4}$$

This is the generalization of (11.3.15) that we wanted to obtain.

The correlations now become functionals of the *one-particle distribution function.*[1]

5. Bogoliubov's Theory

In 1946, Bogoliubov published a most interesting approach to a general dynamic theory of gases. His main assumptions are the following (see also Uhlenbeck, 1959; Kac, 1959):

a. After a short time, of the order of the duration of a collision, the singlet distribution function f_1 satisfies a separate equation of the form

$$\partial f_1/\partial t = A(\boldsymbol{x}, f_1) \tag{1}$$

b. After this time the higher order distribution functions become functionals of f_1,

$$f_s(\boldsymbol{x}_1 \cdots \boldsymbol{x}_s, t) = F_s(\boldsymbol{x}_1 \cdots \boldsymbol{x}_s, f_1) \tag{2}$$

These two equations are precisely of the form of (11.4.1) and (11.4.4).

We see that Bogoliubov's assumptions are indeed correct for times $t \gg t_{\text{int}}$ in the "thermodynamic case" in which all characteristic times in the diagrams are of the same order t_{int} and in which the hydrodynamic scale is widely separated from the molecular scale.

To the lowest order in the relevant parameter (see Chapters 4, 6, and 9) the kinetic equations of Bogoliubov are identical to those we have derived. However as the explicit form of Bogoliubov's kinetic equations is not known for higher orders we cannot compare the two theories in more general situations at present.

6. Stationary Non-Equilibrium States — Kinetic Equations and Time Scales

The separation of the hydrodynamic scale from the molecular ones implies the existence of quasi-stationary non-equilibrium states in which the distribution function remains practically

[1] For an example of the application of (11.4.4.) see Balescu and de Gottal (1961).

constant over times of the order of the duration of the interaction (or even the relaxation time). The value of the distribution function is then determined by a compensation of the flow term in the transport equation by the collision term. Let us for example consider equation (11.4.1). We may then suppress the term $\partial f_1/\partial t$ in the left-hand side. On the other hand in the right-hand side the term

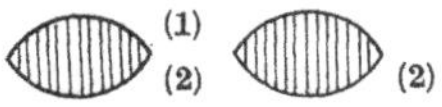 (1)

comes, as we have seen, from the variation of the distribution function over times of the order of the duration of the interaction. Such terms have to be suppressed in the quasi-stationary state too and (11.4.1) reduces to

$\boldsymbol{v}_\alpha \cdot \partial f_1^{st}/\partial \boldsymbol{x}_\alpha = 8\pi^4 \int d\boldsymbol{x}_j \, d\boldsymbol{v}_j$  $\delta(\boldsymbol{x}_\alpha - \boldsymbol{x}_j) \, f_1^{st}(\alpha) \, f_1^{st}(j)$

$+ \, 8\pi^4 C^{-1} \int d\boldsymbol{x}_j \, d\boldsymbol{v}_j \, d\boldsymbol{x}_n \, d\boldsymbol{v}_n$  $\delta(\boldsymbol{x}_\alpha - \boldsymbol{x}_j) \, \delta(\boldsymbol{x}_\alpha - \boldsymbol{x}_n)$

$\times \, f_1^{st}(\alpha) \, f_1^{st}(j) \, f_1^{st}(n)$ (2)

The important point is that only diagonal fragments calculated in an asymptotic way appear in (11.6.2). All terms like (11.6.1), which are due to the finite duration of the collision, have disappeared. This conclusion is quite general and can also be obtained by combining the general formalism for transport coefficients due to Kubo with our method (see Balescu, 1961, 1962). The quasi-stationary distribution function f^{st} and therefore all properties which can be expressed in terms of it depends only on the asymptotic diagonal fragments. The effect of the finite duration of the collision disappears completely.

We see therefore that the form of the kinetic equation depends essentially on the time scale in which one is interested:

a. In the general case the fundamental kinetic equation (11.1.9) is of a non-Markowian type.

b. For times long with respect to all characteristic times involved in the diagrams it reduces to the Markowian equation (11.2.5), corrected for effects of the duration of the interactions.

c. For quasi-stationary situations the operator Ω in (11.2.5) drops out and it is only the asymptotic form of the diagonal fragment that enters the collision operator.

These conditions show in a striking way the direct link between the time scale and the "precision" of the mechanical description that is required.

CHAPTER 12

General H-Theorem

1. Introduction

We shall now show that the kinetic equations we established in Chapter 11 drive the system to equilibrium. The proof of this "general H-theorem", valid to all orders of the strength of the interaction or of the concentration, will involve the following steps (Prigogine et Henin, 1960; Henin, Résibois, and Andrews, 1961; Andrews, 1960, 1961):

a. the proof that ρ_0 approaches the equilibrium velocity distribution function ρ_0^{eq};

b. the proof that ρ'_γ vanishes for large t [see (11.3.4)];

c. the proof that whenever ρ_0 has reached its equilibrium value the correlations ρ_γ given by (11.3.13) correctly describe the correlations given by equilibrium statistical mechanics.

Let us consider these steps in succession.

2. Approach to Equilibrium of the Velocity Distribution Function

We shall constantly make use of the following basic theorem: The sum of all diagonal fragments containing η vertices and involving ν particles yields zero when acting on a function of H_0 alone, where H_0 is as usual the unperturbed Hamiltonian. We shall write this theorem as

 $$f(H_0) = 0 \qquad (1)$$

In order to avoid duplication we shall postpone the proof of (12.2.1) as well as the discussion of its physical meaning to § 4.

With the help of (12.2.1) we may easily follow the approach to equilibrium of the velocity distribution. We may immediately start with equation (11.2.5) which is valid for $t \gg t_{\text{int}}$. As in (11.2.5), the diagonal fragments in (12.2.1) are taken asymptotically. The finite duration of the collisions is taken into account in (11.2.5) by the operator Ω.

Let us decompose the operator Ω as well as the diagonal fragments in (11.2.5) according to the number of particles they involve

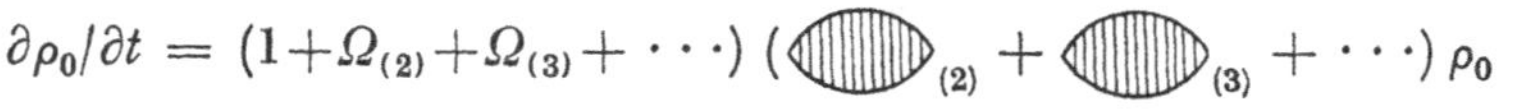

$$\partial\rho_0/\partial t = (1+\Omega_{(2)}+\Omega_{(3)}+\cdots)\,(\langle\!\langle\,\rangle\!\rangle_{(2)} + \langle\!\langle\,\rangle\!\rangle_{(3)} + \cdots)\,\rho_0 \qquad (2)$$

In conjunction with the calculation of reduced distribution functions each "dummy" particle in the diagram introduces a supplementary power of the concentration. We could of course have ordered the diagrams according to the number of vertices (that is according to the powers of λ instead of the number of particles involved) as well.

Let us also expand $\rho_0(t)$ in powers of the concentration

$$\rho_0 = \sum_{k=0}^{\infty} \rho_0^{(k)} \qquad (3)$$

Substituting this into (12.2.2) and ordering the terms in powers of the concentration we obtain the system of equations

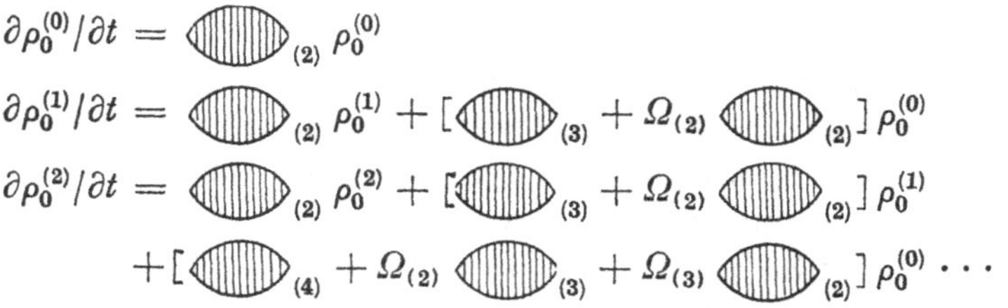

$$\partial\rho_0^{(0)}/\partial t = \langle\!\langle\,\rangle\!\rangle_{(2)}\,\rho_0^{(0)}$$

$$\partial\rho_0^{(1)}/\partial t = \langle\!\langle\,\rangle\!\rangle_{(2)}\,\rho_0^{(1)} + [\langle\!\langle\,\rangle\!\rangle_{(3)} + \Omega_{(2)}\,\langle\!\langle\,\rangle\!\rangle_{(2)}]\,\rho_0^{(0)} \qquad (4)$$

$$\partial\rho_0^{(2)}/\partial t = \langle\!\langle\,\rangle\!\rangle_{(2)}\,\rho_0^{(2)} + [\langle\!\langle\,\rangle\!\rangle_{(3)} + \Omega_{(2)}\,\langle\!\langle\,\rangle\!\rangle_{(2)}]\,\rho_0^{(1)}$$

$$+ [\langle\!\langle\,\rangle\!\rangle_{(4)} + \Omega_{(2)}\,\langle\!\langle\,\rangle\!\rangle_{(3)} + \Omega_{(3)}\,\langle\!\langle\,\rangle\!\rangle_{(2)}]\,\rho_0^{(0)}\cdots$$

In the ordering of the terms we have to consider only semi-connected diagrams. For example in the product of the two operators $\Omega_{(3)}$ and $\langle\!\langle\,\rangle\!\rangle_{(2)}$, there has to be one particle in common and we have therefore three dummy particles.

Using the classical H-theorem for dilute gases (or weakly coupled gases, see Chapter 4, § 3; Chapter 5, § 2) the first equation (12.2.4) shows that

$$\rho_0^{(0)} \to f(H_0) \qquad (5)$$

but then theorem (12.2.1) permits us to reduce the second

equation to

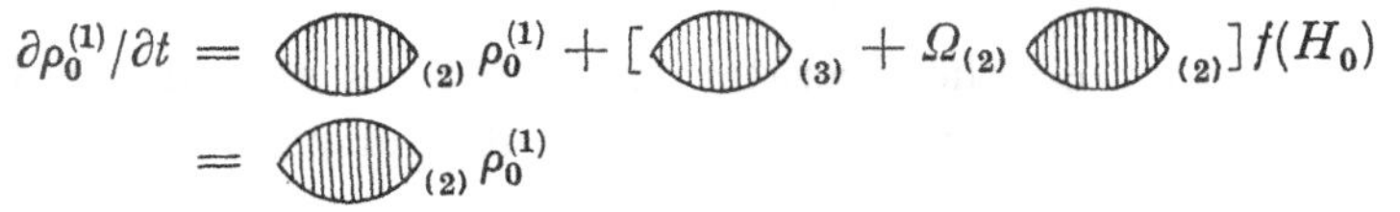

$$\partial\rho_0^{(1)}/\partial t = \text{⬭}_{(2)}\,\rho_0^{(1)} + [\,\text{⬭}_{(3)} + \Omega_{(2)}\,\text{⬭}_{(2)}]\,f(H_0)$$

$$= \text{⬭}_{(2)}\,\rho_0^{(1)} \tag{6}$$

The equation for $\rho_0^{(1)}$ therefore becomes identical to that for $\rho_0^{(0)}$ and we have again

$$\rho_0^{(1)} \rightarrow f(H_0) \tag{7}$$

The iteration of this procedure shows that correct to all orders in concentration

$$\rho_0 \rightarrow f(H_0) \tag{8}$$

We may choose a normalized $\rho_0^{(0)}$ such that

$$\rho_0^{(0)} = \rho_0 = f(H_0)$$
$$\rho_0^{(n)} = 0 \qquad (n \neq 0) \tag{9}$$

The important feature of this proof is that it does not depend of the actual form of the cross section of the collision processes involved (whose calculation would require complicated sommations of diagrams in the general case of many-body collisions). It depends only on the general property of diagonal fragments expressed by (12.2.1). We see also how important the fact that the Ω acts at the left of the diagonal fragment is. This permits us to apply (12.2.1) without specifying the form of Ω.

Exactly the same method may be used to prove (11.3.4). We start with (11.3.3'), expand ρ_γ' in powers of the concentration (or of the coupling parameter), and use the fact that for weakly coupled systems (see 5.3.3)

$$\rho_\gamma'^{(0)} \xrightarrow[t\rightarrow\infty]{} 0 \tag{10}$$

By iteration this result is then proved to all orders in λ.

3. Approach to Equilibrium of the Correlations

Let us consider the canonical distribution function

$$\rho^{eq} = \exp\left[-(1/kT)(H_0+\lambda V)\right]/\int dx\,dp\,\exp\left[-(1/kT)(H_0+\lambda V)\right] \tag{1}$$

Because the potential energy V depends only on the intermolecular distances we obtain the velocity distribution function

$$\rho_0^{\mathrm{eq}} = \int d\boldsymbol{x}\, \rho^{\mathrm{eq}} = (\pi \boldsymbol{k}T)^{3N/2} \exp(-H_0/\boldsymbol{k}T) \tag{2}$$

We want to show that when this is introduced into (11.3.13) the correct equilibrium correlations are indeed derived.

Now $C^+_{\gamma 0}(0)$ is by definition the matrix element of an arbitrary creation fragment in the resolvent $\langle \boldsymbol{k}|R(z)|0\rangle$ with z replaced by $i\varepsilon$, $\varepsilon > 0$ [see (8.3.7), (8.3.11), and (8.4.18)]. We may therefore write (11.3.13) more explicitly

$$\rho_\gamma = \sum_{\eta=0}^{\infty} (-\lambda)^\eta \langle\{\boldsymbol{k}\}\,|(L_0 - i\varepsilon)^{-1}\delta L (L_0 - i\varepsilon)^{-1}\delta L \cdots \delta L|\{0\}\rangle\, \rho_0^{\mathrm{eq}} \tag{3}$$

where the set $\{\boldsymbol{k}\}$ contains γ non-vanishing $\boldsymbol{k}$ vectors and where the matrix element contains the operator δL η times. Let us order the creation fragments involving η vertices according to the number of particles ν that participate in the process. Clearly we have $\nu \geqq \gamma$ and (12.3.3) becomes[1]

$$\rho_\gamma = \sum_{\eta=\gamma-1}^{\infty} \sum_{\nu=\gamma}^{\theta} (-\lambda)^\eta [N^{(\nu-\gamma)}/(\nu-\gamma)!] \langle\{\boldsymbol{k}\}|[(L_0 - i\varepsilon)^{-1}\delta L \cdots \delta L]_\nu^\eta\,|\{0\}\rangle \rho_0^{\mathrm{eq}} \tag{4}$$

We have defined

$$\theta = (\eta + \gamma)/2 \qquad \text{or} \qquad \theta = (\eta + \gamma + 1)/2$$

whichever is integral. This is the maximum number of particles that may appear in the creation fragment corresponding to η vertices.

We have also inserted the factor $N^{\nu-\gamma}/(\nu-\gamma)!$ because the number of ways of choosing the group of $\nu-\gamma$ "extra" or "dummy" particles from N is

$$\lim_{N\to\infty} N!/(N-\nu+\gamma)!\,(\nu-\gamma)! = N^{(\nu-\gamma)}/(\nu-\gamma)! \tag{5}$$

We now replace the matrix element in (12.3.4) by its definition [see (4.1.4)]. We then obtain [see also (12.3.2)]

[1] The notation $[\ ]_\nu^\eta$ in (12.3.4) indicates an interaction involving η vertices and ν particles. Similarly the notation $(\ )_\nu$ in (12.3.7) indicates an interaction involving ν particles.

$$\rho_\gamma = \sum_{\eta=\gamma-1}^{\infty} \sum_{\nu=\gamma}^{\theta} [(-\lambda)^\eta/(\nu-\gamma)!] \times [N^{\nu-\gamma}/L^{3(\nu-\gamma)}] (1/L^3) \int d\boldsymbol{x}_1 \cdots d\boldsymbol{x}_\nu$$
$$\times e^{-i\sum_1^\gamma \boldsymbol{k}_j \cdot \boldsymbol{x}_j} \{(L_0 - i\varepsilon)^{-1} \delta L (L_0 - i\varepsilon)^{-1} \cdots \delta L\}_\nu^\eta (\pi kT)^{3N/2} e^{-H_0/kT} \tag{6}$$

The analysis of this equation is much simplified by using the identity

$$(1/s!)\,(\sum_{i<j=1}^{\nu} V_{ij})^s (\delta L)_\nu \partial^s \rho_0(H_0)/\partial H_0^s$$
$$= (i/s!)\,(\sum_{i<j=1}^{\nu} V_{ij})^s \sum_{m<n=1}^{\nu} (\partial V_{mn}/\partial \boldsymbol{x}_m) \cdot \boldsymbol{D}_{mn} (\partial^s \rho_0(H_0)/\partial H_0^s)$$
$$= (i/s!)(\partial^{s+1} \rho_0(H_0)/\partial H_0^{s+1})(\sum_{i<j=1}^{\nu} V_{ij})^s \sum_{m<n=1}^{\nu} (\partial V_{mn}/\partial \boldsymbol{x}_m) \cdot (\boldsymbol{v}_m - \boldsymbol{v}_n)$$
$$= (i/s!)\,(\partial^{s+1} \rho_0(H_0)/\partial H_0^{s+1})(\sum_{i<j}^{\nu} V_{ij})^s \sum_{t=1}^{\nu} \boldsymbol{v}_t \cdot (\partial/\partial \boldsymbol{x}_t)(\sum_{i<j=1}^{\nu} V_{ij})$$
$$= [-1/(s+1)!]\,(\partial^{s+1} \rho_0(H_0)/\partial H_0^{s+1})(L_0)_\nu (\sum_{i<j=1}^{\nu} V_{ij})^{s+1} \tag{7}$$

Briefly speaking the operator δL acting on a function of H_0 alone is replaced by the operator L_0 acting on a function of the positions. Because of the obvious identity

$$\lim_{\varepsilon \to 0} (L_0 - i\varepsilon)^{-1} L_0 = 1 \tag{8}$$

we may also write (12.3.7) in the form

$$(1/s!)\,(\sum_{i<j=1}^{\nu} V_{ij})^s [(L_0 - i\varepsilon)^{-1} \delta L]_\nu (\partial^s \rho_0(H_0)/\partial H_0^s)$$
$$= -\,[1/(s+1)!]\,[\partial^{s+1} \rho_0(H_0)/\partial H_0^{s+1}]\,(\sum_{i<j=1}^{\nu} V_{ij})^{s+1} \tag{9}$$

Using this relation, (12.3.6) reduces to

$$\rho_\gamma = \sum_{\eta=\gamma-1}^{\infty} \sum_{\nu=\gamma}^{\theta} [(-\lambda/kT)^\eta/(\nu-\gamma)!\,\eta!]\, C^{(\nu-\gamma)} (\pi kT)^{3N/2}$$
$$\times \exp(-H_0/kT)(1/L^3) \int d\boldsymbol{x}_1 \cdots d\boldsymbol{x}_\nu\, e^{-i\sum_{j=1}^\gamma \boldsymbol{k}_j \cdot \boldsymbol{x}_j} (\sum_{i<j=1}^{\nu} V_{ij})^\eta \tag{10}$$

However the sommation over η and ν is no longer unrestricted because certain classes of terms in (12.3.6) give a vanishing contribution. This restriction will be studied in detail in § 4.

Formula (12.3.10) provides us with the concentration expansion of the Fourier coefficient ρ_γ. We could compare it directly to the expansion obtained from the canonical distribution (12.3.1). However in order to derive the standard expressions of equilibrium statistical mechanics we shall make use of the correlation functions $u(1, 2, \ldots, s)$ introduced in Chapter 7, § 4, which are Fourier transforms of the ρ_γ. Moreover as we are interested, in this part of the chapter, only in space correlations we shall integrate (12.3.10) over the velocities. The correlation functions used here depend therefore only on the relative distances between the particles.

We obtain in this way from (12.3.10)

$$u(1, 2, \ldots, \gamma) = \sum_{\eta=\gamma-1}^{\infty} \sum_{\nu=\gamma}^{\theta} [(-\lambda/kT)^\eta/(\nu-\gamma)!\,\eta!]\, C^{(\nu-\gamma)} \times \int dx_{\gamma+1} \cdots dx_\nu \Big(\sum_{i<j=1}^{\nu} V_{ij}\Big)^\eta \tag{11}$$

But we have the multinomial expansion

$$\Big(\sum_{i<j=1}^{\nu} V_{ij}\Big)^\eta = \sum_{\{m_{ij}\}} \eta! \prod_{i<j=1}^{\nu} (V_{ij})^{m_{ij}}/m_{ij}! \tag{12}$$

with

$$\sum_{i<j=1}^{\nu} m_{ij} = \eta \tag{13}$$

Substituting this into (12.3.11) and summing over η, we obtain

$$\begin{aligned} u(1, 2, \ldots, \gamma) &= \sum_{\nu=\gamma}^{\infty} [C^{(\nu-\gamma)}/(\nu-\gamma)!] \int dx_{\gamma+1} \cdots dx_\nu \sum_{\{m_{ij}\}} \prod_{i<j=1}^{\nu} (-\lambda V_{ij}/kT)^{m_{ij}}/m_{ij}! \\ &= \sum_{n=0}^{\infty} (C^n/n!) \int dx_1 \cdots dx_n \sum_{\{m_{ij}\}} \prod_{i<j=1}^{\gamma+n} (-\lambda V_{ij}/kT)^{m_{ij}}/m_{ij}! \end{aligned} \tag{14}$$

In the summation over $\{m_{ij}\}$ one has to take account of (12.3.13). Now a diagram involving ν particles ending with γ lines may contain not less than $(\nu-1)$ vertices; therefore the requirement (12.3.13) has to be replaced by

$$\sum_{i<j=1}^{\nu} m_{ij} \geqq \nu-1 \qquad \text{or} \qquad \sum_{i<j=1}^{\gamma+n} m_{ij} \geqq \gamma+n-1 \tag{15}$$

Expression (12.3.14) is precisely the concentration expansion of the distribution function, which is discussed in many textbooks on equilibrium statistical mechanics (see especially Münster, 1956; Hill, 1956; also among the original papers on this subject we want to note Montroll and Mayer, 1941; de Boer, 1948–1949; Meeron, 1958; Salpeter, 1958; and Van Leeuwen, Groeneveld, and de Boer, 1959). To show the identity between (12.3.14) and the concentration expansion as derived by the equilibrium methods we first have to analyze in more detail the nature of the terms that vanish in the transition from (12.3.6) to (12.3.10).

4. Articulation Points and Principle of Detailed Balance

We shall now show that every term in (12.3.6) that contains an "articulation point" is zero. A diagram with an articulation point is characterized by a group of σ particles, which are not part of the finally correlated γ particles, coupled by intermolecular potentials V_{ij} to only *one* other particle. The last vertex where a σ particle is involved is called an *articulation point*. Two examples are

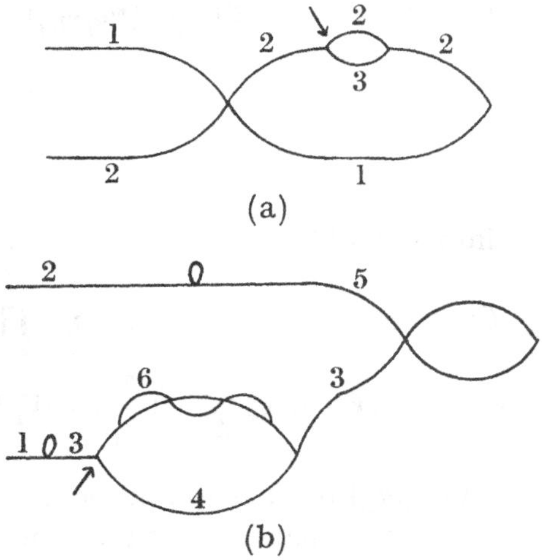

Fig. 12.4.1 Examples of diagrams with an articulation point. The arrow indicates the articulation point.

given in Fig. 12.4.1. In example (a) the group of σ particles reduces to particle 3 which is coupled to 2. In example (b) the group σ is formed by particles 4 and 6 coupled to 3. The terminology "articulation point", which is somewhat inappropriate here, is taken over from equilibrium statistical mechanics. We shall come back to its meaning in terms of Mayer diagrams in § 5.

To prove that such contributions vanish, we apply relation (12.3.9) to formula (12.3.6) until we come to the articulation point. This vertex is indicated by an arrow in Fig. 12.4.1. At this vertex we have to deal with an integral of the form (we shall use example (b) of Fig. 12.4.1):

$$V_{53}^{a_1} V_{52}^{a_2} (\mathbf{v}_3 - \mathbf{v}_4) \int d\mathbf{x}_4 \, d\mathbf{x}_6 \, (\partial V_{34}/\partial \mathbf{x}_4) \, V_{34}^{a_3} V_{36}^{a_4} V_{46}^{a_5} \tag{1}$$

where the powers $a_1, \ldots, a_5$ are arbitrary. The important point is that we may integrate over the positions of the "σ particles", 4 and 6 because they will no longer appear at the left of the vertex. To evaluate (12.4.1) we take the position of the molecule 3 as the new origin in configuration space. The integration is over $d\mathbf{x}_{43} \, d\mathbf{x}_{63}$. Now the force $\partial V_{34}/\partial \mathbf{x}_{43}$ changes sign when were place $\mathbf{x}_{43}$ by $-\mathbf{x}_{43}$ while the intermolecular potentials remain unaltered. Therefore the integral vanishes.

As a special case we may now prove the important theorem (12.2.1). Indeed a diagonal fragment is a special case of a diagram containing an articulation point in which all particles involved are "σ particles." We have to calculate an integral of the form [see (12.3.6)]

$$\int d\mathbf{x}_1 \cdots d\mathbf{x}_\nu \{\delta L (L_0 - i\varepsilon)^{-1} \delta L \cdots \delta L\}_\nu^\eta f(H_0) \tag{2}$$

We again apply (12.3.9) and (12.3.7). At the last vertex we are left with

$$\int d\mathbf{x}_1 \cdots d\mathbf{x}_\nu L_0 \Big(\sum_{i<j=1}^{\nu} V_{ij} \Big)^\eta \tag{3}$$

where as usual L_0 is the unperturbed Liouville operator. This is a surface integral and vanishes identically.

It is very interesting to understand clearly the physical meaning of the two important results we have established: the vanishing contribution of both diagonal fragments and creation fragments containing an articulation point when applied to a function of the unperturbed Hamiltonian.

Let us go back to the scattering theory studied in Chapter 6. We have seen that to a contribution of order λ^m in Fourier space corresponds the sum of the contributions to scattering due to m successive interactions at arbitrary points $\mathbf{x}_1, \mathbf{x}_2, \ldots, \mathbf{x}_m$ in ordinary space [see (6.3.3) and Fig. 6.3.1]. The same is true for the many-body problems we are now studying. A diagram like  (3) involving 3 particles and say η vertices corresponds to the sum of all possible three-body collisions involving η successive interactions, which may occur at arbitrary points inside the range of mutual interaction of the particles. Formula (12.2.1) states therefore, that the sum of all ν-body collisions involving a given number of vertices does not alter the velocity distribution function once it has reached its equilibrium value.

This may be considered as an application of the classical Principle of Detailed Balance to the velocity distribution. Our theorem indeed says that the number of particles that, as the result of these processes, undergo the transition $\{v\}^0 \rightarrow \{v\}$, from some values of the velocities to some other values, is equal to the number that undergo the inverse transition $\{v\} \rightarrow \{v\}^0$. In this way the distribution function remains unaltered. The elimination of the articulation points may also be understood from the same point of view. Indeed, the integral which appears in (12.4.1) is similar to the integral that would correspond to a diagonal fragment involving the σ particles for a given position of the articulation point. As the consequence of the existence of an articulation point there is a part of the diagram that corresponds to the sum of all possible collision processes involving the σ particles. Again, when the velocity distribution has reached its equilibrium value, the number of transitions, involving the σ particles, from a given state to another is equal to the number of inverse transitions. Therefore such diagrams contain a factor that is vanishing.

5. Two-Particle Correlation Function

As an example let us study the two-particle correlation function $u(1, 2)$ in some detail, starting with formula (12.3.14). In the summation over the $\{m_{ij}\}$ we have to take account of the following restrictions: (a) we have to consider only diagrams without articulation points; (b) we have the condition (12.3.15) for the sum of the $\{m_{ij}\}$.

Let us consider the contributions to (12.3.14) arising from $n = 0, 1, 2 \ldots$ in succession. For $n = 0$ we have simply

$$\sum_{m=1}^{\infty} (-\lambda V_{12}/kT)^m/m! = \exp(-\lambda V_{12}/kT) - 1 = f_{12} \tag{1}$$

This expression is generally denoted by f_{12} in equilibrium statistical mechanics. It corresponds to the "direct" correlation between particles 1 and 2 due to their interaction. It vanishes outside the range of intermolecular forces. Two particles, ij, corresponding to a non-vanishing factor f are, in the usual notation of equilibrium statistical mechanics, connected by a line which is also called a "bond" (see Chapter 7, § 5).

The next contribution ($n = 1$) involves one dummy particle, which we shall call 3:

$$C \exp[-(\lambda/kT) V_{12}] \int dx_3 f_{13} f_{23} = C \exp[-(\lambda/kT) V_{12}] \beta_1(1, 2) \tag{2}$$

No contribution arising from V_{13} or V_{23} alone exists because this would correspond to an articulation point. Also a term

$\beta_1(1, 2) =$ 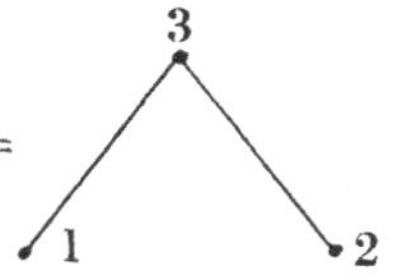

Fig. 12.5.1 Contribution to the two-particle correlation function, to first order in the concentration.

of at least second order in the intermolecular potential is required by (12.3.15). Expression (12.5.2) defines the cluster integral $\beta_1(1, 2)$. It is represented in terms of the Mayer diagrams in Fig. 12.5.1. Diagrams like those represented in Fig. 12.5.2 do not contribute to the two-particle distribution function because they contain an articulation point.

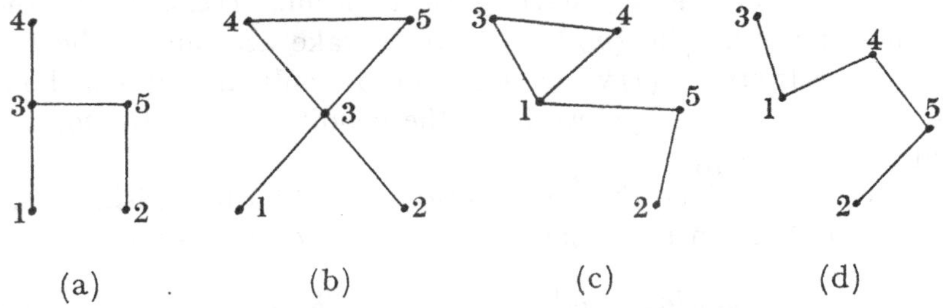

Fig. 12.5.2 Examples of diagrams with an articulation point. In (a), (b) 3 is an articulation point. In (c), (d) 1 is an articulation point.

We see that the definition of the articulation point, as implied in Fig. 12.5.2, coincides with that used in § 4. For example in Fig. 12.5.2 (b) particles 4 and 5 are the "σ particles" as defined in § 4. However the origin of the terminology "articulation point" is clearer in the equilibrium diagrams: an articulation point is a point which forms the only connection between the main part of the diagram (containing the points 1 and 2) and an appended diagram, such that cutting the diagram at the articulation point disconnects the appendage from the main part of the diagram.

Let us now consider the higher order contributions ($n > 1$) to (12.3.14). We obtain in the same way

$$u(1, 2) = f_{12} + \sum_{n=1}^{\infty} (C^n/n!) \exp(-\lambda V_{12}/kT)\, \beta_n(1, 2) \qquad (3)$$

$\beta_2(1, 2) =$ 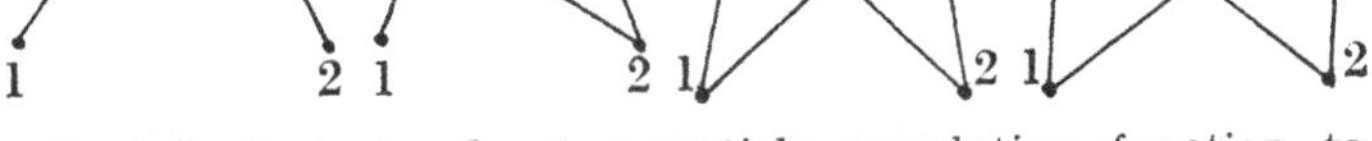

Fig. 12.5.3 Contributions to the two-particle correlation function to second order in concentration.

where the sum is over all diagrams satisfying the above requirements. In Fig. 12.5.3 we show diagrams that contribute to β_2.

Equation (12.5.3) is precisely the equilibrium concentration expansion of the two-particle correlation function; s-particle correlation functions introduce no new problems and will therefore not be considered. We have thus completed the proof of the generalized H-theorem.

6. Mechanism of Irreversibility

The kinetic equations obtained in Chapter 11 as well as the H-theorem permit us to understand in a very intuitive way the mechanism of irreversibility, which may be described as a "cascade mechanism."

Let us go back to Fig. 7.6.4 and consider the evolution in a homogeneous system of the reduced distribution function corresponding to an arbitrary number of momenta. We shall not explicitly show diagrams corresponding to the propagation of correlations. We obtain the set of equations described in Fig. 12.6.1.

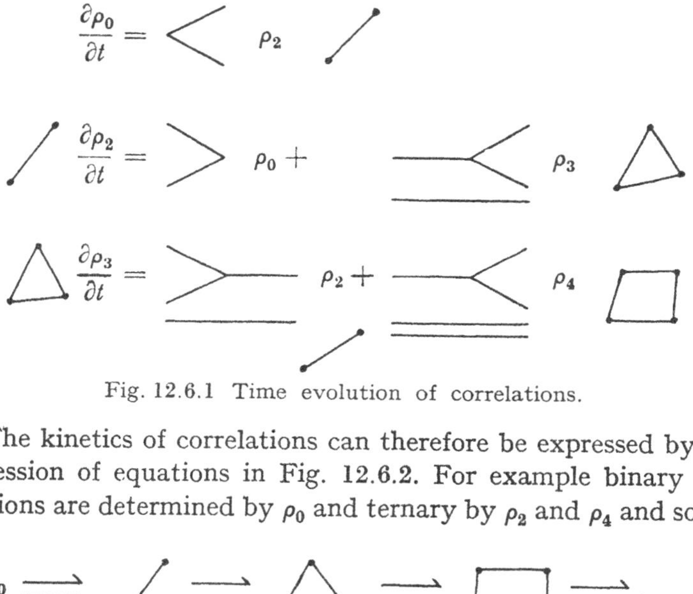

Fig. 12.6.1 Time evolution of correlations.

The kinetics of correlations can therefore be expressed by the succession of equations in Fig. 12.6.2. For example binary correlations are determined by ρ_0 and ternary by ρ_2 and ρ_4 and so on.

$$\rho_0 \rightleftharpoons \diagup \rightleftharpoons \triangle \rightleftharpoons \square \rightleftharpoons \cdots\cdots$$

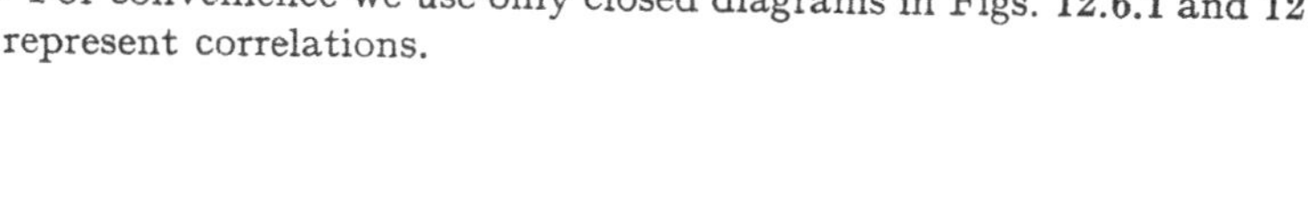

Fig. 12.6.2 Kinetics of correlations.[1]

[1] For convenience we use only closed diagrams in Figs. 12.6.1 and 12.6.2 to represent correlations.

Now the existence, for sufficiently long times, of a separate equation for ρ_0 and of equation (11.3.9), which expresses the correlations as functionals of ρ_0 shows that the "information" contained in ρ_0 is transmitted in succession to higher and higher correlations.

Instead of the "reversible" kinetics represented in Fig. 12.6.2 we then have the directed flow of correlations represented in Fig. 12.6.3. This flow finally disappears in the "sea" of highly multiple, incoherent correlations.

Fig. 12.6.3 Directed flow of correlations.

While the very meaning of irreversibility is difficult to express in terms of the usual dynamics of particles, it acquires a direct intuitive sense in terms of the dynamics of correlations, which we have developed in a systematic way in this monograph.

CHAPTER 13

Quantum Mechanics

1. Quantum Mechanical Density Matrix

This monograph is specifically devoted to the study of irreversible processes in the frame of classical mechanics. It is however interesting, for a better understanding of the concepts we have been using, to consider the transition to quantum mechanics briefly. The role of the classical density in phase space $\rho(\boldsymbol{x}_1 \cdots \boldsymbol{x}_N, \boldsymbol{p}_1 \cdots \boldsymbol{p}_N)$ is played in quantum mechanics by the *statistical operator* ρ due to von Neumann (see for example Tolman, 1938). Its time evolution is given by the quantum mechanical Liouville equation [1]

$$\partial\rho/\partial t = (1/i\hbar)(H\rho - \rho H) \tag{1}$$

where $\hbar = h/2\pi$ is Planck's constant divided by 2π. This equation replaces equation (1.2.5) which is valid in classical mechanics. In the right-hand side of (13.1.1) we have the commutator of ρ and the Hamiltonian. The average value of any physical quantity L is given by

$$\bar{L} = \sum_{mn} \rho_{mn} L_{nm} = \text{Tr } \rho L \tag{2}$$

Here Tr ρL is the usual notation for the trace of an operator, that is, the sum of its diagonal elements. [For more details about the statistical operator ρ, see especially Fano, 1957; also Tolman, 1938; and ter Haar, 1954]. We shall also call ρ the density operator. To each operator in an given representation corresponds a matrix in quantum mechanics. In this way the density matrix ρ_{mn} corresponds to the density operator ρ.

We shall show that the Liouville equation in interaction representation (1.8.7), which is the basis of our whole method,

[1] In (13.1.1) both ρ and H are operators. We generally use no special symbol for operators.

remains valid for the density matrix in the quantum mechanical case. The only difference between the two lies in the definition of the vertex. In the quantum mechanical case to each vertex corresponds a finite transfer of momentum, while in the classical case we had an infinitesimal transfer.

As a consequence the methods we have developed in Chapter 7–12 can easily be transposed to quantum systems. This is a very great simplification of the quantum problem (Prigogine and Ono, 1959; Prigogine, 1959), especially from the point of view of the asymptotic time integrations which are much more complicated when one starts directly from (13.1.1) (see Van Hove, 1955; Prigogine and Résibois, 1958).

A few interesting applications have already been worked out in detail (see for example the quantum mechanical Brownian motion problem, studied by George, 1960; the quantum mechanical plasma, Balescu, 1961, 1962; the quantum mechanical general H-theorem, Résibois, 1961). In this monograph however we shall not give a systematic exposition of the quantum mechanical version of the theory (we hope to do this in a special monograph); we shall only consider a simple example. We shall first consider the two-body problem corresponding to the Hamiltonian (6.1.1) and derive the quantum mechanical form of the Liouville equation (6.1.3). By a straightforward extension to the N-body problem we shall then derive the quantum mechanical transport equation for the weakly coupled case (the Pauli equation).

2. Quantum Mechanical Liouville Equation in Interaction Representation

As in Chapter 1, § 8, we first want to eliminate the unperturbed Hamiltonian. This is done in quantum mechanics by introducing the interaction representation through the relations

$$\begin{aligned} \rho_{\text{int}} &= e^{iH_0 t/\hbar} \rho\, e^{-iH_0 t/\hbar} \\ V_{\text{int}} &= e^{iH_0 t/\hbar} V e^{-iH_0 t/\hbar} \end{aligned} \tag{1}$$

We then obtain, instead of (13.1.1),

$$\partial \rho_{\text{int}}/\partial t = (\lambda/i\hbar)(V_{\text{int}}\,\rho_{\text{int}} - \rho_{\text{int}}\,V_{\text{int}}) \tag{2}$$

Let us write this relation explicitly in the momentum representation. We get [1]

$$i\hbar(\partial/\partial t)\langle \boldsymbol{p}|\rho_{\text{int}}|\boldsymbol{p}'\rangle = (\lambda/L^3)\sum_{\boldsymbol{p}''}[\langle \boldsymbol{p}|V_{\text{int}}|\boldsymbol{p}''\rangle\langle \boldsymbol{p}''|\rho_{\text{int}}|\boldsymbol{p}'\rangle - \langle \boldsymbol{p}|\rho_{\text{int}}|\boldsymbol{p}''\rangle\langle \boldsymbol{p}''|V_{\text{int}}|\boldsymbol{p}'\rangle] \quad (3)$$

Instead of $\boldsymbol{p}$ and $\boldsymbol{p}'$ we shall introduce new variables $\boldsymbol{P}$ and $\boldsymbol{k}$ defined by

$$\begin{aligned} \boldsymbol{p} &= \boldsymbol{P} + (\hbar/2)\boldsymbol{k} \\ \boldsymbol{p}' &= \boldsymbol{P} - (\hbar/2)\boldsymbol{k} \end{aligned} \quad (4)$$

or inversely

$$\begin{aligned} \boldsymbol{P} &= (1/2)(\boldsymbol{p}+\boldsymbol{p}') \\ \boldsymbol{k} &= (1/\hbar)(\boldsymbol{p}-\boldsymbol{p}') \end{aligned} \quad (5)$$

We shall see that the momentum $\boldsymbol{P}$ and the wave length $\boldsymbol{k}$ play precisely the role of the variables $\boldsymbol{p}$ and $\boldsymbol{k}$ in equation (6.1.3), which is valid in classical mechanics. Let us also use the notation

$$\begin{aligned} \langle \boldsymbol{p}|\rho_{\text{int}}|\boldsymbol{p}'\rangle &= \langle \boldsymbol{P}+(\hbar/2)\boldsymbol{k}|\rho_{\text{int}}|\boldsymbol{P}-(\hbar/2)\boldsymbol{k}\rangle \\ &= \rho_k(\boldsymbol{P}) \end{aligned} \quad (6)$$

Equation (13.2.3) can now be written in the form

$$\begin{aligned} i\hbar(\partial\rho_k(\boldsymbol{P})/\partial t) &= (\lambda/L^3)\sum_{l}\{\langle \boldsymbol{P}+(\hbar/2)\boldsymbol{k}|V_{\text{int}}|\boldsymbol{P}+(\hbar/2)\boldsymbol{k}+\hbar\boldsymbol{l}\rangle \\ &\quad\times\langle \boldsymbol{P}+(\hbar/2)\boldsymbol{k}+\hbar\boldsymbol{l}|\rho_{\text{int}}|\boldsymbol{P}-(\hbar/2)\boldsymbol{k}\rangle \\ &\quad-\langle \boldsymbol{P}+(\hbar/2)\boldsymbol{k}|\rho_{\text{int}}|\boldsymbol{P}-(\hbar/2)\boldsymbol{k}-\hbar\boldsymbol{l}\rangle\langle \boldsymbol{P}-(\hbar/2)\boldsymbol{k}-\hbar\boldsymbol{l}|V_{\text{int}}|\boldsymbol{P}-(\hbar/2)\boldsymbol{k}\rangle\} \\ &= \lambda(2\pi/L)^3\sum_{l}V_{-l}[\exp\{(i/\hbar)[E(|\boldsymbol{P}+(\hbar/2)\boldsymbol{k}|)-E(|\boldsymbol{P}+(\hbar/2)\boldsymbol{k}+\hbar\boldsymbol{l}|)]t\} \\ &\quad\times\rho_{k+l}(\boldsymbol{P}+(\hbar/2)\boldsymbol{l})-\exp\{(i/\hbar)[E(|\boldsymbol{P}-(\hbar/2)\boldsymbol{k}-\hbar\boldsymbol{l}|)-E(|\boldsymbol{P}-(\hbar/2)\boldsymbol{k}|)]t\} \\ &\quad\times\rho_{k+l}(\boldsymbol{P}-(\hbar/2)\boldsymbol{l})] \end{aligned} \quad (7)$$

We replaced the potential energy in the interaction representation by (13.2.1), and used the Fourier expansion (6.1.2). We now

[1] We use Dirac's notation (1947). For example $\langle \boldsymbol{p}'|V_{\text{int}}|\boldsymbol{p}''\rangle = \int d\boldsymbol{x}\, e^{-i\boldsymbol{p}'\cdot\boldsymbol{x}/\hbar}\, V_{\text{int}}\, e^{i\boldsymbol{p}''\cdot\boldsymbol{x}/\hbar}$. In general, a matrix element is written $\langle m|L|n\rangle$.

introduce the explicit value of the unperturbed energy

$$E(|\boldsymbol{p}|) = p^2/2m \tag{8}$$

and displacement operators defined by (see Chapter 6, § 1)

$$e^{\alpha\,\partial/\partial x} f(x) = f(x+\alpha) \tag{9}$$

We may then write (13.2.7) as follows:

$$i\hbar(\partial\rho_k(\boldsymbol{P})/\partial t) = \lambda(2\pi/L)^3 \sum_l V_l\{\exp[-(i/m)((\hbar/2)l^2 - \boldsymbol{P}\cdot\boldsymbol{l} - (\hbar/2)\boldsymbol{k}\cdot\boldsymbol{l})t]$$
$$\times \exp[-(\hbar/2)\,\boldsymbol{l}\cdot(\partial/\partial\boldsymbol{P})] - \exp[(i/m)((\hbar/2)\,l^2 + \boldsymbol{P}\cdot\boldsymbol{l} - (\hbar/2)\,\boldsymbol{k}\cdot\boldsymbol{l})\,t]$$
$$\times \exp[(\hbar/2)\,\boldsymbol{l}\cdot(\partial/\partial\boldsymbol{P})]\}\,\rho_{k-l}(\boldsymbol{P}) \tag{10}$$

Using (13.2.9) again we may also put (13.2.10) into the form

$$\partial\rho_k(\boldsymbol{P})/\partial t = \lambda(2\pi/L)^3 \sum_l e^{(i/m)\,\boldsymbol{P}\cdot\boldsymbol{k}t}\,V_l$$
$$\times (1/i\hbar)[e^{-(\hbar/2)\,\boldsymbol{l}\cdot(\partial/\partial\boldsymbol{P})} - e^{(\hbar/2)\,\boldsymbol{l}\cdot(\partial/\partial\boldsymbol{P})}]\,e^{-(i/m)\,\boldsymbol{P}\cdot(\boldsymbol{k}-\boldsymbol{l})\,t}\,\rho_{k-l}(\boldsymbol{P}) \tag{11}$$

This equation is indeed the quantum analogue of (6.1.3). The only difference is that the *differential* operator in (6.1.3)

$$\boldsymbol{l}\cdot(\partial/\partial\boldsymbol{p}) \tag{12}$$

is now replaced by the *displacement* operator

$$(1/\hbar)[e^{(\hbar/2)\,\boldsymbol{l}\cdot(\partial/\partial\boldsymbol{P})} - e^{-(\hbar/2)\,\boldsymbol{l}\cdot(\partial/\partial\boldsymbol{P})}] \tag{13}$$

which reduces to (13.2.12) in the limit $\hbar \to 0$.

The appearance of the displacement operator (13.2.13) expresses the fact that a *finite momentum change occurs* in a single step. This is a well known feature of quantum scattering theory (see, e.g., Bohm, 1951).

One may consider the case of anharmonic solids as well. In that case the displacement operators would act on the energy of a normal mode and tell us that the energy of this normal mode can only change by a multiple of $h\nu_k$ in a single interaction. Indeed only an *integral* number of phonons can be destroyed or created.

Inversely we could even use the substitution of (13.2.12) by (13.2.13) as the starting point for the transition from classical to quantum mechanics. Going "backwards" then we may verify that this implies the validity of (13.2.2) and (13.1.1). In other words the substitution of the differential operator (13.2.12) by the displacement operator (13.2.13) is equivalent to the replacement of the Poisson bracket by the commutator (13.1.1).

3. Wave Vector Conservation

We see that the variables we used in the classical mechanical description keep a simple meaning in the quantum mechanical case; the wave vector $\boldsymbol{k}$ is given by (13.2.5). It is equal to the difference between the two values of the momenta that specify a matrix element of ρ in the momentum representation. To the value $\boldsymbol{k} = 0$ correspond the *diagonal* elements of the statistical matrix.

For the N-body problem a similar derivation yields, instead of (13.2.11) [see (4.1.9)],

$$(\partial/\partial t)\rho_{\{k\}}(\boldsymbol{P}_1\cdots \boldsymbol{P}_N, t) = \lambda(2\pi/L)^3 \sum_{jn}\sum_{l} \exp[(i/m)(\boldsymbol{k}_j\cdot \boldsymbol{P}_j+\boldsymbol{k}_n\cdot \boldsymbol{P}_n)t]$$
$$\times V_l(1/i\hbar)\{\exp\{(\hbar/2)\,\boldsymbol{l}\cdot[(\partial/\partial \boldsymbol{P}_n)-(\partial/\partial \boldsymbol{P}_j)]\}$$
$$-\exp\{-(\hbar/2)\,\boldsymbol{l}\cdot[(\partial/\partial \boldsymbol{P}_n)-(\partial/\partial \boldsymbol{P}_j)]\}\}$$
$$\times \exp\{-(i/m)[(\boldsymbol{k}_j-\boldsymbol{l})\cdot \boldsymbol{P}_j+(\boldsymbol{k}_n+\boldsymbol{l})\cdot \boldsymbol{P}_n]t\}\,\rho_{k_j-l,\,k_n+l,\,\{k'\}}(\boldsymbol{P}_1\cdots \boldsymbol{P}_N, t) \tag{1}$$

As in the classical case we have the wave vector conservation law ([see (4.1.8)]

$$\boldsymbol{k}_n'+\boldsymbol{k}_j' = \boldsymbol{k}_n+\boldsymbol{k}_j \tag{2}$$

It is interesting to consider the origin of this relation more closely. We have in the momentum representation

$$\langle \boldsymbol{p}_n \boldsymbol{p}_j\,|V_{nj}|\,\boldsymbol{p}_n' \boldsymbol{p}_j'\rangle = \int d\boldsymbol{x}_n\, d\boldsymbol{x}_j\, e^{-i(\boldsymbol{p}_n\cdot \boldsymbol{x}_n+\boldsymbol{p}_j\cdot \boldsymbol{x}_j)/\hbar}$$
$$\times V(|\boldsymbol{x}_n-\boldsymbol{x}_j|)\, e^{i(\boldsymbol{p}_n'\cdot \boldsymbol{x}_n+\boldsymbol{p}_j'\cdot \boldsymbol{x}_j)/\hbar} \tag{3}$$

As a consequence of the *invariance* of V with *respect to translation* [see Chapter 4, § 1; $V(|\boldsymbol{x}_n-\boldsymbol{x}_j|) = V(|(\boldsymbol{x}_n+\boldsymbol{a})-(\boldsymbol{x}_j+\boldsymbol{a})|)$] this expression vanishes except when the momentum conservation condition $\boldsymbol{p}_n+\boldsymbol{p}_j = \boldsymbol{p}_n'+\boldsymbol{p}_j'$ is satisfied. A matrix element

$\langle \boldsymbol{p}_j \boldsymbol{p}_n | \rho_{\text{int}} | \boldsymbol{p}'_j \boldsymbol{p}'_n \rangle$ on the left-hand side of (13.2.3) can only be connected to an element $\langle \boldsymbol{p}''_j \boldsymbol{p}''_n | \rho_{\text{int}} | \boldsymbol{p}'_j \boldsymbol{p}'_n \rangle$ of the right-hand side in a single step if

$$\boldsymbol{p}_j + \boldsymbol{p}_n = \boldsymbol{p}''_j + \boldsymbol{p}''_n \tag{4}$$

Using the variables $\boldsymbol{P}$, $\boldsymbol{k}$ defined by (13.2.5) we have

$$\begin{aligned}&\langle \boldsymbol{p}_j \boldsymbol{p}_n | \rho_{\text{int}} | \boldsymbol{p}'_j \boldsymbol{p}'_n \rangle \\ &= \langle \boldsymbol{P}_j + (\hbar/2)\boldsymbol{k}_j, \boldsymbol{P}_n + (\hbar/2)\boldsymbol{k}_n | \rho_{\text{int}} | \boldsymbol{P}_j - (\hbar/2)\boldsymbol{k}_j, \boldsymbol{P}_n - (\hbar/2)\boldsymbol{k}_n \rangle\end{aligned} \tag{5}$$

$$\begin{aligned}&\langle \boldsymbol{p}''_j \boldsymbol{p}''_n | \rho_{\text{int}} | \boldsymbol{p}'_j \boldsymbol{p}'_n \rangle \\ &= \langle \boldsymbol{P}'_j + (\hbar/2)\boldsymbol{k}'_j, \boldsymbol{P}'_n + (\hbar/2)\boldsymbol{k}'_n | \rho_{\text{int}} | \boldsymbol{P}'_j - (\hbar/2)\boldsymbol{k}'_j, \boldsymbol{P}'_n - (\hbar/2)\boldsymbol{k}'_n \rangle\end{aligned} \tag{6}$$

Relations (13.3.5) and (13.3.6) define the sets of variables $\boldsymbol{P}$, $\boldsymbol{P}'$, $\boldsymbol{k}$, $\boldsymbol{k}'$; we have

$$\begin{aligned}\boldsymbol{k}'_j &= (1/\hbar)(\boldsymbol{p}''_j - \boldsymbol{p}'_j), \qquad & \boldsymbol{k}_j &= (1/\hbar)(\boldsymbol{p}_j - \boldsymbol{p}'_j) \\ \boldsymbol{k}'_n &= (1/\hbar)(\boldsymbol{p}''_n - \boldsymbol{p}'_n), \qquad & \boldsymbol{k}_n &= (1/\hbar)(\boldsymbol{p}_n - \boldsymbol{p}'_n)\end{aligned} \tag{7}$$

Therefore taking account of (13.3.4),

$$\boldsymbol{k}'_j + \boldsymbol{k}'_n = \boldsymbol{k}_j + \boldsymbol{k}_n \tag{8}$$

This is precisley the wave vector conservation law.

4. Pauli Equation

As the simplest possible example we shall study the evolution of the diagonal elements of the statistical matrix ρ and derive the quantum mechanical form of the master equation (4.2.3).

The basic closed set of equations is again given by Fig. 8.11.2, and the evolution to equilibrium will again be represented by the succession of the cycles of Fig. 8.11.1. When written explicitly, using (13.3.1), the two equations of Fig. 8.11.2 become [see (6.5.2)]

$$\begin{aligned}\partial\rho_0(\boldsymbol{P}_1 \cdots \boldsymbol{P}_N, t)/\partial t = \lambda(2\pi/L)^3 \sum_{jn}\sum_{l} V_l (1/i\hbar)(e^{(\hbar/2)\boldsymbol{l}\cdot\boldsymbol{D}_{nj}} - e^{-(\hbar/2)\boldsymbol{l}\cdot\boldsymbol{D}_{nj}}) \\ \times e^{-i\boldsymbol{l}\cdot\boldsymbol{g}_{nj}t} \rho(j_{-l}, n_l, \boldsymbol{P}_1 \cdots \boldsymbol{P}_N, t)\end{aligned} \tag{1}$$

$$\begin{aligned}\partial\rho(j_{-l}, n_l, \boldsymbol{P}_1 \cdots \boldsymbol{P}_N, t)/\partial t = \lambda(2\pi/L)^3 V_{-l} e^{i\boldsymbol{l}\cdot\boldsymbol{g}_{nj}t}(1/i\hbar) \\ \times (e^{-(\hbar/2)\boldsymbol{l}\cdot\boldsymbol{D}_{nj}} - e^{(\hbar/2)\boldsymbol{l}\cdot\boldsymbol{D}_{nj}})\rho_0(\boldsymbol{P}_1 \cdots \boldsymbol{P}_N, t)\end{aligned} \tag{2}$$

By elimination of the matrix element $\rho(j_{-l}, n_l, \boldsymbol{P}_1 \cdots \boldsymbol{P}_N, t)$ between (13.4.1) and (13.4.2) we obtain, as in the classical case, the master equation for the velocity distribution function

$$\begin{aligned}\partial\rho_0(\boldsymbol{P}_1 \cdots \boldsymbol{P}_N, t)/\partial t \\ &= \lambda^2(2\pi/L)^6(1/\hbar^2)\sum_{jn}\sum_{l}|V_l|^2(e^{(\hbar/2)\,\boldsymbol{l}\cdot\boldsymbol{D}_{nj}} - e^{-(\hbar/2)\,\boldsymbol{l}\cdot\boldsymbol{D}_{nj}}) \\ &\quad\times \pi\delta(\boldsymbol{l}\cdot\boldsymbol{g}_{nj})(e^{-(\hbar/2)\,\boldsymbol{l}\cdot\boldsymbol{D}_{nj}} - e^{(\hbar/2)\,\boldsymbol{l}\cdot\boldsymbol{D}_{nj}})\,\rho_0(\boldsymbol{P}_1 \cdots \boldsymbol{P}_N, t) \qquad (3)\end{aligned}$$

It differs from the classical master equation only by the appearance of the displacement operators. If we now commute the δ function with the displacement operator and use (13.2.9) we obtain

$$\begin{aligned}\partial\rho_0(\boldsymbol{P}_1 \cdots \boldsymbol{P}_N, t)/\partial t \\ &= -\lambda^2[(2\pi)^6/L^6\hbar^2]\sum_{jn}\sum_{l}|V_l|^2\{\pi\delta[\boldsymbol{l}\cdot(\boldsymbol{P}_n + (\hbar/2)\,\boldsymbol{l}) - \boldsymbol{l}\cdot(\boldsymbol{P}_j - (\hbar/2)\boldsymbol{l})] \\ &\quad\times [\rho_0(\boldsymbol{P}_1 \cdots \boldsymbol{P}_N, t) - \rho_0(\boldsymbol{P}_1, \ldots, \boldsymbol{P}_j - \hbar\boldsymbol{l}, \ldots, \boldsymbol{P}_n + \hbar\boldsymbol{l}, \ldots, \boldsymbol{P}_N, t)] \\ &\quad -\pi\delta[\boldsymbol{l}\cdot(\boldsymbol{P}_n - (\hbar/2)\,\boldsymbol{l}) - \boldsymbol{l}\cdot(\boldsymbol{P}_j + (\hbar/2)\,\boldsymbol{l})] \\ &\quad\times [\rho_0(\boldsymbol{P}_1, \ldots \boldsymbol{P}_j + \hbar\boldsymbol{l}, \ldots, \boldsymbol{P}_n - \hbar\boldsymbol{l}, \ldots, \boldsymbol{P}_N, t) - \rho_0(\boldsymbol{P}_1 \cdots \boldsymbol{P}_N, t)]\} \qquad (4)\end{aligned}$$

or [see (13.2.8)], exchanging $\boldsymbol{l}$ and $-\boldsymbol{l}$ in the second part of the right-hand side:

$$\begin{aligned}\partial\rho_0(\boldsymbol{P}_1 \cdots \boldsymbol{P}_N, t)/\partial t \\ &= -(2\pi\lambda^2(2\pi)^6/L^6\hbar^2)\sum_{jn}\sum_{l}|V_l|^2 \\ &\quad\times \delta[E(\boldsymbol{P}_n) + E(\boldsymbol{P}_j) - E(|\boldsymbol{P}_n + \boldsymbol{l}\hbar|) - E(|\boldsymbol{P}_j - \boldsymbol{l}\hbar|)] \\ &\quad\times [\rho_0(\boldsymbol{P}_1 \cdots \boldsymbol{P}_N, t) - \rho_0(\boldsymbol{P}_1, \ldots, \boldsymbol{P}_j - \boldsymbol{l}\hbar, \ldots, \boldsymbol{P}_n + \boldsymbol{l}\hbar, \ldots, \boldsymbol{P}_N, t)] \qquad (5)\end{aligned}$$

It is very interesting to see how the δ function expressing the conservation of "frequencies" in (13.4.3) is transformed, by commutation with the displacement operator, into the usual condition of conservation of energy as it appears in (13.4.4) or (13.4.5).

This equation may be called the Pauli equation because it was first derived by Pauli (1928). More precisely, Pauli derived the corresponding "Boltzmann equation" in which the molecular chaos condition (4.3.3) is used to factorize $\rho_0(\boldsymbol{P}_1 \cdots \boldsymbol{P}_N, t)$. In his derivation Pauli used assumptions essentially similar to those introduced in the standard derivation of the classical Boltzmann equation (see also Chapter 6, § 7 and Tolman, 1938). This includes the repeated use of the "random phase approximation." In other words it is *assumed* that the density matrix remains diagonal at every moment. The first derivation of the Pauli equation in which the repeated use of the random phase approximation was avoided is due to Van Hove (1955), (see also Appendix IV).

For t sufficiently large we shall have

$$\rho_0(\boldsymbol{P}_1 \cdots \boldsymbol{P}_N) = \text{constant} \tag{6}$$

for all momenta satisfying the conservation of energy as expressed by the δ function. In other words, ρ_0 will then depend only on H_0 and we have the quantum mechanical H-theorem

$$\underset{t\to\infty}{\rho_0(\boldsymbol{P}_1 \cdots \boldsymbol{P}_N)} = \rho_0(H_0) \tag{7}$$

The domain of validity of (13.4.5) is the same as that of the corresponding classical equation which has been discussed in Chapter 8, § 11. It implies that only terms of the order $(\lambda^2 t)^n$ are retained.

5. Discussion

In the example of the Pauli equation we have just treated, we clearly see the great similarity between the classical and the quantum mechanical situations when one uses the "Liouville" formalism (see Prigogine, Balescu, Henin, and Résibois, 1960).

Our method may also be extended to include the effect of quantum statistics (Prigogine, 1959; Résibois, 1961; Balescu, 1962). Therefore it seems at present that the methods used here give to statistical physics a remarkable unity and coherence. Indeed the same formalism remains valid for both the quantum mechanical and the classical cases, and includes both nonequilibrium and equilibrium situations.

CHAPTER 14

Irreversibility and Invariants of Motion

1. The Condition of Dissipativity

The approach to equilibrium we described in Chapter 11 and 12 is closely related to the destruction of the invariants of the unperturbed motion by the interactions between the different degrees of freedom of the system. It is clear for example that if the momenta of the individual molecules would remain invariants of the motion, the velocity distribution could not evolve towards the Maxwell distribution.

In this chapter we want to analyze more closely the problem of the invariants of motion for large systems in the sense of (7.1.12), for which boundary effects may be neglected.

We shall first discuss the condition under which such a system approaches statistical equilibrium. This condition may be called the "*condition of dissipativity.*" We shall then discuss the relation between this condition and the existence of invariants of motion, especially in connection with Poincaré's fundamental theorem about invariants which are analytic in the coupling constant.

We show that for systems which satisfy the dissipativity condition there exist no invariants of motion (except the energy, total momentum, . . .) which are analytic in the coupling constant and have non-singular Fourier coefficients. However as a consequence of the existence of a continuous spectrum for the large systems in which we are interested, there exist invariants which are analytic in the coupling constants but have singular Fourier transforms (involving δ functions).

This rather unexpected conclusion (Résibois and Prigogine, 1960) shows that the systems we are studying are *not* "metrically undecomposable" in the sense of modern ergodic theory (see § 2). There exist invariants of motion but they do not prevent the approach to equilibrium in the way described in Chapter 11 and 12. This situation will be discussed in more detail in § 4.

Let us first study the condition of dissipativity. As we have seen, the basic role in the approach to equilibrium is played by the velocity distribution ρ_0. As we are interested in this chapter only in the long time behaviour (and not in the precise form of the kinetic equation for arbitrary values of t) we may use the Markowian form (11.2.5) of the evolution equation for ρ_0

$$\partial \rho_0 / \partial t = \Omega i \Psi^+(0)\, \rho_0(t) = \Omega$$  $$\rho_0(t) \qquad (1)$$

The condition of dissipativity is simply that the operator $i\Psi^+(0)$ corresponding to the sum of arbitrary diagonal fragments does not vanish identically.

Let us first consider weakly coupled systems. The only diagonal fragments we have to consider then are the two-vertex diagrams studied in Chapters 2–5. In the case of three phonon processes, which we studied in Chapter 2 (see 2.7.9), the dissipativity condition requires that $V_{kk'-k''}$ shall not vanish for wave vectors such that the resonance condition

$$\omega_k + \omega_{k'} - \omega_{k''} = 0 \qquad (2)$$

is satisfied. Now we have seen [see (2.1.25) and (2.1.27)] that $V_{kk'-k''}$ is only different from zero if (apart from a vector of the reciprocal lattice)

$$\boldsymbol{k} + \boldsymbol{k}' = \boldsymbol{k}'' \qquad (3)$$

We have already stressed, in Chapter 2, the analogy between (14.1.3) and the conservation of *momentum* in quantum mechanics. A similar analogy exists between (14.1.2) and the conservation of *energy* in quantum mechanics (see also Chapter 13, § 4).

The conditions of dissipativity (14.1.2) and (14.1.3), expressed in quantum mechanical language, mean therefore that *energy and momentum may be simultaneously conserved.*

We may easily generalize (14.1.2) and (14.1.3) to an arbitrary weakly coupled system using action-angle variables (see Chapter 1, §§ 6, 7). Let us write the potential energy in the general form

$$V = \sum_{\{n\}} V_{\{n\}} e^{i \Sigma n_k \alpha_k} \qquad (4)$$

with

$$V_{\{-n\}} = V^*_{\{n\}} \qquad (5)$$

The summation in (14.1.4) over $\{n\}$ includes all interacting degrees of freedom. Then the dissipativity condition requires that the resonance condition

$$\sum n_k \omega_k = 0 \tag{6}$$

has non-trivial solutions (for which not all n_k are zero) and that for these solutions the corresponding Fourier coefficients of the potential energy are different from zero

$$V_{\{n\}} \neq 0 \tag{7}$$

The situation is similar in gases. There we have to require [see (4.2.3)] that the force

$$\boldsymbol{l} V_l \neq 0 \tag{8}$$

shall not vanish when the resonance condition

$$\boldsymbol{l} \cdot (\boldsymbol{v}_j - \boldsymbol{v}_n) = 0 \tag{9}$$

is satisfied (with $\boldsymbol{l} \neq 0$).

For most physical systems the condition of dissipativity to order λ^2 is indeed satisfied. A simple exception is a system of particles placed on a line ("one-dimensional" models). Then the wave vector $\boldsymbol{l}$ and the velocity $\boldsymbol{v}$ have a single component, and (14.1.9) requires that

$$l = 0 \tag{10}$$

The resonance condition has only trivial solutions and the operator corresponding to the diagonal fragments vanishes.

Another more physical example is the case of a free particle (for example, an electron) coupled to lattice waves. Here the conservation of frequencies requires (for more details see Bak, Henin, and Goche, 1959)

$$\omega_l - \boldsymbol{l} \cdot \boldsymbol{v} = 0 \tag{11}$$

Using the Debye approximation for the frequency ω_l this gives

$$cl - \boldsymbol{l} \cdot \boldsymbol{v} = 0 \tag{12}$$

where c is the velocity of sound. Clearly this can only be satisfied if

$$v > c \tag{12'}$$

Therefore there will be dissipation only if the velocity v is larger than the velocity of sound (as long as the particle remains free). In the case of a free particle coupled with an electromagnetic field one again obtains condition (14.1.12), where c is now the velocity of light (see Prigogine and Leaf, 1959). In this case no dissipation can occur at all because of the theory of relativity which requires $v < c$. However, if the electron propagates in a dielectric the velocity of light is reduced and (14.1.12′) may be satisfied. This is the so called Cerenkov radiation.

But the most interesting case in which the dissipation condition cannot be satisfied corresponds to a sufficiently slow motion of a particle in a superfluid liquid like He^4 (see for example Landau and Lifschitz, 1958). This situation is very similar to that of a particle coupled to lattice waves [(14.1.11) through (14.1.12′)].

Since this monograph is specifically devoted to irreversible processes in the frame of classical mechanics we shall not discuss quantum fluids in detail.

Even if the λ^2 contribution to $\Psi^{+}(0)$ vanishes, this does not imply that the system will not approach equilibrium. There may be higher order terms in λ which are not vanishing, so that the condition of dissipativity is still satisfied.

For example, in one-dimensional systems the simplest diagonal fragment that does not vanish is represented in Fig. 14.1.1.

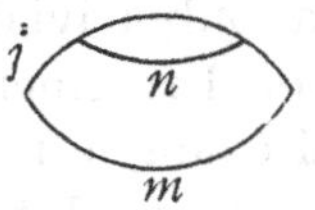

Fig. 14.1.1 Simplest non-vanishing diagonal fragment for one-dimensional systems.

This fragment is of order λ^4 and involves three particles. This diagram now contains the factor

$$lV_l\,\delta[kv_n-(k+l)v_j+lv_m] \tag{13}$$

The corresponding resonance condition is

$$kv_n-(k+l)v_j+lv_m=0 \tag{14}$$

It has non-trivial solutions corresponding to k, l different from zero.

In "weakly coupled" one-dimensional systems the relaxation time is therefore of order

$$t_r \sim 1/\lambda^4 \tag{15}$$

Here, triple collisions determine the approach to equilibrium (see F. Henin, 1961).

It is possible to give a more compact form to the condition of dissipativity using the theory of Cauchy integrals developed in Chapter 8, § 4. Let us write the operator $\Psi(z)$ which appears in (14.1.1) in the form of a Cauchy integral [see (8.4.5)]

$$\Psi(z) = (1/2\pi i) \int_{-\infty}^{+\infty} d\omega\, f(\omega)/(\omega - z) \tag{16}$$

Therefore [see (8.4.8)]

$$\begin{aligned} i\Psi^+(0) &= (1/2\pi i) \int_{-\infty}^{+\infty} d\omega [i\pi\delta(\omega) + \mathscr{P}(1/\omega)] f(\omega) \\ &= (1/2) f(0) + (1/2\pi i) \int_{-\infty}^{+\infty} d\omega\, f(\omega)\, \mathscr{P}(1/\omega) \end{aligned} \tag{17}$$

The operator $i\Psi^+(0)$ which acts on the velocity distribution is real (see Chapter 9, § 3). The imaginary part in (14.1.17) has to vanish for reasons of symmetry when we take account of all diagonal fragments.

The condition of dissipativity means therefore the non-vanishing of $f(0)$. But as we have seen in formula (8.4.10), $f(x)$ is equal to the discontinuity of the resolvent when one crosses the real axis. Therefore the condition of dissipativity means that the operator $\Psi(z)$ presents a discontinuity at the origin $z = 0$. A similar condition was first formulated by Van Hove (1957).

2. Dissipativity and Poincaré's Theorem

We shall now show that the validity of the dissipativity conditions, (14.1.6) and (14.1.7), implies important consequences about the invariants of the motion.[1]

[1] We shall follow the method due to Poincaré (1892) very closely to prove his fundamental theorem about the invariants of motion. This theorem will be stated and discussed in this as well as in the next paragraph.

Let us consider the case of interacting particles characterized by the Hamiltonian (4.1.1). As usual the interaction potential will be Fourier analyzed [see (4.1.2)]. As long as we neglect λV there exist a large number of invariants of motion. Indeed, all momenta $\boldsymbol{p}_i$ are then invariants of the motion and therefore any arbitrary function of them is also an invariant of the motion. There may also exist invariants which are both functions of the momenta and the coordinates (see § 3).

Let us consider some function Φ^0 which is an invariant of the motion with respect to the unperturbed Hamiltonian H_0. Using the notation of Chapter 1 this means [see (1.1.3) and (1.3.3)]

$$[H_0, \Phi^0] = 0 \qquad \text{or} \qquad L_0 \Phi^0 = 0 \tag{1}$$

On the contrary Φ^0 is not an invariant for the complete Hamiltonian H,

$$[H, \Phi^0] \neq 0 \qquad \text{or} \qquad L\Phi^0 \neq 0 \tag{2}$$

The basic question then is the following: Is it possible to find a function Φ which reduces to Φ^0 for $\lambda \to 0$ and which is an invariant of the motion? Such a function would have to satisfy the relations

$$[H, \Phi] = 0 \qquad \text{or} \qquad L\Phi = 0 \tag{3}$$

We can give a precise answer to this question if we assume that Φ is *analytic* in the coupling constant λ. We may then expand Φ in a Taylor series in λ,

$$\Phi = \sum_{r=0}^{\infty} \lambda^r \Phi^{(r)}(\boldsymbol{p}_1 \cdots \boldsymbol{p}_N, \boldsymbol{x}_1 \cdots \boldsymbol{x}_N) \tag{4}$$

Moreover we assume periodic boundary conditions as in Chapter 1, § 5 and expand each $\Phi^{(r)}$ in a Fourier series

$$\Phi^{(r)} = (2\pi/L)^{3N} \sum_{\{k\}} \varphi^{(r)}_{\{k\}}(\boldsymbol{p}_1 \cdots \boldsymbol{p}_N)\, e^{i\Sigma \boldsymbol{k}_j \cdot \boldsymbol{x}_j} \tag{5}$$

We must study the possibility of determining the coefficients $\varphi^{(r)}_{\{k\}}$ in such a way that they satisfy (14.2.3). From (14.2.3) and the series expansion (14.2.4) we have, by equating powers of λ, the succession of equations

$$[H_0, \Phi^0] = 0$$
$$[H_0, \Phi^{(r)}] + [V, \Phi^{(r-1)}] = 0 \qquad r = 1, 2, \ldots \tag{6}$$

We first prove that the leading term, Φ^0, may be chosen independent of H_0. Suppose

$$\Phi^0 = \psi(H_0) \tag{7}$$

Then form the invariant $\Phi - \psi(H)$. For $\lambda \to 0$ this invariant vanishes, by definition. We may then write

$$\Phi - \psi(H) = \lambda[\Phi'^{(0)} + \lambda\Phi'^{(1)} + \cdots]$$
$$= \lambda\Phi' \tag{8}$$

where Φ' is again an invariant. Either $\Phi'^{(0)}$ is no longer a function of H_0 or we may repeat the operation.

Thus if Φ^0 is a function of H_0, we may always find another invariant whose leading term is not a function of H_0, except of course in the trivial case in which Φ is a function of H.

We shall now show that as long as we consider only invariants of the motion which are regular in the wave vector $\boldsymbol{k}$ we may also exclude the case in which Φ^0 depends on the coordinates. Indeed, using (14.2.5), condition (14.2.6) gives

$$\varphi^0_{\{k\}} \sum_j \boldsymbol{k}_j \cdot \boldsymbol{p}_j = 0 \tag{9}$$

This requires

$$\varphi^0_{\{k\}} = 0 \qquad \text{or} \qquad \sum_j \boldsymbol{k}_j \cdot \boldsymbol{p}_j = 0 \tag{10}$$

But the second equation can only be satisfied identically (that is, for all values of $\boldsymbol{p}$) if

$$\boldsymbol{k}_1 = \boldsymbol{k}_2 = \cdots = \boldsymbol{k}_N = 0 \tag{11}$$

We may therefore conclude that

$$\varphi^0_{\{k\}} = 0 \qquad \text{except for} \qquad \{\boldsymbol{k}\} = 0 \tag{12}$$

We shall see in § 3 that this is no longer so when we permit $\varphi^0_{\{k\}}$ to become infinite at resonance. But in this paragraph we shall treat only the case of invariants with regular Fourier transforms.

We now consider equation (14.2.6), corresponding to first order in λ. We have, taking account of (14.2.12),

$$\varphi^{(1)}_{\{k\}} \sum_j \boldsymbol{p}_j \cdot \boldsymbol{k}_j = V_{\{k\}} \sum_j \boldsymbol{k}_j \cdot (\partial \Phi^0 / \partial \boldsymbol{p}_j) \tag{13}$$

Here $V_{\{k\}}$ is the Fourier coefficient of the potential energy [see (4.1.2)]

$$V = (2\pi/L)^{3N} \sum_{\{k\}} V_{\{k\}}\, e^{i \Sigma \boldsymbol{k}_j \cdot \boldsymbol{x}_j} \tag{14}$$

We see from the condition of dissipativity [see (14.1.6)–(14.1.9)] that $V_{\{k\}}$ does not vanish when the resonance conditions

$$\sum_j \boldsymbol{k}_j \cdot \boldsymbol{p}_j = 0 \tag{15}$$

are satisfied. It follows that at every point where the resonance conditions are satisfied, we also have

$$\sum_j \boldsymbol{k}_j \cdot (\partial \Phi^0 / \partial \boldsymbol{p}_j) = 0 \tag{16}$$

This equation is therefore a consequence of the validity of the resonance conditions (14.2.15) and for all points for which (14.2.15) is satisfied we have

$$\boldsymbol{p}_1 / (\partial \Phi^0 / \partial \boldsymbol{p}_1) = \boldsymbol{p}_2 / (\partial \Phi^0 / \partial \boldsymbol{p}_2) = \cdots = \boldsymbol{p}_N / (\partial \Phi^0 / \partial \boldsymbol{p}_N) \tag{17}$$

The crucial point in Poincaré's proof is to show that the resonance conditions have so "many" solutions that (14.2.17) has to be valid *identically*.

To understand this, let us consider the case of two degrees of freedom and replace the $\boldsymbol{k}$'s by their values (1.5.9) using periodic boundary conditions:

$$\boldsymbol{k}_1 = (2\pi/L)\boldsymbol{n}_1 \qquad \boldsymbol{k}_2 = (2\pi/L)\boldsymbol{n}_2 \tag{18}$$

The resonance condition (14.2.15) then becomes

$$\boldsymbol{p}_1 \cdot \boldsymbol{n}_1 + \boldsymbol{p}_2 \cdot \boldsymbol{n}_2 = 0 \tag{19}$$

and simply expresses that $\boldsymbol{p}_1$ and $\boldsymbol{p}_2$ are commensurable with each other. Now in each domain of variation of $\boldsymbol{p}_1$ and $\boldsymbol{p}_2$, however small, there are an infinite number of values of $\boldsymbol{p}_1$ and $\boldsymbol{p}_2$ for which this condition is satisfied.

For all these points (14.2.17) is valid. Therefore if we restrict ourselves to Φ^0's that are continuous functions, these relations have to be satisfied *identically*.

But this means that all determinants

$$\begin{vmatrix} \boldsymbol{p}_i & \boldsymbol{p}_j \\ \partial\Phi^0/\partial\boldsymbol{p}_i & \partial\Phi^0/\partial\boldsymbol{p}_j \end{vmatrix} = 0 \tag{20}$$

vanish. This is precisely the condition that Φ^0 be a function of H_0.

$$F(\Phi^0, H_0) = 0 \tag{21}$$

Indeed, from the differentiation of (14.2.21) with respect to $\boldsymbol{p}_i$ and $\boldsymbol{p}_j$ it follows that

$$\begin{aligned} (\partial F/\partial\Phi^0)(\partial\Phi^0/\partial\boldsymbol{p}_i) + (\partial F/\partial H_0)\boldsymbol{p}_i/m &= 0 \\ (\partial F/\partial\Phi^0)(\partial\Phi^0/\partial\boldsymbol{p}_j) + (\partial F/\partial H_0)\boldsymbol{p}_j/m &= 0 \end{aligned} \tag{22}$$

These relations are equivalent to (14.2.20).

We have therefore shown that Φ^0 has to be a function of H_0, in contradiction to our previous result — that we may restrict ourselves to invariants which are *not* functions of H_0.

We may conclude that for systems that satisfy the dissipativity condition there exist no invariants of the motion, except the energy, that are analytic in the coupling constant and have non-singular Fourier coefficients. *All such invariants are destroyed by the interactions.* Now this is precisely a formulation of Poincaré's fundamental theorem (1892). Indeed Poincaré has shown (see also Fermi, 1923) that for a large class of mechanical systems there exist no invariants analytic in the coupling constant except the trivial ones. As long as we restrict ourselves to invariants with regular Fourier transforms this theorem applies to the systems we are considering here. The case of analytic invariants with singular Fourier transforms is treated in § 3.

It is now important to recognize that the validity of Poincaré's theorem (in our sense) is a much less restricted condition than the condition of "*metrical transitivity*" used in modern ergodic theory (see for example Hopf, 1948; Kinchine, 1949; or ter Haar 1954). This condition means that the energy surface is "metrically indecomposable" in the following sense: It is impossible to divide this surface into two parts (both of non-vanishing measure) such that the trajectories are confined to one of these regions.

This condition is tied up with Poincaré's theorem in an obvious way. If the system admits additional invariants, the trajectories are confined to certain subsurfaces of the energy surface, which is inconsistent with metric indecomposibility.

Thus Poincaré's theorem is a necessary condition for "metrical transitivity". It is not however sufficient, because Poincaré's theorem is only concerned with the restricted class of invariants that are analytic in the coupling constant λ.

3. Analytic Invariants with Singular Fourier Transforms

We now permit the volume to go to infinity in order to treat the wave vectors as *continuous variables.*

Let us first consider (14.2.9). We may now satisfy it by invariants of motion of the form

$$\varphi^0_{\{k\}} = \delta(\sum_j \boldsymbol{k}_j \cdot \boldsymbol{p}_j)\, F_{\{k\}}(\{\boldsymbol{p}\}) \tag{1}$$

Indeed (remember that in conjunction with an integration over x one has $x\delta(x) = 0$)

$$\sum_j \boldsymbol{k}_j \cdot \boldsymbol{p}_j\, \varphi^0_{\{k\}} = \sum_j \boldsymbol{k}_j \cdot \boldsymbol{p}_j\, \delta(\sum_j \boldsymbol{k}_j \cdot \boldsymbol{p}_j)\, F_{\{k\}} = 0 \tag{2}$$

These invariants correspond to functions of both the wave vectors $\boldsymbol{k}$ and the momenta $\boldsymbol{p}$ such that the "total" wave vector is orthogonal to the total momentum

$$\sum_j \boldsymbol{k}_j \cdot \boldsymbol{p}_j = 0 \tag{3}$$

From (14.3.1) we derive the invariant $\Phi^0(\boldsymbol{x}_1 \cdots \boldsymbol{x}_N, \boldsymbol{p}_1 \cdots \boldsymbol{p}_N)$ in phase space given by

$$\Phi^0 = \int (d\boldsymbol{k})^N e^{i \Sigma \boldsymbol{k}_j \cdot \boldsymbol{x}_j}\, \delta(\sum_j \boldsymbol{k}_j \cdot \boldsymbol{p}_j)\, F_{\{k\}} \tag{4}$$

This is indeed an invariant because it is a time-independent solution of the unperturbed Liouville equation

$$L_0 \Phi^0 = i \sum (\boldsymbol{p}_j/m) \cdot (\partial \Phi^0 / \partial \boldsymbol{x}_j) = 0 \tag{5}$$

As a very simple example we may consider a single-particle distribution in which only k_x and p_y are different from zero. This corresponds to a distribution which is only x-dependent while the particles move in the y direction (see Fig. 14.3.1). It is clear that such a distribution function will be an invariant of the motion.

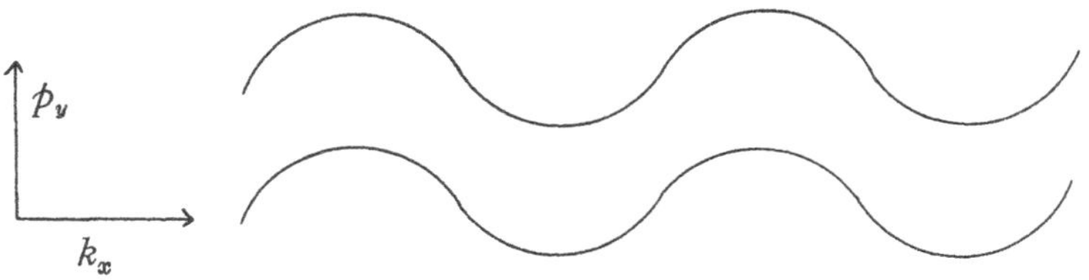

Fig. 14.3.1 Example of a space-dependent invariant of motion.

We see that the consideration of functions with singularities in wave vector space permits us to *extend the class of invariants of the motion.* Moreover these functions, while singular in wave vector space, are perfectly well-behaved in configuration space.

We shall now show, in the same way, that starting with (14.2.13) we obtain a class of invariants of motion which are *analytic* in the coupling constant λ but have singular Fourier transforms.

We could start with an arbitrary invariant of the unperturbed motion Φ^0; but in order to avoid unnecessary complications in notation, we start with a function Φ^0 which depends only on the velocities.

We first solve (14.2.13) formally,

$$\varphi^{(1)}_{\{k\}} = V_{\{k\}} (1/i \sum_j \boldsymbol{p}_j \cdot \boldsymbol{k}_j) \sum_j i\, \boldsymbol{k}_j \cdot (\partial \Phi^0 / \partial \boldsymbol{p}_j) + F_{\{k\}}(\{\boldsymbol{p}\}) \tag{6}$$

The function $F_{\{k\}}$ has to satisfy the condition

$$\sum_j \boldsymbol{p}_j \cdot \boldsymbol{k}_j F_{\{k\}} = 0 \tag{7}$$

This again implies [see (14.3.1) and (14.3.2)] that it contains a δ singularity.

To give a meaning to (14.3.6) in conjunction with an integration over $\{k\}$ we understand the singular denominator to be a *principal part*. Moreover choose $F_{\{k\}}$ to satisfy the requirement of *causality* formulated in Chapter 6, § 3, for scattering theory. We shall verify that this implies [1]

$$F_{\{k\}}(\{p\}) = -\pi\delta(\sum_j \boldsymbol{k}_j\cdot\boldsymbol{p}_j)\, V_{\{k\}} \sum_j i\boldsymbol{k}_j\cdot(\partial\Phi^0/\partial\boldsymbol{p}_j) \tag{8}$$

This choice satisfies conditions (14.3.2). Moreover (14.3.6) then becomes [see (2.5.25)]

$$\begin{aligned}\varphi^{(1)}_{\{k\}} &= V_{\{k\}}[(1/i\sum_j \boldsymbol{k}_j\cdot\boldsymbol{p}_j) - \pi\delta(\sum_j \boldsymbol{k}_j\cdot\boldsymbol{p}_j)] \sum_j i\boldsymbol{k}_j\cdot(\partial\Phi^0/\partial\boldsymbol{p}_j) \\ &= -V_{\{k\}}\pi\delta_+(\sum_j \boldsymbol{k}_j\cdot\boldsymbol{p}_j)\sum_j i\boldsymbol{k}_j\cdot(\partial\Phi^0/\partial\boldsymbol{p}_j)\end{aligned} \tag{9}$$

Now as we have seen in Chapter 6, §§ 2 and 3 the δ_+ function is the free particle propagator in Fourier space. Indeed, as shown by (6.2.1) it is the Fourier transform of the free particle propagator $G(\boldsymbol{r}-\boldsymbol{r}', \boldsymbol{p})$ in phase space. The requirement of causality is expressed by the fact that G connects the distribution function in a given point of phase space to all other points which correspond to *earlier* times on the trajector.

Using (6.2.1) we easily go from (14.3.9) back to phase space by an inverse Fourier transform, and obtain

$$\begin{aligned}\Phi^{(1)} = \lambda \sum_{jn} \int d\boldsymbol{x}'_j\, d\boldsymbol{x}'_n\, G(\boldsymbol{x}_j-\boldsymbol{x}'_j, \boldsymbol{p}_j, \boldsymbol{x}_n-\boldsymbol{x}'_n, \boldsymbol{p}_n) \\ \times\ (\partial V_{jn}/\partial\boldsymbol{x}'_j)\cdot[(\partial/\partial\boldsymbol{p}_j)-(\partial/\partial\boldsymbol{p}_n)]\,\Phi^0\end{aligned} \tag{10}$$

with [see (6.2.3)]

$$\begin{aligned}G(\boldsymbol{x}_j-\boldsymbol{x}'_j, \boldsymbol{p}_j, \boldsymbol{x}_n-\boldsymbol{x}'_n, \boldsymbol{p}_n) = \int_0^\infty d\tau\, \delta(\boldsymbol{x}_j-\boldsymbol{x}'_j-\boldsymbol{p}_j\tau/m) \\ \times\ \delta(\boldsymbol{x}_n-\boldsymbol{x}'_n-\boldsymbol{p}_n\tau/m)\end{aligned}$$

Because of the formal procedure we followed, it is interesting to verify that (14.3.10) is really an invariant of the motion. This is a straightforward consequence of (6.2.8)

[1] The method we follow is identical to that used in quantum mechanical scattering theory (see Dirac, 1947).

$$
\begin{aligned}
[H^0, \Phi^{(1)}] &= \lambda \sum_{jn} \int dx'_j\, dx'_n\, \delta(x_j - x'_j)\, \delta(x_n - x'_n) \\
&\quad \times (\partial V_{jn}/\partial x'_j) \cdot [(\partial/\partial p_j) - (\partial/\partial p_n)]\, \Phi^0 \\
&= \lambda \sum_{jn} (\partial V/\partial x_j) \cdot [(\partial/\partial p_j) - (\partial/\partial p_n)]\, \Phi^0 \\
&= -[V, \Phi^0]
\end{aligned}
\tag{11}
$$

We could extend our discussion to an arbitrary order in λ. But it is more convenient to observe that any function Φ which satisfies the integral equation

$$
\begin{aligned}
\Phi = \Phi^0 + \lambda \sum_{jn} \int dx'_j\, dx'_n\, G(x_j - x'_j, p_j, x_n - x'_n, p_n) \\
\times (\partial V/\partial x'_j) \cdot [(\partial/\partial p_j) - (\partial/\partial p_n)]\, \Phi
\end{aligned}
\tag{12}
$$

is an invariant of the motion which for $\lambda \to 0$ reduces to Φ^0. Indeed using (6.2.8) once more we have

$$[H, \Phi] = 0 \qquad \text{or} \qquad L\Phi = 0 \tag{13}$$

We see therefore that to each invariant of motion Φ^0 of the unperturbed Hamiltonian H_0, the equation (14.3.12) associates the invariant Φ of the complete Hamiltonian $H_0 + \lambda V$. The invariant Φ^0 may be an arbitrary function of the momenta of the particles.

The actual calculation of the Φ associated with a given Φ_0 proceeds exactly in the same way as the calculation of the stationary distribution function $f(x, v)$ associated with a given incident flow $\rho_0(v, t = 0)$ in scattering problems.

Indeed equation (6.3.6) which gives $f(x, v)$ and equation (14.3.12) for Φ are identical. We may therefore visualize Φ as the stationary phase space distribution which for $\lambda \to 0$ goes over to Φ_0 (See for example Chapter 6, § 4).

Let us now discuss the meaning of our results.

4. Approach to Equilibrium and Invariants

We see from the results we have derived in § 3 that the systems we have considered are not metrically indecomposable in the sence defined in § 2. At least as long as we neglect boundary effects there exists a class of invariants analytic in the coupling constant with singular Fourier transforms.

In a more restricted sense however the Hamiltonian (as well as the other usual invariants, like the total momentum, etc.) plays a special role because it remains the only invariant analytic in λ with *regular* Fourier transforms.

The existence of this new class of invariants explains why time-independent perturbation methods can be applied to the Liouville equation (see the example given in Appendix II). We have seen, indeed, that in accordance with (1.4.14) each eigenfunction of the Liouville operator gives an invariant of motion. Now time-independent perturbation methods are based on an expansion of the eigenfunctions as powers of λ and presuppose, therefore, the existence of *invariants of the motion which are analytic in* λ.

The existence of analytic invariants of motion precludes any attempt to prove that the complete phase distribution function ρ approaches the distribution function, $\rho^{eq}(H)$, of the equilibrium theory of statistical mechanics with time. It is only for certain classes of properties that the asymptotic distribution function $\rho(t \to \infty)$ and the equilibrium distribution ρ^{eq} can give equivalent results. This is a different formulation of a classical argument (P. and T. Ehrenfest, 1911): Because of the conservation of the *"fine-grained"* distribution function, ρ, the approach to equilibrium can only be conceived in a *"coarse-grained"* sense.

This conclusion is in complete agreement with the mechanism of irreversibility that we have described in this book. We have only shown that properties depending on a finite number of molecules exhibit, in the limit $N \to \infty$, $\Omega \to \infty$, $C =$ finite, an irreversible behavior. Our method permits no conclusion whatever about the behavior of functions explicitly involving all or an appreciable fraction of the molecules.

Therefore our derivation of equilibrium statistical mechanics as a long time result of interactions between the degrees of freedom, has to be understood precisely in this sense. All properties involving a finite number of degrees of freedom have a long time behavior identical to that predicted by equilibrium statistical mechanics. It is in this sense that our method leads to a justification for the use of equilibrium statistical mechanics.

APPENDIX I

Laguerre Functions and Bessel Functions

For convenience we shall go here into more detail about the properties of Laguerre and Bessel functions.[1]

1. Laguerre's Polynomials

These polynomials can be obtained as solution of Laguerre's differential equation

$$x\frac{d^2y}{dx^2}+(1-x)\frac{dy}{dx}+my = 0 \qquad m = \text{positive integer} \tag{1}$$

Let us write the solution of this equation formally as a series

$$y = \sum_{\lambda=0}^{\infty} a_\lambda x^{\kappa+\lambda} \tag{2}$$

and introduce it into (A.I.1); we obtain the following:

$$a_0\kappa^2 x^{\kappa-1}+\sum_{\lambda=0}^{\infty}\{a_{\lambda+1}(\kappa+\lambda+1)^2-a_\lambda(\kappa+\lambda+m)\}x^{\kappa+\lambda} = 0 \tag{3}$$

This identity must hold whatever the value of x. Hence

$$\begin{aligned} a_0\kappa^2 &= 0 \\ a_{\lambda+1}(\kappa+\lambda+1)^2 &= a_\lambda(\kappa+\lambda-m) \end{aligned} \tag{3'}$$

The first equation (usually called the indicial equation) leads to

$$\kappa = 0 \tag{4}$$

The other equations form a set of recurrence relations which

[1] An excellent introduction to this subject may be found in Sneddon (1956).

determine the coefficients a_λ in terms of a_0. The solution is:

$$a_\lambda = (-1)^\lambda a_0 \frac{m(m-1)\cdots(m-\lambda+1)}{(\lambda!)^2} \tag{5}$$

As m is a positive integer,

$$a_\lambda = 0 \qquad \text{for} \qquad \lambda > m$$

and therefore (A.I.2) is a polynomial of the mth degree. The Laguerre polynomials as used in Chapter 3 correspond to the special choice:

$$a_0 = m!$$

and therefore we have:

$$L_m(x) = \sum_{\lambda=0}^{m} (-1)^\lambda \left(\frac{m!}{\lambda!}\right)^2 \frac{1}{(m-\lambda)!} x^\lambda \tag{6}$$

This expression can be written in a more compact form

$$L_m(x) = e^x \frac{d^m}{dx^m} (x^m e^{-x}) \tag{7}$$

as can be verified using the following well-known formula:

$$\frac{d^m}{dx^m} u(x)\, v(x) = \sum_{p=0}^{m} \frac{m!}{p!(m-p)!} \frac{d^p u}{dx^p} \frac{d^{m-p} v}{dx^{m-p}} \tag{8}$$

2. Associated Laguerre Polynomials

The associated Laguerre polynomial $L_m^{(k)}(x)$ is defined as the kth ($k \leqq m$) derivative of $L_m(x)$ with respect to x:

$$L_m^{(k)}(x) = \frac{d^k}{dx^k} L_m(x) \qquad k \leqq m \tag{9}$$

Using (A.I.6), we obtain

$$L_m^{(k)}(x) = \sum_{\lambda=k}^{m} (-1)^\lambda \frac{(m!)^2}{\lambda!(\lambda-k)!(m-\lambda)!} x^{\lambda-k} \tag{10}$$

We also have, using (A.I.7):

$$L_m^{(k)}(x) = \frac{d^k}{dx^k}\left\{e^x \frac{d^m}{dx^m}\left[x^m e^{-x}\right]\right\} \tag{11}$$

We can easily obtain the differential equation satisfied by $L_m^{(k)}(x)$ by differentiating (A.I.1) k times with respect to x:

$$\begin{aligned}&\frac{d^k}{dx^k}\left\{x\frac{d^2 L_m(x)}{dx^2} + (1-x)\frac{dL_m(x)}{dx} + mL_m(x)\right\}\\ &= x\frac{d^2 L_m^{(k)}}{dx^2} + (k+1-x)\frac{dL_m^{(k)}}{dx} + (m-k)L_m^{(k)} = 0\end{aligned} \tag{12}$$

3. Generating Function for Laguerre Polynomials

The generating function $F^{(k)}(z, x)$ for Laguerre polynomials is defined by:

$$F^{(k)}(z, x) = \sum_{m=k}^{\infty} \frac{L_m^{(k)}(x)}{m!} z^m \tag{13}$$

Let us first consider the case $k = 0$. Taking into account (A.I.6), we have

$$F^{(0)}(z, x) = \sum_{m=0}^{\infty} \frac{L_m(x)}{m!} z^m = \sum_{m=0}^{\infty} \sum_{\lambda=0}^{m} (-)^\lambda \frac{m!\, x^\lambda z^m}{(m-\lambda)!\,(\lambda!)^2} \tag{14}$$

We change the order of the summations:

$$\sum_{m=0}^{\infty} \sum_{\lambda=0}^{m} = \sum_{\lambda=0}^{\infty} \sum_{m=\lambda}^{\infty} \tag{15}$$

and take as a new variable $m-\lambda$,

$$n = m - \lambda \tag{16}$$

We then obtain

$$F^{(0)}(z, x) = \sum_{\lambda=0}^{\infty} \sum_{n=0}^{\infty} (-1)^\lambda \frac{(n+\lambda)!}{n!\,(\lambda!)^2} (xz)^\lambda z^n \tag{17}$$

By series expansion, we have

$$\left(\frac{1}{1-z}\right)^{\lambda+1} = \sum_{n=0}^{\infty} \frac{(\lambda+n)!}{\lambda!\,n!} z^n \tag{18}$$

and therefore (A.I.17) can be written

$$F^{(0)}(z, x) = \sum_{\lambda=0}^{\infty} (-1)^{\lambda} \frac{(xz)^{\lambda}}{\lambda!} \frac{1}{(1-z)^{\lambda+1}} \tag{19}$$

or

$$F^{(0)}(z, x) = \sum_{m=0}^{\infty} \frac{L_m(x)}{m!} z^m = \frac{1}{1-z} \exp\left[-\frac{xz}{1-z}\right] \tag{20}$$

Differentiating k times with respect to x, we obtain the generating function for the associated Laguerre polynomials $L_m^{(k)}(x)$:

$$F^{(k)}(z, x) = \sum_{m=k}^{\infty} \frac{L_m^{(k)}(x)}{m!} z^m = (-1)^k \frac{z^k}{(1-z)^{k+1}} \exp\left[-\frac{xz}{1-z}\right] \tag{21}$$

4. Orthonormalized Laguerre Functions

The Laguerre polynomials themselves do not form an orthogonal set. However the Laguerre functions defined as

$$y_n^{(k)} = e^{-x/2} x^{k/2} L_n^{(k)}(x) \tag{22}$$

form an orthogonal set for a given value of k, i.e.,

$$\begin{aligned} I_{n,m}^{(k)} = \int_0^{\infty} dx\, e^{-x} x^k L_n^{(k)}(x) L_m^{(k)}(x) &\neq 0 \qquad \text{for} \qquad n = m \\ &= 0 \qquad\qquad\qquad n \neq m \end{aligned} \tag{23}$$

In order to establish this, let us consider the function $g^{(k)}(z_1, z_2)$, defined as

$$\begin{aligned} g^{(k)}(z_1, z_2) &= \int_0^{\infty} dx\, F^{(k)}(z_1, x) F^{(k)}(z_2, x) e^{-x} x^k \\ &= \sum_{n=k}^{\infty} \sum_{m=k}^{\infty} \frac{z_1^n z_2^m}{n!\, m!} I_{n,m}^{(k)} \end{aligned} \tag{24}$$

where we have used (A.I.13).

Taking (A.I.21) into account, we have

$$g^{(k)}(z_1, z_2) = \frac{(z_1 z_2)^k}{[(1-z_1)(1-z_2)]^{k+1}} \int_0^{\infty} dx\, x^k \exp\left[-x \frac{1-z_1 z_2}{(1-z_1)(1-z_2)}\right] \tag{25}$$

Using the following well-known formula:

$$\int_0^\infty dx\, x^k e^{-ax} = \frac{k!}{a^{k+1}} \tag{26}$$

we obtain

$$g^{(k)}(z_1, z_2) = k!\frac{(z_1 z_2)^k}{(1-z_1 z_2)^{k+1}} \tag{27}$$

We can expand the denominator in a power series [see (A.I.18)]. Then we have

$$\begin{aligned} g^{(k)}(z_1, z_2) &= \sum_{n=0}^{\infty} \frac{(k+n)!}{n!} (z_1 z_2)^{n+k} \\ &= \sum_{m=k}^{\infty} \frac{m!}{(m-k)!} (z_1 z_2)^m \end{aligned} \tag{28}$$

If we compare this with (A.I.24) we obtain

$$I^{(k)}_{n,m} = \frac{(m!)^3}{(m-k)!} \delta^{\text{Kr}}_{n,m} \tag{29}$$

where $\delta^{\text{Kr}}_{n,m}$ is the Kronecker delta function. Therefore, the functions

$$\left[\frac{(m-k)!}{(m!)^3}\right]^{1/2} e^{-x/2} x^{k/2} L^{(k)}_m(x) \tag{30}$$

form an orthonormal set.

5. Bessel Functions of Integral Order

These functions are solutions of the differential equation

$$x^2 \frac{d^2 y}{dx^2} + x\frac{dy}{dx} + (x^2-n^2)y = 0 \qquad n = \text{integer} \tag{31}$$

Following the same method as in § 1, we can write

$$y = \sum_{\lambda=0}^{\infty} a_\lambda x^{\kappa+\lambda} \tag{32}$$

which leads to the indicial equation [see (A.I.3′)]

$$a_0(\kappa^2-n^2) = 0 \tag{33}$$

This equation has the two roots

$$\kappa = \pm n \tag{34}$$

We now obtain the recurrence relations

$$a_\lambda\{(\kappa+\lambda)^2-n^2\}+a_{\lambda-2} = 0 \tag{35}$$

For $\kappa = n$ the solution of (A.I.35) is:

$$a_{2\lambda} = (-1)^\lambda a_0 \frac{n!}{2^{2\lambda}\lambda!(n+\lambda)!} \tag{36}$$
$$a_{2\lambda+1} = 0$$

The Bessel function of integral order $J_n(x)$ corresponds to the choice

$$a_0 = \frac{1}{2^n n!} \tag{37}$$

and we have therefore

$$J_n(x) = \sum_{\lambda=0}^{\infty} (-1)^\lambda \frac{1}{\lambda!(n+\lambda)!}\left(\frac{x}{2}\right)^{n+2\lambda} \tag{38}$$

6. Bessel Functions of Integral Order and Purely Imaginary Argument

This function is defined as

$$I_n(x) = i^{-n} J_n(ix) \tag{39}$$

or using (A.I.38)

$$I_n(x) = \sum_{\lambda=0}^{\infty} \frac{1}{\lambda!(n+\lambda)!}\left(\frac{x}{2}\right)^{n+2\lambda} \tag{40}$$

Taking (A.I.39) and equation (A.I.31) for the function $J_n(x)$ into account, it is easily verified that the differential equation for the functions $I_n(x)$ is

$$x^2\frac{d^2y}{dx^2}+x\frac{dy}{dx}-(x^2+n^2)y = 0 \tag{41}$$

It is immediately verified from (A.I.40) that

$$I_n(x) > 0 \qquad \text{for } x > 0 \tag{42}$$

$$I_0(0) = 1 \tag{43}$$

$$I_n(0) = 0 \qquad n \neq 0 \tag{44}$$

$$I_n(x) \simeq \frac{1}{n!}\left(\frac{x}{2}\right)^n \qquad \text{for small } x \tag{45}$$

7. Proof of Formula (3.3.34)

We now want to prove

$$\tfrac{1}{2} e^{z \cos u} = \sum_{n=1}^{\infty} I_n(z) \cos nu + \tfrac{1}{2} I_0(z) \tag{46}$$

Let us first expand the left-hand side in a power series

$$\begin{aligned}\tfrac{1}{2} e^{z \cos u} &= \tfrac{1}{2} \sum_{n=0}^{\infty} \frac{1}{n!} (z \cos u)^n \\ &= \tfrac{1}{2} \sum_{n=0}^{\infty} \frac{1}{n!} \left(\frac{z}{2}\right)^n (e^{iu}+e^{-iu})^n\end{aligned} \tag{47}$$

Using the binomial theorem

$$(a+b)^n = \sum_{p=0}^{n} \frac{n!}{p!(n-p)!} a^p b^{n-p} \tag{48}$$

we obtain

$$\tfrac{1}{2} e^{z \cos u} = \tfrac{1}{2} \sum_{n=0}^{\infty} \sum_{p=0}^{n} \frac{1}{p!(n-p)!} \left(\frac{z}{2}\right)^n e^{i(2p-n)u} \tag{49}$$

Changing the order of the summations [see (A.I.15)] and taking as a new variable $n = m+2p$, we obtain

$$\begin{aligned}\tfrac{1}{2} e^{z \cos u} &= \tfrac{1}{2} \sum_{p=0}^{\infty} \sum_{m=-p}^{\infty} \frac{e^{-imu}}{p!(m+p)!} \left(\frac{z}{2}\right)^{m+2p} \\ &= \tfrac{1}{2} \sum_{p=0}^{\infty} \sum_{m=0}^{\infty} \frac{e^{-imu}}{p!(m+p)!} \left(\frac{z}{2}\right)^{m+2p} \\ &\quad + \tfrac{1}{2} \sum_{p=0}^{\infty} \sum_{m=-p}^{-1} \frac{e^{-imu}}{p!(m+p)!} \left(\frac{z}{2}\right)^{m+2p}\end{aligned} \tag{50}$$

Taking (A.I.40) into account we have

$$\tfrac{1}{2}\sum_{m=0}^{\infty} e^{-imu} \sum_{p=0}^{\infty} \frac{1}{p!\,(m+p)!}\left(\frac{z}{2}\right)^{m+2p} = \tfrac{1}{2}\sum_{m=0}^{\infty} e^{-imu} I_m(z) \tag{51}$$

On the other hand, if we replace m by $-\lambda$ in the second term in the right-hand side of (A.I.50) and change the order of the summations over p and λ, we obtain

$$\begin{aligned} &\tfrac{1}{2}\sum_{p=0}^{\infty} \sum_{m=-p}^{-1} \frac{e^{-imu}}{p!\,(m+p)!}\left(\frac{z}{2}\right)^{m+2p} \\ &= \tfrac{1}{2}\sum_{\lambda=1}^{\infty} \sum_{p=\lambda}^{\infty} \frac{e^{i\lambda u}}{p!\,(p-\lambda)!}\left(\frac{z}{2}\right)^{2p-\lambda} \\ &= \tfrac{1}{2}\sum_{\lambda=1}^{\infty} e^{i\lambda u} I_\lambda(z) \end{aligned} \tag{52}$$

where we have taken account of (A.I.40).

Adding (A.I.51) and (A.I.52) we indeed obtain (A.I.46).

8. Bessel Functions in Terms of Laguerre Polynomials

We shall now show that

$$(-1)^\alpha e^w w^{\alpha/2} J_\alpha\{2(xw)^{1/2}\} = x^{\alpha/2} \sum_{m=\alpha}^{\infty} w^m \frac{L_m^{(\alpha)}(x)}{(m!)^2} \tag{53}$$

Using (A.I.10), the right-hand side can be written

$$x^{\alpha/2} \sum_{m=\alpha}^{\infty} w^m \frac{L_m^{(\alpha)}(x)}{(m!)^2} = x^{\alpha/2} \sum_{m=\alpha}^{\infty} \sum_{\lambda=\alpha}^{m} (-1)^\lambda \frac{w^m x^{\lambda-\alpha}}{\lambda!\,(\lambda-\alpha)!\,(m-\lambda)!} \tag{54}$$

If we take as variables

$$n = m-\alpha \qquad l = \lambda-\alpha$$

we have

$$x^{\alpha/2} \sum_{m=\alpha}^{\infty} w^m \frac{L_m^{(\alpha)}(x)}{(m!)^2} = \sum_{n=0}^{\infty} \sum_{l=0}^{n} (-1)^{l+\alpha} \frac{w^{n+\alpha} x^{l+(\alpha/2)}}{(n-l)!\,l!\,(l+\alpha)!} \tag{55}$$

Let us change the order of the summations again and introduce a new variable $\rho = n-l$. Then we have

$$x^{\alpha/2} \sum_{m=\alpha}^{\infty} w^m \frac{L_m^{(\alpha)}(x)}{(m!)^2} = (-1)^\alpha w^\alpha x^{\alpha/2} \sum_{l=0}^{\infty} \frac{(-1)^l (xw)^l}{l!\,(l+\alpha)!} \sum_{\rho=0}^{\infty} \frac{w^\rho}{\rho!} \tag{56}$$

Taking (A.I.38) into account, the right-hand side of this expression is identical to the left-hand side of (A.I.53).

9. Laguerre Polynomials as Integrals Over Bessel Functions

We now want to show that the Laguerre polynomials can be written as

$$(-1)^{\alpha}\frac{L_m^{(\alpha)}(x)}{m!} = \frac{e^{x}x^{-\alpha/2}}{(m-\alpha)!}\int_0^{\infty} dt\, e^{-t}t^{m-(\alpha/2)} J_{\alpha}\{2(tx)^{1/2}\} \tag{57}$$

Let us calculate the generating function for the right-hand side of (A.I.57),

$$F^{(\alpha)}(w,x) = \sum_{m\geqq\alpha}^{\infty} w^m \frac{e^{x}x^{-\alpha/2}}{(m-\alpha)!}\int_0^{\infty} dt\, e^{-t}t^{m-(\alpha/2)} J_{\alpha}\{2(tx)^{1/2}\} \tag{58}$$

We replace J_α by its expression (A.I.38). We then obtain

$$F^{(\alpha)}(w,x) = \sum_{m=\alpha}^{\infty}\sum_{\lambda=0}^{\infty}\frac{(-1)^{\lambda}e^{x}x^{\lambda}w^{m}}{(m-\alpha)!\,\lambda!\,(\lambda+\alpha)!}\int_0^{\infty} dt\, e^{-t}t^{m+\lambda} \tag{59}$$

Taking (A.I.26) into account and expanding e^x in a power series, we have

$$F^{(\alpha)}(w,x) = \sum_{m=\alpha}^{\infty}\sum_{\lambda=0}^{\infty}\sum_{n=0}^{\infty}\frac{(-1)^{\lambda}x^{\lambda+n}w^{m}(m+\lambda)!}{n!\,(m-\alpha)!\,\lambda!\,(\lambda+\alpha)!} \tag{60}$$

If we take as variables $l = n+\lambda$ and $\mu = m-\alpha$, this can also be written

$$F^{(\alpha)}(w,x) = \sum_{\mu=0}^{\infty}\sum_{\lambda=0}^{\infty}\sum_{l=\lambda}^{\infty}\frac{(-1)^{\lambda}x^{l}w^{\mu+\alpha}(\mu+\alpha+\lambda)!}{(l-\lambda)!\,\mu!\,\lambda!\,(\lambda+\alpha)!} \tag{61}$$

The summation over μ can be performed taking (A.I.18) into account. In this way we obtain

$$F^{(\alpha)}(w,x) = \frac{w^{\alpha}}{(1-w)^{\alpha+1}}\sum_{\lambda=0}^{\infty}\sum_{l=\lambda}^{\infty}\frac{(-1)^{\lambda}x^{l}}{(l-\lambda)!\,\lambda!\,(1-w)^{\lambda}} \tag{62}$$

Changing the order of the summations and introducing $\rho = l-\lambda$

this can be written

$$F^{(\alpha)}(w, x) = \frac{w^\alpha}{(1-w)^{\alpha+1}} \sum_{l=0}^{\infty} \frac{(-1)^l x^l}{(1-w)^l} \sum_{\rho=0}^{l} \frac{(-1)^\rho (1-w)^\rho}{\rho!(l-\rho)!} \tag{63}$$

Taking (A.I.48) into account, we see that

$$\sum_{\rho=0}^{l} \frac{(-1)^\rho (1-w)^\rho}{\rho!(l-\rho)!} = \frac{(-1)^l}{l!}[(1-w)-1]^l = \frac{w^l}{l!} \tag{64}$$

and therefore

$$\begin{aligned} F^{(\alpha)}(w, x) &= \frac{w^\alpha}{(1-w)^{\alpha+1}} \exp\left[-\frac{xw}{1-w}\right] \\ &= (-1)^\alpha \sum_{m=\alpha}^{\infty} \frac{L_m^{(\alpha)}(x)}{m!} w^m \end{aligned} \tag{65}$$

where in the last step we have taken (A.I.21) into account. One easily obtains (A.I.57) by comparing (A.I.58) to (A.I.65).

10. Myller-Lebedeff Formula

This formula is

$$\begin{aligned} &\exp\left[-\frac{x+x_0}{2}\right] (xx_0)^{n/2} a^{-n/2} \sum_{m=n}^{\infty} \frac{(m-n)!}{(m!)^3} L_m^{(n)}(x) L_m^{(n)}(x_0) a^m \\ &= \frac{1}{(1-a)} \exp\left[-\left(\frac{x+x_0}{2}\right)\left(\frac{1+a}{1-a}\right)\right] I_n\left(2\frac{\sqrt{xx_0 a}}{1-a}\right) \end{aligned} \tag{66}$$

Using (A.I.57) to express $L_m^{(n)}(x_0)$ in terms of Bessel functions, the left-hand side can be written

$$\begin{aligned} &\exp\left[-\frac{x-x_0}{2}\right] x^{n/2} a^{-n/2} \sum_{m=n}^{\infty} \frac{L_m^{(n)}(x)}{(m!)^2} a^m \\ &\qquad \times (-1)^n \int_0^\infty dt\, e^{-t} t^{m-(n/2)} J_n\{2(tx_0)^{1/2}\} \end{aligned} \tag{67}$$

or, taking account of (A.I.53) for $w = at$, to express the sum over the Laguerre functions, we obtain

$$\exp\left[-\frac{x+x_0}{2}\right](xx_0)^{n/2}a^{-n/2}\sum_{m=n}^{\infty}\frac{(m-n)!}{(m!)^3}L_m^{(n)}(x)\,L_m^{(n)}(x_0)\,a^m$$
$$=\exp\left[-\frac{x-x_0}{2}\right]\int_0^{\infty}dt\,J_n\{2(tx_0)^{1/2}\}J_n\{2(axt)^{1/2}\}e^{(a-1)t} \tag{68}$$

This integral is known. It is given by the formula (its proof is given in Watson, 1952, p. 395):

$$\int_0^{\infty}\exp(-p^2t^2)\,J_n(at)\,J_n(bt)\,t\,dt$$
$$=(1/2)\int_0^{\infty}\exp(-p^2u)\,J_n(a\sqrt{u})\,J_n(b\sqrt{u})\,du \tag{69}$$
$$=\frac{1}{2p^2}\exp\left[-\frac{a^2+b^2}{4p^2}\right]I_n\left(\frac{ab}{2p^2}\right)$$

Substituting in (A.I.68) we obtain the Myller-Lebedeff's formula (A.I.66).

11. Moments of the Energy Distribution Function for the Brownian Motion of a Normal Mode

This distribution function is [see (3.3.31) for $n=0$]

$$f_0(xt|x_0t=0)=e^{-x}\sum_{m=0}^{\infty}e^{-m\tau}\frac{1}{(m!)^2}L_m(x)\,L_m(x_0) \tag{70}$$

Its moments are [see (3.4.6)]

$$\overline{x^r}^{x_0}=\int_0^{\infty}dx\,x^r f_0(xt|x_0t=0)$$
$$=\sum_{m=0}^{\infty}\frac{1}{(m!)^2}e^{-m\tau}L_m(x_0)\int_0^{\infty}dx\,x^r\frac{d^m}{dx^m}(x^m e^{-x}) \tag{71}$$

where we have taken account of (A.I.7).

The integral on the right-hand side can easily be performed by integration by parts. All terms for which $m>r$ are equal to zero. Therefore

$$\overline{x^r}^{x_0}=\sum_{m=0}^{r}(-1)^m\frac{r!\,e^{-m\tau}}{(m!)^2(r-m)!}L_m(x_0)\int_0^{\infty}dx\,e^{-x}x^r \tag{72}$$

or, using (A.I.26)

$$\overline{x^r}^{x_0} = \sum_{m=0}^{r} (-1)^m \left(\frac{r!}{m!}\right)^2 \frac{e^{-m\tau}}{(r-m)!} L_m(x_0) \tag{73}$$

For $r = 1$ and $r = 2$, we obtain

$$\overline{x}^{x_0} = L_0(x_0) - e^{-\tau} L_1(x_0) = 1 - e^{-\tau} + x_0 e^{-\tau} \tag{74}$$

$$\begin{aligned} \overline{x^2}^{x_0} &= 2L_0(x_0) - 4e^{-\tau} L_1(x_0) + e^{-2\tau} L_2(x_0) \\ &= 2(1-e^{-\tau})^2 + 4x_0 e^{-\tau}(1-e^{-\tau}) + x_0^2 e^{-2\tau} \end{aligned} \tag{75}$$

The energy fluctuations are then given by

$$\begin{aligned} \overline{(x - \overline{x}^{x_0})^2}^{x_0} &= [\overline{x^2}^{x_0} - (\overline{x}^{x_0})^2] \\ &= (1-e^{-\tau})(1-e^{-\tau}+2x_0 e^{-\tau}) \end{aligned} \tag{76}$$

APPENDIX II

Derivation of the Master Equation by Time-Independent Perturbation Theory

As an example of the use of the time-independent perturbation theory (1.11.1), (1.11.2), we shall derive the master equation (4.2.3) for weakly coupled gases (P. Résibois and B. Leaf, 1960).

We have here [see (1.5.3) and (4.1.3)]

$$L\varphi_\mu = \lambda_\mu \varphi_\mu \tag{1}$$

$$L = L_0 + \lambda \delta L \tag{2}$$

$$L_0 = -i \sum_\alpha \mathbf{v}_\alpha \cdot \frac{\partial}{\partial \mathbf{x}_\alpha}, \qquad \delta L = i \sum_{\alpha \neq \beta} \frac{\partial V}{\partial \mathbf{x}_\alpha} \cdot \left(\frac{\partial}{\partial \mathbf{v}_\alpha} - \frac{\partial}{\partial \mathbf{v}_\beta} \right) \tag{3}$$

It is now convenient to use eigenfunctions in the *complete* phase space (and not only in coordinate space as in Chapter 1, § 5). For the unperturbed operator L_0 these eigenfunctions satisfy the equations

$$L_0 \varphi_\mu^0 = \lambda_\mu^0 \varphi_\mu^0 \tag{4}$$

In the case of a single particle the eigenfunctions and the eigenvalues are therefore given by [see (1.5.11), (1.5.6) and (1.5.9)]

$$\varphi_\mu^0 = \left(\frac{1}{L}\right)^{3/2} e^{i\mathbf{k}\cdot\mathbf{x}} \delta(\mathbf{v} - \mathbf{v}_0) \tag{5}$$

$$\lambda_\mu^0 = \mathbf{k} \cdot \mathbf{v}_0 \tag{6}$$

The single index μ corresponds, therefore, to the *two* "quantum numbers" $\mathbf{k}$ and $\mathbf{v}_0$. It is immediately clear that the eigenfunctions (A.II.5) form an orthonormal set

$$\left(\frac{1}{L}\right)^3 \int d\mathbf{x}\, d\mathbf{v}\, e^{-i\mathbf{k}\cdot\mathbf{x}} \delta(\mathbf{v} - \mathbf{v}_0)\, e^{i\mathbf{k}'\cdot\mathbf{x}} \delta(\mathbf{v} - \mathbf{v}_0') = \delta_{kk'}^{\mathrm{Kr}} \delta(\mathbf{v}_0 - \mathbf{v}_0') \tag{7}$$

For a finite volume L^3, the spectrum is discrete in the index $\mathbf{k}$, while it is always continuous in the index $\mathbf{v}_0$.

In the case of an arbitrary number of particles we have to take the product of the eigenfunctions (A.II.5) and the sum of the eigenvalues (A.II.6) [see (1.5.18) and (1.5.19)].

We now introduce the expansion (1.11.2) into (A.II.1) and obtain the set of equations

$$L_0 \varphi_\mu^0 = \lambda_\mu^0 \varphi_\mu^0$$
$$L_0 \varphi_\mu^{(1)} + \delta L \varphi_\mu^0 = \lambda_\mu^{(1)} \varphi_\mu^0 + \lambda_\mu^0 \varphi_\mu^{(1)} \tag{8}$$

$$L_0 \varphi_\mu^{(2)} + \delta L \varphi_\mu^{(1)} = \lambda_\mu^{(2)} \varphi_\mu^0 + \lambda_\mu^{(1)} \varphi_\mu^{(1)} + \lambda_\mu^0 \varphi_\mu^{(2)} \tag{9}$$

We will follow the standard procedure of quantum mechanics closely (see for example Bohm, 1951). We expand $\varphi_\mu^{(1)}$, $\varphi_\mu^{(2)}$, ... in the eigenfunctions of the unperturbed problem

$$\varphi_\mu^{(j)} = \sum_{\mu'} \gamma_{\mu\mu'}^{(j)} \varphi_{\mu'}^0 \tag{10}$$

and we substitute this in (A.II.8) and (A.II.9).

In this way, the first order equations (A.II.8) become

$$\sum_{\mu'} \gamma_{\mu\mu'}^{(1)} \lambda_{\mu'}^0 \varphi_{\mu'}^0 + \delta L \varphi_\mu^0 = \lambda_\mu^{(1)} \varphi_\mu^0 + \lambda_\mu^0 \sum_{\mu'} \gamma_{\mu\mu'}^{(1)} \varphi_{\mu'}^0 \tag{11}$$

We then multiply by φ_ν^{0*} and integrate over phase space.

Using (A.II.7) we obtain

$$\lambda_\nu^0 \gamma_{\mu\nu}^{(1)} + \langle \nu | \delta L | \mu \rangle = \lambda_\mu^{(1)} \delta_{\nu\mu} + \lambda_\mu^0 \gamma_{\mu\nu}^{(1)} \tag{12}$$

with

$$\delta_{\mu\nu} = \delta_{k_\mu k_\nu}^{\mathrm{Kr}} \delta(\mathbf{v}_\mu^0 - \mathbf{v}_\nu^0) \qquad \begin{array}{l} \mu \equiv \mathbf{k}_\mu, \mathbf{v}_\mu^0 \\ \nu \equiv \mathbf{k}_\nu, \mathbf{v}_\nu^0 \end{array} \tag{13}$$

Note that the matrix element

$$\langle \nu | \delta L | \mu \rangle = \int d\mathbf{v}\, d\mathbf{x}\, \varphi_\nu^{0*} \delta L \varphi_\mu^0 \tag{14}$$

is an ordinary number here (and not an operator) because we use a set of eigenfunctions in the *complete* phase space. The elements of (A.II.14) diagonal in $\mathbf{k}_\mu$ vanish [see (4.1.3) with $\mathbf{l} = 0$]

$$\langle \mathbf{k}_\mu \mathbf{v}_\mu^0 | \delta L | \mathbf{k}_\mu \mathbf{v}_\nu^0 \rangle = 0 \tag{15}$$

We obtain, from (A.II.12), for $\mu \neq \nu$,

$$\gamma_{\mu\nu}^{(1)} = \frac{1}{\lambda_\mu^0 - \lambda_\nu^0} \langle \nu | \delta L | \mu \rangle \tag{16}$$

As in quantum mechanics the normalization condition for $\varphi_\mu^{(1)}$ implies that $\gamma_{\mu\mu}^{(1)}$ is a purely imaginary factor (see Bohm, 1951, p.455)

$$\gamma_{\mu\mu}^{(1)} = -\gamma_{\mu\mu}^{(1)*} \tag{17}$$

which can be taken equal to zero.

In order to obtain $\lambda_\mu^{(1)}$ we have to take $\mu = \nu$ in (A.II.12). We are then however left with $\lambda_\mu^{(1)}\ \delta_{\mu\mu}$ which contains the infinite factor $\delta(0)$ [see (A.II.13)]. We therefore proceed in the following way: we first take $\boldsymbol{k}_\mu = \boldsymbol{k}_\nu$ [see (A.II.13)] and then integrate both sides of (A.II.12) over $\boldsymbol{v}_\nu^0$. This integral is taken over a small interval Δ containing $\boldsymbol{v}_\mu^0$. In this way we get away with the δ function in (A.II.13) and obtain

$$\lambda_\mu^{(1)} = \int_\Delta d\boldsymbol{v}_\nu^0 \langle \boldsymbol{k}_\mu \boldsymbol{v}_\nu^0 | \delta L | \boldsymbol{k}_\mu \boldsymbol{v}_\mu^0 \rangle \tag{18}$$

However, because of (A.II.15) the first-order correction to the eigenvalues vanish

$$\lambda_\mu^{(1)} = 0 \tag{19}$$

Let us now consider the second-order equation (A.II.9). We obtain in a similar way

$$\gamma_{\mu\nu}^{(2)} = \frac{1}{\lambda_\mu^0 - \lambda_\nu^0} \sum_{\mu'} \langle \nu | \delta L | \mu' \rangle \frac{1}{\lambda_\mu^0 - \lambda_{\mu'}^0} \langle \mu' | \delta L | \mu \rangle \qquad (\mu \neq \nu) \tag{20}$$

$$\lambda_\mu^{(2)} = \int_\Delta d\boldsymbol{v}_\nu^0 \sum_{\mu'} \langle \boldsymbol{k}_\mu \boldsymbol{v}_\mu^0 | \delta L | \mu' \rangle \frac{1}{\lambda_\mu^0 - \lambda_{\mu'}^0} \langle \mu' | \delta L | \boldsymbol{k}_\mu \boldsymbol{v}_\nu^0 \rangle \tag{21}$$

This expression for the second-order correction to the eigenvalue λ_μ^0 is closely related to the master equation operator as we shall see now.

Let us expand the phase density ρ in terms of the eigenfunctions φ_μ. According to equations (1.3.14) through (1.3.17) we have

$$\rho(p, q, t) = \sum_\mu a_\mu(t) \varphi_\mu(p, q) \tag{22}$$

$$a_\mu(t) = \int dp\, dq\, \varphi_\mu^*(p, q)\, \rho(p, q, t) \tag{23}$$

$$i \frac{da_\mu}{dt} = \lambda_\mu a_\mu \tag{24}$$

In integrated form

$$a_\mu(t) = e^{-i\lambda_\mu t} a_\mu(0) \tag{25}$$

We shall be interested in the evolution of the velocity distribution. Let us therefore take $\boldsymbol{k}_\mu = 0$ or $\mu = (0, \boldsymbol{v})$. Now using (1.11.2),

$$\lambda_{0,v} = \lambda^0_{0,v} + \lambda\lambda^{(1)}_{0,v} + \lambda^2 \lambda^{(2)}_{0,v} + \cdots \tag{26}$$

However $\lambda^0_{0,v}$ vanishes because of (A.II.6) and $\lambda^{(1)}_{0,v}$ because of (A.II.19). We have, therefore, using (A.II.21) and (A.II.14)

$$\lambda_{0,v} = -\lambda^2 \int_\Delta d\boldsymbol{v}^0_\nu \sum_{\mu'=(k',v')} \langle 0\, \boldsymbol{v} | \delta L | \boldsymbol{k}'\, \boldsymbol{v}' \rangle \frac{1}{\boldsymbol{k}' \cdot \boldsymbol{v}'} \langle \boldsymbol{k}' \cdot \boldsymbol{v}' | \delta L | 0\, \boldsymbol{v}^0_\nu \rangle \tag{27}$$

$$= \lambda^2 \int_\Delta d\boldsymbol{v}^0_\nu \int d\boldsymbol{v}' \sum_{k'} \left\langle 0 \left| \frac{\partial V}{\partial \boldsymbol{x}} \right| \boldsymbol{k}' \right\rangle \int d\boldsymbol{v}_1 \delta(\boldsymbol{v}_1 - \boldsymbol{v})$$
$$\times \frac{\partial}{\partial \boldsymbol{v}_1} \delta(\boldsymbol{v}_1 - \boldsymbol{v}') \frac{1}{\boldsymbol{k}' \cdot \boldsymbol{v}'} \left\langle \boldsymbol{k}' \left| \frac{\partial V}{\partial \boldsymbol{x}} \right| 0 \right\rangle \int d\boldsymbol{v}_2 \delta(\boldsymbol{v}_2 - \boldsymbol{v}')$$
$$\times \frac{\partial}{\partial \boldsymbol{v}_2} \delta(\boldsymbol{v}_2 - \boldsymbol{v}^0_\nu) \tag{28}$$

We may immediately perform the integrations over $\boldsymbol{v}_1$, $\boldsymbol{v}_2$, and $\boldsymbol{v}^0_\nu$. This expression then reduces to

$$\lambda_{0,v} = \lambda^2 \sum_{k'} \int d\boldsymbol{v}' \left\langle 0 \left| \frac{\partial V}{\partial \boldsymbol{x}} \right| \boldsymbol{k}' \right\rangle \frac{\partial}{\partial \boldsymbol{v}} \delta(\boldsymbol{v} - \boldsymbol{v}') \frac{1}{\boldsymbol{k}' \cdot \boldsymbol{v}'}$$
$$\times \left\langle \boldsymbol{k}' \left| \frac{\partial V}{\partial \boldsymbol{x}} \right| 0 \right\rangle \frac{\partial}{\partial \boldsymbol{v}'} \tag{29}$$
$$= \lambda^2 \sum_{k'} \left\langle 0 \left| \frac{\partial V}{\partial \boldsymbol{x}} \right| \boldsymbol{k}' \right\rangle \frac{\partial}{\partial \boldsymbol{v}} \frac{1}{\boldsymbol{k}' \cdot \boldsymbol{v}} \left\langle \boldsymbol{k}' \left| \frac{\partial V}{\partial \boldsymbol{x}} \right| 0 \right\rangle \frac{\partial}{\partial \boldsymbol{v}}$$

We see, therefore, that by performing the integration over the velocities the eigenvalue becomes an *operator*. Using a notation that explicitly takes account of the different particles present, we obtain (see Chapter IV, §§ 1 and 2)

$$\lambda_{0,v} = -i\lambda^2 \left(\frac{2\pi}{L}\right)^6 \sum_{jn} \sum_{k'} V_{k'}\, i\boldsymbol{k}' \cdot \left(\frac{\partial}{\partial \boldsymbol{v}_j} - \frac{\partial}{\partial \boldsymbol{v}_n}\right) \frac{1}{i\boldsymbol{k}' \cdot \boldsymbol{g}_{jn}}$$
$$\times V_{-k'}(-i\boldsymbol{k}') \cdot \left(\frac{\partial}{\partial \boldsymbol{v}_j} - \frac{\partial}{\partial \boldsymbol{v}_n}\right) \tag{30}$$

We have still to give a meaning to the singular denominator $i\boldsymbol{k}' \cdot \boldsymbol{g}_{jn}$ which appears here as well as in the time-dependent theory. We have seen that $1/i\boldsymbol{k}' \cdot \boldsymbol{g}_{jn}$ is the Fourier transform of the propagator $G(\boldsymbol{r}-\boldsymbol{r}', \boldsymbol{v})$ in ordinary space [see (6.2.1)]. It has to be understood as the limit [see (2.5.25) and (2.5.26)]

$$-\lim_{\varepsilon \to 0} \frac{1}{i(\boldsymbol{k}' \cdot \boldsymbol{g} + i\varepsilon)} = \pi \delta_+(\boldsymbol{k}' \cdot \boldsymbol{g}) \tag{31}$$

However as in Chapter 4 the principal part does not contribute to (A.II.30) and we are left precisely with the operator which appears in the master equation for weakly coupled gases

$$\lambda_{0,v} = i\lambda^2 \left(\frac{2\pi}{L}\right)^6 \sum_{jn} \sum_{k'} V_{k'} i\boldsymbol{k}' \cdot \left(\frac{\partial}{\partial \boldsymbol{v}_j} - \frac{\partial}{\partial \boldsymbol{v}_n}\right) \pi \delta(\boldsymbol{k}' \cdot \boldsymbol{g}_{jn})$$
$$\times V_{k'}(-i\boldsymbol{k}') \cdot \left(\frac{\partial}{\partial \boldsymbol{v}_j} - \frac{\partial}{\partial \boldsymbol{v}_n}\right) \tag{32}$$

To derive the master equation itself we observe that the coefficient $a_{0,v}$ in the expansion of the density in phase space in terms of the eigenfunctions φ_μ is given by [see (A.II.23)]

$$\begin{aligned} a_{0,v} &= \int d\boldsymbol{v}'\, d\boldsymbol{r}\, \varphi^*_{0,v}(\boldsymbol{v}', \boldsymbol{r})\, \rho(\boldsymbol{v}', \boldsymbol{r}, t) \\ &= \int d\boldsymbol{v}'\, d\boldsymbol{r}\, \varphi^{0*}_{0,v}(\boldsymbol{v}', \boldsymbol{r})\, \rho(\boldsymbol{v}', \boldsymbol{r}, t) + O(\lambda) \\ &= \left(\frac{1}{L}\right)^{3N/2} \int d\boldsymbol{r}\, \delta(\boldsymbol{v}' - \boldsymbol{v})\, \rho(\boldsymbol{v}', \boldsymbol{r}, t) + O(\lambda) \\ &= L^{3N/2} \rho_0(\boldsymbol{v}, t) + O(\lambda) \end{aligned} \tag{33}$$

In the usual limit $\lambda \to 0$, $t \to \infty$, $\lambda^2 t =$ finite, the equation (A.II.25) therefore reduces to the master equation in integrated form [see (4.2.3)]

$$\rho_0(\mathbf{v}, t) = \exp\left\{\lambda^2 t\left(\frac{2\pi}{L}\right)^6 \sum_{jn}\sum_{k'} V_{k'} i\mathbf{k}' \cdot \left(\frac{\partial}{\partial \mathbf{v}_j} - \frac{\partial}{\partial \mathbf{v}_n}\right)\right.$$
$$\left. \times\, \pi\delta(\mathbf{k}'\cdot\mathbf{g}_{jn}) V_{k'}(-i\mathbf{k}')\cdot\left(\frac{\partial}{\partial \mathbf{v}_j} - \frac{\partial}{\partial \mathbf{v}_n}\right)\right\} \rho_0(\mathbf{v}, 0) \tag{34}$$

It is interesting to stress the fact that the "dissipative form" of the master equation [the operator (A.II.32) is purely imaginary] is essentially a consequence of the interpretation of the propagator $1/i\mathbf{k}'\cdot\mathbf{g}'$ as a δ_+ function, in accordance with (A.II.31).

Now this interpretation is forced upon us by causality considerations (see Chapter 6, § 2). Therefore one may say that *irreversibility appears as a special aspect of the physical causality requirement*, which states that the distribution function at a given point is influenced only by the distribution function at points which corresponds to *earlier* times on the trajectory.

But the most unexpected aspect of this time-independent perturbation theory is that it shows the existence of analytic invariants. Indeed we have seen in Chapter 1, § 4 (see also Chapter 1, § 11) that each eigenfunction of the Liouville operator gives us an invariant of the motion. The existence of eigenfunctions which are power series in the coupling constant λ indicates, according to (1.11.3) the existence of invariants of the motion which are also expressible as power series in λ.

The significance of this result and its relation to Poincaré's theorem on analytic invariants and on ergodic theory is discussed in Chapter 14.

APPENDIX III

General Theory of Anharmonic Oscillators

In Chapters 7, 8, and 11 we considered the general theory of the approach to equilibrium of interacting particles. With minor modifications it applies as well to interacting normal modes (Prigogine and Henin, 1960).

The essential point in Chapter 7 was to introduce a class of initial conditions such that in the limit of a large system the evolution equations for reduced distribution functions no longer depend on the size of the system. In this appendix we shall discuss the corresponding conditions for the case of interacting normal modes.

A difference from the case considered in Chapter 7 is that here the ratio N/L^3 is automatically fixed by the lattice constant. Therefore, instead of (7.1.12) we now have to use the limit

$$N \to \infty \tag{1}$$

This limit simply corresponds to the elimination of surface effects. We may then again make a clear distinction between extensive and intensive quantities.

For example, in thermodynamic equilibrium we have for the correlation between the displacement of three particles from their *equilibrium* positions

$$\overline{u_n u_{n'} u_{n''}} \underset{N\to\infty}{\to} \text{independent of } N \tag{2}$$

and for the energy

$$\bar{E}/N \underset{N\to\infty}{\to} \text{independent of } N \tag{3}$$

We shall, as in the case of gases (see Chapter 7, § 1), again explicitly restrict our considerations to situations for which the distinction between extensive and intensive variables is maintained and for which (A.III.2) and (A.III.3) will still be valid. Th introduces precise statements about the N dependence of the Fourier coefficients of the distribution function, which we shall study now. Let us start with the Fourier expansion (1.8.5), but for convenience we shall include the complete time dependence in $\rho_{\{n\}}$. We shall first derive the conditions on the Fourier coefficients which express that a solid is homogeneous. We then have

$$\overline{u_{n_1} u_{n_2} \cdots u_{n_r}} = \overline{u_{n_1+p} u_{n_2+p} \cdots u_{n_r+p}} \tag{4}$$

This equation tells us that the average $\overline{u_{n_1} \cdots u_{n_r}}$ is invariant with respect to translation.

Let us use (2.1.5) and (2.1.19). We obtain[1]

$$\overline{u_{n_1} \cdots u_{n_r}} = \sum_{k_1 \cdots k_r} \frac{1}{(Nm)^{r/2}} e^{i(k_1 n_1 + \cdots + k_r n_r) a} \times \overline{[J_{k_1} \cdots J_{k_r}/\omega_{k_1} \cdots \omega_{k_r}]^{1/2} e^{i(\alpha_{k_1} + \cdots + \alpha_{k_r})}} \tag{5}$$

$$\overline{u_{n_1} \cdots u_{n_r}} = \sum_{k_1 \cdots k_r} \frac{1}{(Nm)^{r/2}} e^{i(k_1 n_1 + \cdots + k_r n_r) a} \times \int dJ\, \rho_{-k_1 \cdots -k_r} (J_{k_1} \cdots J_{k_r}/\omega_{k_1} \cdots \omega_{k_r})^{1/2} \tag{6}$$

The condition of invariance with respect to translation then implies

$$\rho_{\{n_i\}} = 0 \text{ except for } \pm k_1 \pm k_2 \pm k_3 \pm \cdots = 0 \quad \text{(modulo reciprocal vector)} \tag{7}$$

This condition is of course similar to (7.2.6). Here again, if once valid, (A.III.7) will persist in time because of the conservation of the wave vectors in each interaction [see (2.1.25), (2.1.27)].

We shall now consider the Fourier expansion (1.8.5) in the equilibrium case more closely. We then have

$$\rho^{\text{eq}}(J_1 \cdots J_N, \alpha_1 \cdots \alpha_N) = \frac{f(H_0 + \lambda V)}{\int dJ\, d\alpha\, f(H_0 + \lambda V)} \tag{8}$$

[1] As shown by (2.1.19) each normal coordinate when expressed in terms of action-angle variables gives two terms, one in k, the other in $-k$. To avoid expressions that are too cumbersome in (A.II.5) and (A.II.6) we have written only the contribution of the terms in k explicitly.

where $f(H_0+\lambda V)$ is an arbitrary function of the total Hamiltonian. We shall assume the normalization

$$\int dJ\, f(H_0) = 1 \tag{9}$$

Moreover

$$\int dJ\, d\alpha\, V f(H_0) = \int dJ\, f(H_0) \int d\alpha\, V = 0 \tag{10}$$

[see (2.1.26), note that the coefficient $V_{000} = 1/6(2Nm)^{-3/2} \times \sum_{nn'n''} B_{nn'n''}$ vanishes for the same reason as $\sum_{n'} A_{nn'}$ in (2.1.4)].

We then have

$$\rho^{\text{eq}} = \left[1+\lambda V \frac{\partial}{\partial H_0} + \frac{\lambda^2}{2} V^2 \frac{\partial^2}{\partial H_0^2} - \frac{\lambda^2}{2}\left(\int dJ\, d\alpha\, V^2 \frac{\partial^2 f(H_0)}{\partial H_0^2}\right) + \lambda^3 \cdots\right] f(H_0) \tag{11}$$

We see that the only non-vanishing Fourier coefficients will be of the form

$$\rho_{\pm 1_k \pm 1_{k'} \pm 1_{k''} \cdots \pm 1_m \pm 1_{m'} \pm 1_{m''}} \tag{12}$$

with

$$\pm k \pm k' \pm k'' = 0, \qquad \cdots, \qquad \pm m \pm m' \pm m'' = 0 \tag{13}$$

in agreement with the homogeneity condition (A.III.7).

Let us adopt the canonical distribution

$$f(H) = \frac{\prod_k \omega_k}{(kT)^N} e^{-(H_0+\lambda V)/kT} \tag{14}$$

We then obtain from (A.III.11)

$$\begin{aligned}\rho_{\{0\}}^{\text{eq}} = \frac{\prod_k \omega_k}{(kT)^N} \exp\left(-\frac{\sum_k \omega_k J_k}{kT}\right)\Bigg\{1 + \frac{\lambda^2}{(kT)^2} \\ \times \sum_{kk'k''} \frac{|V_{ll'l''}|^2 + 3|V_{ll'-l''}|^2}{\omega_l \omega_{l'} \omega_{l''}} \left[J_l J_{l'} J_{l''} - \frac{(kT)^3}{\omega_l \omega_{l'} \omega_{l''}}\right] + \lambda^4 \cdots\Bigg\}\end{aligned} \tag{15}$$

$$\rho^{eq}_{1_l 1_{l'} 1_{l''}} = \frac{\prod_k \omega_k}{(kT)^N} \exp\left(-\frac{\sum_k \omega_k J_k}{kT}\right) e^{i(\omega_l + \omega_{l'} + \omega_{l''})t} \times \left\{-\frac{\lambda}{kT} V_{ll'l''} \left(\frac{J_l J_{l'} J_{l''}}{\omega_l \omega_{l'} \omega_{l''}}\right)^{1/2} + \lambda^3 \cdots\right\} \tag{16}$$

$$\rho^{eq}_{1_k 1_{k'} 1_{k''} 1_l 1_{l'} 1_{l''}} = \frac{\prod_k \omega_k}{(kT)^N} \exp\left(-\frac{\sum_k \omega_k J_k}{kT}\right) e^{i(\omega_k + \cdots + \omega_{l''})t} \times \left\{\frac{\lambda^2}{(kT)^2} V_{kk'k''} V_{ll'l''} \left(\frac{J_k \cdots J_{l''}}{\omega_k \cdots \omega_{l''}}\right)^{1/2} + \lambda^4 \cdots\right\} \tag{17}$$

As could be expected the minimum order in λ of $\rho^{eq}_{1_k 1_{k'} 1_{k''} \cdots 1_m 1_{m'} 1_{m''}}$ corresponding to m groups of three phonons is λ^m. To specify the order in N we observe that [see (2.1.27), (2.1.24)]

$$V_{kk'k''} = O(N^{-1/2}) \tag{18}$$

therefore

$$\sum_{kk'k''} |V_{kk'k''}|^2 = N^2 \frac{1}{N} = O(N) \tag{19}$$

We see using (A.III.15) through (A.III.17) that the Fourier coefficient ρ^{eq}_{3m} corresponding to m groups of three phonons has the following N dependence.

$$\rho^{eq}_{3m} = \frac{\lambda^m}{N^{m/2}} (1 + \lambda^2 N + \lambda^4 N^2 + \cdots) \tag{20}$$

The equilibrium distribution when arranged according to powers of λ therefore takes the form

$$\rho^{eq} = 1 + \lambda N^{3/2} + \lambda^2 (N^3 + N) + \lambda^3 (N^{9/2} + N^{5/2}) + \cdots \tag{21}$$

The N dependence of the ρ^{eq}_{3m} is complicated. This is essentially due to the N dependence of the potential energy, which in turn originates in the transition from displacements and velocities of particles to normal coordinates. It is clear that the expansion (A.III.21) diverges, in the limit $N \to \infty$, but this divergence is spurious in the sense that if we use (A.III.20) or (A.III.21) to calculate averages of physical quantities like $\overline{u_{n_1} u_{n_2} u_{n_3}}$ or $\bar{E}$, the N

factors cancel completely for intensive variables or it remains exactly a *single* N factor as in the case of $\bar{E}$. Let us verify these statements to order λ^2 in the case of the energy. We have

$$\bar{E} = \int dJ\, d\alpha (H_0+\lambda V)\, f(H_0+\lambda V) \Big/ \int dJ\, d\alpha\, f(H_0+\lambda V)$$

$$= \int dJ\, H_0\, \rho_0^{\text{eq}} + \lambda \sum_{kk'k''} V_{kk'k''} \int dJ \left(\frac{J_k J_{k'} J_{k''}}{\omega_k\, \omega_{k'}\, \omega_{k''}}\right)^{1/2} \rho^{\text{eq}}_{-1_k-1_{k'}-1_{k''}} \tag{22}$$

Using (A.III.15) and (A.III.16) we find

$$\begin{aligned} \bar{E} &= NkT - \lambda^2 (kT)^2 \sum_{kk'k''} \frac{|V_{kk'k''}|^2 + 3|V_{kk'-k''}|^2}{\omega_k^2 \omega_{k'}^2 \omega_{k''}^2} \\ &= O(N) \end{aligned} \tag{23}$$

Similarly, to the lowest order in λ [see (A.III.2)], we have for example

$$\begin{aligned} \overline{u_{n_1} u_{n_2} u_{n_3}} &\sim \sum_{k_1 k_2 k_3} \frac{1}{N^{3/2}}\, \rho_{1_{k_1} 1_{k_2} 1_{k_3}} \sim N^2 \frac{1}{N^{3/2}} \frac{\lambda}{N^{1/2}} \\ &= \lambda O(N^0) \end{aligned} \tag{24}$$

The basic assumption we have to introduce in order to make the evolution equations for distribution functions of a finite number of normal modes independent of the size of the system, is that the Fourier coefficients already have, the *same N dependence as in equilibrium*. In this way, reduced distribution functions involving a finite number of normal modes exist in the limit $N \to \infty$, even out of equilibrium.

Once this restriction is introduced the general theory for anharmonic oscillators is essentially identical to that of Chapters 7, 8, and 11, taking into account the difference in the nature of the diagrams (compare Chapter 2, § 2 to Chapter 7, § 6). Therefore we shall not repeat it and refer for more detail to our original publication (Prigogine and Henin, 1960).

APPENDIX IV

Scattering by Randomly Distributed Centers and Van Hove's Diagonal Singularity Condition

In this monograph we are essentially concerned with the problem of the approach to equilibrium of systems with many degrees of freedom as the result of the purely mechanical interactions between these degrees of freedom.

A different and much simpler situation occurs if a random element is introduced into the Hamiltonian itself. In this case even a system which has only a few degrees of freedom may undergo an irreversible approach to equilibrium. An example is the scattering of a particle by randomly distributed centers. Such a situation occurs in the scattering of an electron in a solid by randomly distributed impurities or in the scattering of phonons as the effect of random isotopic substitution (see Kohn and Luttinger, **1957**, **1958**; George, **1959**; Kac, **1959**; Brocas and Résibois, **1961**).

It is interesting to compare the situation for the scattering problem to that of the general many-body problem in some detail (see Philippot, **1961**).

Let us consider this problem using the Liouville formalism developed in Chapter **6**. We may immediately apply equation (**6.1.3**), but we do not make any assumption about the homogeneity of the system. We then have by iteration [see (**6.1.4**)]

$$\rho_k(t) = \rho_k(0) + \lambda \left(\frac{2\pi}{L}\right)^3 \sum_l \int_0^t dt_1\, e^{i\mathbf{k}\cdot\mathbf{v}t_1} V_l\, i\mathbf{l}\cdot\frac{\partial}{\partial \mathbf{v}} e^{-i(\mathbf{k}-\mathbf{l})\cdot\mathbf{v}t_1} \rho_{k-l}(0)$$

$$+ \lambda^2 \left(\frac{2\pi}{L}\right)^6 \sum_{ll'} \int_0^t dt_1 \int_0^{t_1} dt_2\, e^{i\mathbf{k}\cdot\mathbf{v}t_1} V_l\, i\mathbf{l}\cdot\frac{\partial}{\partial \mathbf{v}} e^{-i(\mathbf{k}-\mathbf{l})\cdot\mathbf{v}(t_1-t_2)}$$

$$\times V_{l'}\, i\mathbf{l}'\cdot\frac{\partial}{\partial \mathbf{v}} e^{-i(\mathbf{k}-\mathbf{l}-\mathbf{l}')\cdot\mathbf{v}t_2} \rho_{k-l-l'}(0) + \frac{\lambda^3}{L^9}\cdots \tag{1}$$

Let us analyze these different contributions. We call the potential energy corresponding to the interaction of the particle with the nth scattering center $v(|\mathbf{x}-\mathbf{x}_n|)$. By our usual Fourier expansion [see (6.1.2)]

$$v(|\mathbf{x}-\mathbf{x}_n|) = \left(\frac{2\pi}{L}\right)^3 \sum_l v_l \, e^{i\mathbf{l}\cdot(\mathbf{x}-\mathbf{x}_n)} \tag{2}$$

The total potential energy is therefore

$$V(\mathbf{x}) = \sum_n v(|\mathbf{x}-\mathbf{x}_n|) = \left(\frac{2\pi}{L}\right)^3 \sum_l \sum_n v_l \, e^{-i\mathbf{l}\cdot\mathbf{x}_n} e^{i\mathbf{l}\cdot\mathbf{x}} \tag{3}$$

Comparing this with (6.1.2) we see that the Fourier coefficient V_l here is

$$V_l = \sum_n v_l \, e^{-i\mathbf{l}\cdot\mathbf{x}_n} \tag{4}$$

The essential point is that this Fourier coefficient is multiplied by a sum of randomly distributed phase factors.

As in Chapter 4 let us consider the case of weak forces and retain only powers of $(\lambda^2 t)$ in (A.IV.1).

In the first-order term we have the sum

$$\sum_l V_l = \sum_l \sum_n v_l \, e^{-i\mathbf{l}\cdot\mathbf{x}_n} \tag{5}$$

This sum contains

$$L^3 N \tag{6}$$

terms, but because of the random phase factor, its order of magnitude may be expected to be

$$(L^3 N)^{1/2} \tag{7}$$

This term, therefore, gives a contribution of order

$$\lambda \frac{(L^3 N)^{1/2}}{L^3} \sim \lambda \left(\frac{N}{L^3}\right)^{1/2} \tag{8}$$

where N/L^3 is the concentration of the scatterers. In the limit of weak coupling this contribution may be neglected.

Let us now consider the second-order terms. We first take $\boldsymbol{l}' = -\boldsymbol{l}$. We then have the sum

$$\begin{aligned}\sum_l V_l V_{-l} &= \sum_l |v_l|^2 \sum_n \sum_{n'} e^{-i\boldsymbol{l}\cdot(\boldsymbol{x}_n - \boldsymbol{x}_{n'})}\\ &= \sum_l |v_l|^2 \sum_n 1 + \sum_l |v_l|^2 \sum_n \sum_{n'\neq n} e^{-i\boldsymbol{l}\cdot(\boldsymbol{x}_n - \boldsymbol{x}_{n'})}\end{aligned} \tag{9}$$

The first term when introduced into (A.IV.1) gives, by our usual arguments,

$$\frac{\lambda^2 N}{L^6}\sum_l \int_0^t dt_1 \int_0^{t_1} dt_2 \cdots |v_l|^2 \cdots = \lambda^2 t \frac{N}{L^3} \tag{10}$$

This term is indeed of order $\lambda^2 t$ and has to be retained. It corresponds to a diagonal transition relating $\rho_k(t)$ to $\rho_k(0)$. On the contrary, the second term contains a rapidly oscillating function of $\boldsymbol{l}$. The asymptotic time integration (see Chapter 2, § 5) can no longer be performed. We therefore lose the factor of t of (A.IV.10). Moreover the N^2 oscillating terms in (A.IV.9) give a sum of order N. As a consequence we obtain a contribution of order

$$\lambda^2 \frac{N}{L^3} \tag{11}$$

which is again negligible for weakly coupled systems.

Finally for terms for which $\boldsymbol{l}' \neq -\boldsymbol{l}$, we have $(NL^3)^2$ oscillating terms in the sum

$$\sum_l \sum_{l'\neq l} \sum_n \sum_{n'} v_l v_{l'} e^{-i\boldsymbol{l}\cdot\boldsymbol{x}_n} e^{-i\boldsymbol{l}'\cdot\boldsymbol{x}_{n'}} \tag{12}$$

This sum will therefore be of order

$$NL^3 \tag{13}$$

and give a contribution of order

$$\lambda^2 \frac{N}{L^3} \tag{14}$$

in (A.IV.1). This may again be neglected.

This is a very interesting result. Indeed we see that the *sum* of all non-diagonal transitions $\rho_{k'}(0) \to \rho_k(t)$ may be neglected in comparison with the *single* diagonal transition $\rho_k(0) \to \rho_k(t)$. Of course this discussion may be extended to the higher-order terms in λ. We therefore obtain from (A.IV.1) evolution equations identical to the equations for weakly coupled systems which we studied in Chapters 4 or 5.

The essential point in the random scattering problem is that the sum

$$\sum_l V_l V_{-l} \tag{15}$$

corresponding to a diagonal transition in $\boldsymbol{k}$, gives a contribution of the same order in N and L^3 as

$$\sum_l \sum_{l' \neq -l} V_l V_{l'} \tag{16}$$

corresponding to the *sum* of all possible non-diagonal transitions from $\boldsymbol{k}'$ to $\boldsymbol{k}$ (with $\boldsymbol{k}' \neq \boldsymbol{k}$).

This may be considered as a special case of the diagonal singularity property discovered by Van Hove (1955). Its general formulation is the following: Let us consider the plane wave representation $\{\boldsymbol{k}\}$ and the matrix element

$$\langle \{\boldsymbol{k}\} | V A_1 V \cdots A_n V | \{\boldsymbol{k}'\} \rangle \tag{17}$$

where $A_1, \ldots, A_n$ are diagonal operators in the plane wave representation (one may take for example $A_1 = \cdots = A_n = 1$). We have at least two V factors in (A.IV.17). Van Hove has observed that for $\{\boldsymbol{k}\} = \{\boldsymbol{k}'\}$ such a matrix element is larger than for $\{\boldsymbol{k}\} \neq \{\boldsymbol{k}'\}$ by a factor which increases with the size of the system.

The scattering by randomly distributed centers provides us with a simple example. Indeed, as we have shown, the single term (A.IV.15), which may also be written

$$\langle \boldsymbol{k} | V^2 | \boldsymbol{k} \rangle$$

is of the same order as the sum (A.IV.16)

$$\sum_{\boldsymbol{k}' \neq \boldsymbol{k}} \langle \boldsymbol{k} | V^2 | \boldsymbol{k}' \rangle$$

Considered as a function of $\boldsymbol{k}'$ the matrix element $\langle \boldsymbol{k}|V^2|\boldsymbol{k}'\rangle$ has a δ singularity for $\boldsymbol{k}' = \boldsymbol{k}$.[1] Such a situation may be expected to hold for wide classes of problems in which a random element is incorporated into the Hamiltonian. Let us now consider the general N-body problem studied in this monograph. For example, in the case of gases, we know that the diagonal transition corresponding to the cycle (see Chapter 7, § 7)

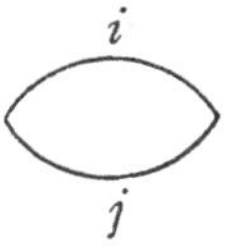

Fig. A. IV.1.

is of order

$$N^2/L^3 = N(N/L^3) \tag{18}$$

while a *given* non-diagonal transition

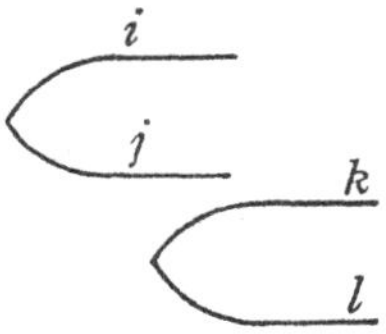

Fig. A.IV.2.

is of order

$$1/L^6 \tag{19}$$

The ratio of these two expressions is

$$N^2 L^3 = N^3 (1/C) \tag{20}$$

which increases with the size of the system. This again expresses Van Hove's diagonal singularity condition.

However the sum of all possible non-diagonal transitions of this type is

$$N^4/L^6 = N^2(N/L^3)^2 \tag{21}$$

[1] This means that the contribution of the single point $\boldsymbol{k}' = \boldsymbol{k}$ is of the same order as the contribution of the integral over a finite range of $\boldsymbol{k}'$.

and therefore larger than (A.IV.18) by a factor N. One may say that the diagonal singularity is "weaker" than in the case of random scattering that we studied before. As a consequence, the large value of the diagonal element, by itself, no longer enables us to neglect all non-diagonal transitions.

Van Hove's procedure (1955, 1956, 1957) is to supplement the diagonal singularity property by the assumption of *random phases* in the plane wave representation. In the quantum mechanical case this permits us to consider only the diagonal elements $\langle \mathbf{p}|\rho|\mathbf{p}\rangle$ of the statistical density matrix. In the case of classical mechanics, the random phase approximation corresponds to neglect all Fourier coefficients of $\rho(\mathbf{x}_1 \cdots \mathbf{x}_N, \mathbf{p}_1 \cdots \mathbf{p}_N)$ except $\rho_{\{0\}}$.

This is precisely what we have done in the case of weakly coupled systems (see especially Chapters 2 and 4) However as has been shown by Philippot (1961) the condition of random phases cannot be used for higher order approximations. Indeed the potential energy depends on the off diagonal elements of the Von Neumann density matrix. The use of the random phase approximation would lead to an infinite average energy for hard spheres. In our approach the condition of random phases is replaced by assuming the existence of a singularity in the Fourier transform of the phase distribution function.

Bibliography

F. C. Andrews, (a) *Bull. classe sci., Acad. roy. Belg.* **46**, 475 (1960); (b) Thesis, Harvard University, 1960.

F. C. Andrews, *Physica* **27**, 1054 (1961).

P. Auer and S. Tamor, private communication, (1957).

T. A. Bak, *Contributions to the Theory of Chemical Kinetics*, Munksgaard, Copenhagen (1959).

T. A. Bak and K. Anderson, *Kgl. Danske Videnskab. Selskab, Mat.-fys. Medd.* **33**, No. 7 (1961).

T. A. Bak, F. Henin, and M. Goche, *J. Mol. Phys.* **2**, 181 (1959).

R. Balescu, *Phys. of Fluids* **3**, 62 (1960).

R. Balescu, *Phys. of Fluids* **4**, 94 (1961).

R. Balescu, *Physica,* **27**, 693 (1961).

R. Balescu, *Statistical Mechanics of Charged Particles*, Interscience, New York, to appear (1962).

R. Balescu and Ph. de Gottal, *Bull. classe Sci., Acad. roy. Belg.* **47**, 245 (1961).

R. Balescu and H. S. Taylor, *Phys. of Fluids,* **4**, 85 (1961).

G. K. Batchelor, *The Theory of Homogeneous Turbulence,* Cambridge University Press, Cambridge (1956).

J. de Boer, *Repts. Progr. Phys.* **12**, 305 (1948–1949).

N. N. Bogolioubov, *Problemi Dynamitcheskij Theorie v Statistitcheskey Phisike,* OGIS, Moscow, (1946). English translation by E. K. Gora, *Problems of a Dynamical Theory in Statistical Physics,* ASTIA document No. AD-213317.

D. Bohm, *Quantum Theory,* Prentice-Hall, Englewood Cliffs, New Jersey (1951).

D. Bohm, *The Many Body Problem,* lectures given at the Ecole d'Eté des Houches, Dunod, Paris (1958).

D. Bohm and D. Pines, *Phys. Rev.* **92**, 608 (1953).

L. Boltzmann, *Vorlesungen über Gastheorie, Leipzig* (1896—1898); *Wien. Ber.* **66**, 275, (1872).

M. Born and H. S. Green, *Proc. Roy. Soc.* (*London*), **A 188**, 10 (1946); **A 189**, 103 (1947); **A 190**, 455 (1947); **A 191**, 168 (1947).
The above papers are collected in *A General Kinetic Theory of Liquids,* Cambridge University Press, Cambridge (1949).

J. Brocas and P. Résibois, *Bull. classe Sci., Acad. roy. Belg.* **47**, 226 (1961).

R. Brout, *Physica* **22**, 509 (1956).

R. Brout, *Phys. Rev.* **108,** 515 (1957).

R. Brout and I. Prigogine, *Physica* **22,** 621 (1956).

K. A. Brueckner, *The Many Body Problem*, lectures given at the Ecole d'Eté des Houches, Dunod, Paris (1958).
S. Chandrasekhar, *Principles of Stellar Dynamics*, University of Chicago Press, Chicago (1942).
S. Chandrasekhar, *Astrophys. J.* **97**, 255 (1943).
S. Chandrasekhar, *Revs. Modern Phys.* **15**, 1 (1943).
S. Chapman and T. G. Cowling, *The Mathematical Theory of Non-Uniform Gases*, Cambridge. University Press, Cambridge (1939).
R. Clausius, *Ann. Physik* **100**, 353 (1857).
P. A. M. Dirac, *The Principles of Quantum Mechanics* (third edition), Clarendon Press, Oxford (1947).
H. Dreicer, *Proc. U.N. Intern. Conf. on the Peaceful Uses of Atomic Energy*, Vol. **31**, p. 57. Geneva (1958).
G. Doetsch, *Laplace Transformation*, Dover, New York, (1943).
H. B. Dwight, *Tables of Integrals*, Macmillan, New York (1947).
P. and T. Ehrenfest, "Begriffliche Grundlagen der Statistischen Auffassung der Mechanik," *Encyclopedie der Mathematischen Wissenschaften*, Vol. **4**, p. 4, (1911); reprinted by Cornell University Press, Ithaca, N. Y. (1959).
H. Falkenhagen, *Elektrolyte*, S. Hirzel, Leipzig (1932).
U. Fano, *Revs. Modern Phys.* **29**, 74 (1957).
E. Fermi, *Physik. Z.* **24**, 261 (1923).
R. P. Feynman, *Phys. Rev.* **76**, 749 (1949).
R. Fowler and E. A. Guggenheim, *Statistical Thermodynamics*, Cambridge, University Press, Cambridge (1939).
S. Fujita, *Introduction to the Many-Body Problem in Quantum Statistics*, Interscience, New York, to appear, (1962).
S. Fujita, *Physica* **28**, 281 (1962).
M. Gell-Mann and K. A. Brueckner, *Phys. Rev.* **106**, 364 (1957).
C. George, *Bull. classe sci. Acad. roy. Belg.* **45**, 239 (1959).
C. George, *Physica*, **26**, 453 (1960).
J. W. Gibbs, *The Collected Works*, Vol. 2, Longmans, Green & Co, London, New York (1928).
H. Goldstein, *Classical Mechanics*, Addison-Wesley, Reading, Mass. (1953).
H. Grad, *Handbuch der Physik*, Vol. XII, Springer-Verlag, Berlin (1958)
M. S. Green, *Proceedings of the International Symposium on Transport Processes in Statistical Mechanics, Brussels*, (1956), Interscience, New York (1958).
P. C. Hemmer, L. C. Maximon, and H. Wergeland, *Phys. Rev.* **111**, 689 (1958).
F. Henin, *Approach to Equilibrium in One-Dimensional Systems*, unpublished (1961).
F. Henin, P. Résibois, and F. C. Andrews, *J. Math. Phys.* **2**, 68 (1961).
J. Higgins, *Bull. classe sci., Acad. roy. Belg.* **46**, 385 (1960).

T. L. Hill, *Statistical Mechanics*, McGraw-Hill, New York (1956).

J. O. Hirschfelder, C. F. Curtiss, and R. B. Bird, *The Molecular Theory of Gases and Liquids*, Wiley, New York (1954).

E. Hopf, *Ergodentheorie*, Chelsea Publishing Company, New York, (1948).

N. M. Hugenholtz, *The Many Body Problem*, lectures given at the Ecole d' Eté des Houches, Dunod, Paris (1958).

J. H. Jeans, *Dynamical Theory of Gases*, Cambridge. University Press, Cambridge (1921).

M. Kac, *Foundations of Kinetic Theory, Proceedings of the Berkeley Symposium on Mathematical Statistics*, (1954).

M. Kac, *Probability and Related Topics in Physical Sciences*, Interscience, New York (1959).

J. G. Kirkwood, *J. Chem. Phys.* **14**, 180 (1946); **15**, 72 (1947).

J. G. Kirkwood and J. Ross, *Proceedings of the International Symposium on Transport Processes in Statistical Mechanics*, Brussels (1956), Interscience, New York (1958).

C. Kittel, *Introduction to Solid State Physics*, Wiley, New York (1953).

A. I. Khinchin, *Mathematical Foundations of Statistical Mechanics*, Dover, New York (1949).

G. Klein and I. Prigogine, *Physica* **19**, 1053 (1953).

W. Kohn and M. Luttinger, *Phys. Rev.* **108**, 590 (1957); **109**, 1892 (1958).

H. A. Kramers, *Physica* **7**, 284 (1940).

A. Kronig, *Ann. Physik.* **99**, 315 (1856).

R. Kubo "*Some Aspects of the Statistical Mechanical Theory of Irreversible Processes*," *Lectures in Theoretical Physics*, Vol. I, W. E. Brittin and L. G. Dunham, eds., Interscience, New York (1959).

L. Landau, *Physik. Z.* Sowjet Union **10**, 154 (1936).

L. Landau and E. Lifshitz, *Statistical Physics*, Pergamon Press, London (1958).

B. Leaf, unpublished, (1959).

A. Lenard, *Ann. Physics* **10**, 390 (1962).

J. Light, *J. Chem. Phys.* **36**, 1016 (1962).

M. J. Lighthill, *Introduction to Fourier Analysis and Generalized Functions*, Cambridge University Press, Cambridge (1958).

B. Lippman and J. Schwinger, *Phys. Rev.* **79**, 469 (1949).

J. Loschmidt, *Wien Ber.* **73**, 139 (1876).

H. Margenau and G. M. Murphy, *The Mathematics of Physics & Chemistry*, Van Nostrand, New York (1953).

D. Massignon, *Mécanique Statistique des Fluides*, Dunod, Paris (1957).

J. C. Maxwell, *Phil. Mag.* **19**, 19 (1860).

J. E. Mayer, *J. Chem. Phys.* **5**, 67 (1937).

J. E. Mayer, *J. Chem. Phys.* **18**, 1426 (1950).

J. E. Mayer and M. G. Mayer, *Statistical Mechanics*, Wiley, New York (1940).

P. Mazur, *Physica* **25**, 149 (1959).

P. Mazur and E. W. Montroll, *J. Math. Phys.* **1**, 70, (1960).
E. Meeron, *J. Chem. Phys.* **28**, 630 (1958); *Phys. of Fluids* **1**, 139 (1958).
S. G. Mikhlin, *Integral Equations*, Pergamon Press, London (1957).
E. W. Montroll and J. E. Mayer, *J. Chem. Phys.* **9**, 626 (1941).
E. W. Montroll and M. S. Green, *Ann. Rev. Phys. Chem.* **5**, 449 (1954).
E. W. Montroll and J. C. Ward, *Phys. of Fluids* **1**, 55 (1958).
A. Münster, *Z. Naturforsch.* **6a**, 139 (1951); **7a**, 613 (1952).
A. Münster, *Statistische Thermodynamik*, Springer, Berlin (1958).
N. I. Muskelishwili, *Singular Integral Equations*, Noordhoff, Groningen (1953).
W. Pauli, *Festschrift zum 60. Geburtstage A. Sommerfelds*, Hirzel, Leipzig (1928).
R. E. Peierls, *Ann. Physik* **3**, 1055, (1929).
R. E. Peierls, *The Quantum Theory of Solids*, Clarendon Press, Oxford (1955).
R. E. Peierls, *Selected Topics in Nuclear Theory, Lectures in Theoretical Physics*, Vol. 1, W. E. Brittin and L. G. Dunham, eds, Interscience, New York (1959).
J. Philippot, *Physica*, **27**, 490 (1961).
H. Poincaré, *Méthodes Nouvelles de la Mécanique Céleste*, Gauthier-Villars (1892); reprinted by Dover, New York (1957).
I. Prigogine, ed., *Proceedings of the International Symposium on Transport Processes in Statistical Mechanics*, Brussels (1956), Interscience, New York (1958).
I. Prigogine, *Bull. classe sci., Acad. roy. Belg.* **44**, 106 (1958).
I. Prigogine, *Superfluidité et Equation de Transport Quantique*, Rédigé par P. Résibois, Institut des Sciences Nucléaires, Monographie No. 6, Bruxelles (1959).
I. Prigogine, unpublished (1959).
I. Prigogine and T. A. Bak, *J. Chem. Phys.* **31**, 1368 (1959).
I. Prigogine and R. Balescu, *Physica* **23**, 555 (1957).
I. Prigogine and R. Balescu, *Physica* **25**, 281 (1959); **25**, 302, (1959).
I. Prigogine and R. Balescu, *Physica* **26**, 145, (1960).
I. Prigogine, R. Balescu, F. Henin and P. Résibois, Proceedings of the Congress on Many Particle Problems, *Physica* **26**, 36 (1960).
I. Prigogine, R. Balescu, and I. M. Krieger, *Physica* **26**, 529 (1960).
I. Prigogine and R. Defay, *Chemical Thermodynamics*, translated by D. Everett, Longmans Green & Co., London, New York (1954).
I. Prigogine and F. Henin, *Bull. classe. sci., Acad. roy. Belg.* **43**, 814 (1957).
I. Prigogine and F. Henin, *Physica* **24**, 214 (1958).
I. Prigogine and F. Henin, *J. Math. Phys.* **1**, 349 (1960).
I. Prigogine and B. Leaf, unpublished (1959).
I. Prigogine and B. Leaf, *Physica* **25**, 1067 (1959).
I. Prigogine and S. Ono, *Physica* **25**, 771 (1959).
I. Prigogine and J. Philippot, *Physica* **23**, 569 (1957).

I. Prigogine and R. Résibois, *Physica* **24**, 705 (1958).

I. Prigogine and P. Résibois, *Physica* **27**, 629 (1961).

P. Résibois, *Physica* **25**, 725 (1959).

P. Résibois, *Contribution à la Théorie Formelle du Scattering*, Thèse, Bruxelles (1960).

P. Résibois, *Phys. Rev. Letters* **5**, 411 (1961).

P. Résibois and B. Leaf, *Bull. classe sci., Acad. roy. Belg.* **46**, 148 (1960).

P. Résibois and I. Prigogine, *Bull. classe sci., Acad. roy. Belg.* **46**, 53 (1960).

P. Résibois, *Physica* **27**, 541 (1961).

S. Rice and H. L. Frisch, *Ann. Rev. Phys. Chem.* **11**, 187 (1960).

M. N. Rosenbluth, W. M. Mac Donald, and D. L. Judd, *Phys. Rev.* **107**, 1 (1957).

K. Sawada, *Phys. Rev.* **106**, 372 (1957).

K. Sawada, K. A. Brueckner, N. Fukuda and R. Brout, *Phys. Rev.* **108**, 507 (1957).

S. S. Schweber, H. A. Bethe, and F. de Hoffmann, *Mesons and Fields*, Vol. 1, *Fields*, Row, Peterson & Company, Evanston, Ill. (1956).

I. N. Sneddon, *Special Functions of Mathematical Physics and Chemistry*, Oliver & Boyd, Edinburgh (1956).

L. Spitzer, Jr., *Physics of Fully Ionized Gases*, Interscience, New York (1956).

M. H. Stone, *Linear Transformations in Hilbert Space and their Applications to Analysis*, American Mathematical Society, New York (1932).

D. ter Haar, *Elements of Statistical Mechanics*, Rinehart, New York (1954).

R. C. Tolman, *The Principles of Statistical Mechanics*, Clarendon Press, Oxford (1938).

G. E. Uhlenbeck and B. Kahn, *Physica* **5**, 399 (1938).

G. E. Uhlenbeck, Supplement to M. Kac, *Probability and Related Topics in Physical Sciences*, Interscience, New York (1959).

H. D. Ursell, *Proc. Cambridge Phil. Soc.* **23**, 685 (1928).

L. Van Hove, *Physica* **21**, 512 (1955).

L. Van Hove, *Physica* **22**, 343 (1956).

L. Van Hove, *Physica* **23**, 441 (1957).

J. M. J. Van Leeuwen, J. Groeneveld, and J. de Boer, *Physica* **25**, 792 (1959).

M. C. Wang and G. E. Uhlenbeck, *Revs. Modern Phys.* **17**, 323 (1945).

G. N. Watson, *A Treatise on the Theory of Bessel Functions*, Cambridge University Press, Cambridge (1952).

N. Wax, ed. *Selected Papers on Noise and Stochastic Processes*, Dover, New York (1954).

J. Yvon, *Fluctuations en Densité*, Hermann, Paris (1937).

J. Yvon, *La Théorie Statistique des Fluides et l'Equation d' Etat*, Hermann, Paris (1935).

E. Zermelo, *Ann. Physik* **57**, 485 (1896); **59**, 793 (1896).

LIST OF SYMBOLS

Latin

a	Distance between first neighbours in solid
C	Concentration or number density (4.3.2)
$C_{\gamma\gamma'}(z)$	Contribution to the resolvent of a creation fragment transforming a correlation γ' to a correlation γ
$\boldsymbol{D}_{jn} = \partial/\partial \mathbf{v}_j - \partial/\partial \mathbf{v}_n$	or $\boldsymbol{D}$, differential operator acting on velocities (6.5.2)
E	Energy
e	Electrical charge
$F(1, 2 \ldots s)$	s-particle distribution function (7.4.3)
f_s	Distribution function for coordinates and momenta of s particles (7.1.5)
$f_{s,r}$	Distribution function of s positions, and r momenta (7.1.9)
$f(x, \alpha, t \| x_0, \alpha_0, 0)$	Fundamental solution or Green's function (3.3.23)
G	Propagator (6.2.1)
$\mathbf{g}$	Relative velocity
H	Hamiltonian (1.1.1)
h	Planck's constant
$\hbar$	Planck's constant divided by 2π
$I_n(x)$	Bessel functions of purely imaginary argument (A.I.39)
J	Action variable (1.6.4)
$J_n(x)$	Bessel functions (A.I.38)
k	Wave vector
$\boldsymbol{k}$	Boltzmann's constant
L	Geometrical length
L	Liouville operator (1.3.1)
δL	Part of the Liouville operator due to interactions (1.8.2)

L_0	Unperturbed Liouville operator (1.8.2)
L_h	Hydrodynamical length
$L_m(x)$	Laguerre polynomials (A.I.6)
$L_m^{(k)}(x)$	Associated Laguerre polynomials (A.I.9)
$\mathbf{l}$	Wave vector
m	Mass
N	Number of particles
n_s	Distribution function for coordinates of s particles (7.1.8)
P_i	Momenta corresponding to cyclic coordinates (1.6.5)
$\mathscr{P}(1/x)$	Principal part of $1/x$ (2.5.9)
p_i	Momenta (1.1.1)
Q_i	Cyclic coordinates (1.6.5)
q_i	Coordinates (1.1.1)
q_k	Normal coordinates (2.1.5)
$R(z)$	Resolvent operator (8.3.7)
T	Temperature in °K
t	Time
t_r	Relaxation time
t_h	Hydrodynamical time (10.1.3)
t_{int}	Duration of an interaction
$u(1, 2 \ldots s)$	s-particle correlation (7.4.5)
V	Potential energy
V_l or $V(l)$	Fourier coefficient of intermolecular potential (4.1.2)
$\mathbf{v}$	Velocity
W	Generating function (1.6.6)
x	Coordinates
α	Angle variable (1.6.3)
β	Friction coefficient (3.5.3)
$\beta_n(1, 2)$	Cluster integral (12.6.3)
$\delta(x)$	Dirac's delta function (2.5.27)
$\delta_+(x)$	Delta plus function (2.5.10)
$\delta_-(x)$	Delta minus function (2.5.25)
$\delta_\alpha^{\mathrm{Kr}}$ or $\delta^{\mathrm{Kr}}(\alpha)$	Kronecker's delta function
η	Dynamical friction (4.6.6)

κ	Inverse of the Debye length (4.7.6)
λ	Strength of interaction (1.8.1)
ν	Frequency
ρ	Density in phase space (1.2.1)
ρ	Statistical operator or density operator
ρ_{mn}	Density matrix
ρ_0 or $\rho_{\{0\}}$	Momentum (or velocity) distribution function
ρ_γ	Correlation involving γ particles
ρ'_γ	Part of the correlations involving γ particles (11.3.1)
ρ''_γ	Part of the correlations involving γ particles (11.3.1)
σ	Cross section (6.3.12)
$\Phi(u)$	Error function (4.5.9)
φ	Velocity distribution function for single molecule
φ_s	Velocity distribution function for s molecules (7.1.7)
Ψ	Wave function
$\Psi(z)$	Contribution of a diagonal fragment to the resolvent (8.5.1)
$\Psi^+(z)$	For Im $z > 0$, identical to $\Psi(z)$ For Im $z < 0$, defined as the analytic continuation of the function $\Psi(z)$, see (8.4.11)
$\Psi^{+(i)}(0)$	$(d^i\,\Psi^+/dz^i)_{z=0}$
Ω	Solid angle (6.3.12)
Ω	Operator (11.2.72)
$\Omega = L^3/(2\pi)^3$	Volume divided by $(2\pi)^3$
Ω_γ	Operator (11.3.3')
ω	$2\pi\nu$

SUBJECT INDEX

A CATALOG OF SELECTED

DOVER BOOKS

IN SCIENCE AND MATHEMATICS

Astronomy

CHARIOTS FOR APOLLO: The NASA History of Manned Lunar Spacecraft to 1969, Courtney G. Brooks, James M. Grimwood, and Loyd S. Swenson, Jr. This illustrated history by a trio of experts is the definitive reference on the Apollo spacecraft and lunar modules. It traces the vehicles' design, development, and operation in space. More than 100 photographs and illustrations. 576pp. 6 3/4 x 9 1/4. 0-486-46756-2

EXPLORING THE MOON THROUGH BINOCULARS AND SMALL TELESCOPES, Ernest H. Cherrington, Jr. Informative, profusely illustrated guide to locating and identifying craters, rills, seas, mountains, other lunar features. Newly revised and updated with special section of new photos. Over 100 photos and diagrams. 240pp. 8 1/4 x 11. 0-486-24491-1

WHERE NO MAN HAS GONE BEFORE: A History of NASA's Apollo Lunar Expeditions, William David Compton. Introduction by Paul Dickson. This official NASA history traces behind-the-scenes conflicts and cooperation between scientists and engineers. The first half concerns preparations for the Moon landings, and the second half documents the flights that followed Apollo 11. 1989 edition. 432pp. 7 x 10. 0-486-47888-2

APOLLO EXPEDITIONS TO THE MOON: The NASA History, Edited by Edgar M. Cortright. Official NASA publication marks the 40th anniversary of the first lunar landing and features essays by project participants recalling engineering and administrative challenges. Accessible, jargon-free accounts, highlighted by numerous illustrations. 336pp. 8 3/8 x 10 7/8. 0-486-47175-6

ON MARS: Exploration of the Red Planet, 1958-1978--The NASA History, Edward Clinton Ezell and Linda Neuman Ezell. NASA's official history chronicles the start of our explorations of our planetary neighbor. It recounts cooperation among government, industry, and academia, and it features dozens of photos from Viking cameras. 560pp. 6 3/4 x 9 1/4. 0-486-46757-0

ARISTARCHUS OF SAMOS: The Ancient Copernicus, Sir Thomas Heath. Heath's history of astronomy ranges from Homer and Hesiod to Aristarchus and includes quotes from numerous thinkers, compilers, and scholasticists from Thales and Anaximander through Pythagoras, Plato, Aristotle, and Heraclides. 34 figures. 448pp. 5 3/8 x 8 1/2. 0-486-43886-4

AN INTRODUCTION TO CELESTIAL MECHANICS, Forest Ray Moulton. Classic text still unsurpassed in presentation of fundamental principles. Covers rectilinear motion, central forces, problems of two and three bodies, much more. Includes over 200 problems, some with answers. 437pp. 5 3/8 x 8 1/2. 0-486-64687-4

BEYOND THE ATMOSPHERE: Early Years of Space Science, Homer E. Newell. This exciting survey is the work of a top NASA administrator who chronicles technological advances, the relationship of space science to general science, and the space program's social, political, and economic contexts. 528pp. 6 3/4 x 9 1/4. 0-486-47464-X

STAR LORE: Myths, Legends, and Facts, William Tyler Olcott. Captivating retellings of the origins and histories of ancient star groups include Pegasus, Ursa Major, Pleiades, signs of the zodiac, and other constellations. "Classic." – *Sky & Telescope.* 58 illustrations. 544pp. 5 3/8 x 8 1/2. 0-486-43581-4

A COMPLETE MANUAL OF AMATEUR ASTRONOMY: Tools and Techniques for Astronomical Observations, P. Clay Sherrod with Thomas L. Koed. Concise, highly readable book discusses the selection, set-up, and maintenance of a telescope; amateur studies of the sun; lunar topography and occultations; and more. 124 figures. 26 halftones. 37 tables. 335pp. 6 1/2 x 9 1/4. 0-486-42820-6

Chemistry

MOLECULAR COLLISION THEORY, M. S. Child. This high-level monograph offers an analytical treatment of classical scattering by a central force, quantum scattering by a central force, elastic scattering phase shifts, and semi-classical elastic scattering. 1974 edition. 310pp. 5 3/8 x 8 1/2. 0-486-69437-2

HANDBOOK OF COMPUTATIONAL QUANTUM CHEMISTRY, David B. Cook. This comprehensive text provides upper-level undergraduates and graduate students with an accessible introduction to the implementation of quantum ideas in molecular modeling, exploring practical applications alongside theoretical explanations. 1998 edition. 832pp. 5 3/8 x 8 1/2. 0-486-44307-8

RADIOACTIVE SUBSTANCES, Marie Curie. The celebrated scientist's thesis, which directly preceded her 1903 Nobel Prize, discusses establishing atomic character of radioactivity; extraction from pitchblende of polonium and radium; isolation of pure radium chloride; more. 96pp. 5 3/8 x 8 1/2. 0-486-42550-9

CHEMICAL MAGIC, Leonard A. Ford. Classic guide provides intriguing entertainment while elucidating sound scientific principles, with more than 100 unusual stunts: cold fire, dust explosions, a nylon rope trick, a disappearing beaker, much more. 128pp. 5 3/8 x 8 1/2. 0-486-67628-5

ALCHEMY, E. J. Holmyard. Classic study by noted authority covers 2,000 years of alchemical history: religious, mystical overtones; apparatus; signs, symbols, and secret terms; advent of scientific method, much more. Illustrated. 320pp. 5 3/8 x 8 1/2. 0-486-26298-7

CHEMICAL KINETICS AND REACTION DYNAMICS, Paul L. Houston. This text teaches the principles underlying modern chemical kinetics in a clear, direct fashion, using several examples to enhance basic understanding. Solutions to selected problems. 2001 edition. 352pp. 8 3/8 x 11. 0-486-45334-0

PROBLEMS AND SOLUTIONS IN QUANTUM CHEMISTRY AND PHYSICS, Charles S. Johnson and Lee G. Pedersen. Unusually varied problems, with detailed solutions, cover of quantum mechanics, wave mechanics, angular momentum, molecular spectroscopy, scattering theory, more. 280 problems, plus 139 supplementary exercises. 430pp. 6 1/2 x 9 1/4. 0-486-65236-X

ELEMENTS OF CHEMISTRY, Antoine Lavoisier. Monumental classic by the founder of modern chemistry features first explicit statement of law of conservation of matter in chemical change, and more. Facsimile reprint of original (1790) Kerr translation. 539pp. 5 3/8 x 8 1/2. 0-486-64624-6

MAGNETISM AND TRANSITION METAL COMPLEXES, F. E. Mabbs and D. J. Machin. A detailed view of the calculation methods involved in the magnetic properties of transition metal complexes, this volume offers sufficient background for original work in the field. 1973 edition. 240pp. 5 3/8 x 8 1/2. 0-486-46284-6

GENERAL CHEMISTRY, Linus Pauling. Revised third edition of classic first-year text by Nobel laureate. Atomic and molecular structure, quantum mechanics, statistical mechanics, thermodynamics correlated with descriptive chemistry. Problems. 992pp. 5 3/8 x 8 1/2. 0-486-65622-5

ELECTROLYTE SOLUTIONS: Second Revised Edition, R. A. Robinson and R. H. Stokes. Classic text deals primarily with measurement, interpretation of conductance, chemical potential, and diffusion in electrolyte solutions. Detailed theoretical interpretations, plus extensive tables of thermodynamic and transport properties. 1970 edition. 590pp. 5 3/8 x 8 1/2. 0-486-42225-9

Engineering

FUNDAMENTALS OF ASTRODYNAMICS, Roger R. Bate, Donald D. Mueller, and Jerry E. White. Teaching text developed by U.S. Air Force Academy develops the basic two-body and n-body equations of motion; orbit determination; classical orbital elements, coordinate transformations; differential correction; more. 1971 edition. 455pp. 5 3/8 x 8 1/2. 0-486-60061-0

INTRODUCTION TO CONTINUUM MECHANICS FOR ENGINEERS: Revised Edition, Ray M. Bowen. This self-contained text introduces classical continuum models within a modern framework. Its numerous exercises illustrate the governing principles, linearizations, and other approximations that constitute classical continuum models. 2007 edition. 320pp. 6 1/8 x 9 1/4. 0-486-47460-7

ENGINEERING MECHANICS FOR STRUCTURES, Louis L. Bucciarelli. This text explores the mechanics of solids and statics as well as the strength of materials and elasticity theory. Its many design exercises encourage creative initiative and systems thinking. 2009 edition. 320pp. 6 1/8 x 9 1/4. 0-486-46855-0

FEEDBACK CONTROL THEORY, John C. Doyle, Bruce A. Francis and Allen R. Tannenbaum. This excellent introduction to feedback control system design offers a theoretical approach that captures the essential issues and can be applied to a wide range of practical problems. 1992 edition. 224pp. 6 1/2 x 9 1/4. 0-486-46933-6

THE FORCES OF MATTER, Michael Faraday. These lectures by a famous inventor offer an easy-to-understand introduction to the interactions of the universe's physical forces. Six essays explore gravitation, cohesion, chemical affinity, heat, magnetism, and electricity. 1993 edition. 96pp. 5 3/8 x 8 1/2. 0-486-47482-8

DYNAMICS, Lawrence E. Goodman and William H. Warner. Beginning engineering text introduces calculus of vectors, particle motion, dynamics of particle systems and plane rigid bodies, technical applications in plane motions, and more. Exercises and answers in every chapter. 619pp. 5 3/8 x 8 1/2. 0-486-42006-X

ADAPTIVE FILTERING PREDICTION AND CONTROL, Graham C. Goodwin and Kwai Sang Sin. This unified survey focuses on linear discrete-time systems and explores natural extensions to nonlinear systems. It emphasizes discrete-time systems, summarizing theoretical and practical aspects of a large class of adaptive algorithms. 1984 edition. 560pp. 6 1/2 x 9 1/4. 0-486-46932-8

INDUCTANCE CALCULATIONS, Frederick W. Grover. This authoritative reference enables the design of virtually every type of inductor. It features a single simple formula for each type of inductor, together with tables containing essential numerical factors. 1946 edition. 304pp. 5 3/8 x 8 1/2. 0-486-47440-2

THERMODYNAMICS: Foundations and Applications, Elias P. Gyftopoulos and Gian Paolo Beretta. Designed by two MIT professors, this authoritative text discusses basic concepts and applications in detail, emphasizing generality, definitions, and logical consistency. More than 300 solved problems cover realistic energy systems and processes. 800pp. 6 1/8 x 9 1/4. 0-486-43932-1

THE FINITE ELEMENT METHOD: Linear Static and Dynamic Finite Element Analysis, Thomas J. R. Hughes. Text for students without in-depth mathematical training, this text includes a comprehensive presentation and analysis of algorithms of time-dependent phenomena plus beam, plate, and shell theories. Solution guide available upon request. 672pp. 6 1/2 x 9 1/4. 0-486-41181-8

Mathematics–Bestsellers

HANDBOOK OF MATHEMATICAL FUNCTIONS: with Formulas, Graphs, and Mathematical Tables, Edited by Milton Abramowitz and Irene A. Stegun. A classic resource for working with special functions, standard trig, and exponential logarithmic definitions and extensions, it features 29 sets of tables, some to as high as 20 places. 1046pp. 8 x 10 1/2. 0-486-61272-4

ABSTRACT AND CONCRETE CATEGORIES: The Joy of Cats, Jiri Adamek, Horst Herrlich, and George E. Strecker. This up-to-date introductory treatment employs category theory to explore the theory of structures. Its unique approach stresses concrete categories and presents a systematic view of factorization structures. Numerous examples. 1990 edition, updated 2004. 528pp. 6 1/8 x 9 1/4. 0-486-46934-4

MATHEMATICS: Its Content, Methods and Meaning, A. D. Aleksandrov, A. N. Kolmogorov, and M. A. Lavrent'ev. Major survey offers comprehensive, coherent discussions of analytic geometry, algebra, differential equations, calculus of variations, functions of a complex variable, prime numbers, linear and non-Euclidean geometry, topology, functional analysis, more. 1963 edition. 1120pp. 5 3/8 x 8 1/2. 0-486-40916-3

INTRODUCTION TO VECTORS AND TENSORS: Second Edition--Two Volumes Bound as One, Ray M. Bowen and C.-C. Wang. Convenient single-volume compilation of two texts offers both introduction and in-depth survey. Geared toward engineering and science students rather than mathematicians, it focuses on physics and engineering applications. 1976 edition. 560pp. 6 1/2 x 9 1/4. 0-486-46914-X

AN INTRODUCTION TO ORTHOGONAL POLYNOMIALS, Theodore S. Chihara. Concise introduction covers general elementary theory, including the representation theorem and distribution functions, continued fractions and chain sequences, the recurrence formula, special functions, and some specific systems. 1978 edition. 272pp. 5 3/8 x 8 1/2. 0-486-47929-3

ADVANCED MATHEMATICS FOR ENGINEERS AND SCIENTISTS, Paul DuChateau. This primary text and supplemental reference focuses on linear algebra, calculus, and ordinary differential equations. Additional topics include partial differential equations and approximation methods. Includes solved problems. 1992 edition. 400pp. 7 1/2 x 9 1/4. 0-486-47930-7

PARTIAL DIFFERENTIAL EQUATIONS FOR SCIENTISTS AND ENGINEERS, Stanley J. Farlow. Practical text shows how to formulate and solve partial differential equations. Coverage of diffusion-type problems, hyperbolic-type problems, elliptic-type problems, numerical and approximate methods. Solution guide available upon request. 1982 edition. 414pp. 6 1/8 x 9 1/4. 0-486-67620-X

VARIATIONAL PRINCIPLES AND FREE-BOUNDARY PROBLEMS, Avner Friedman. Advanced graduate-level text examines variational methods in partial differential equations and illustrates their applications to free-boundary problems. Features detailed statements of standard theory of elliptic and parabolic operators. 1982 edition. 720pp. 6 1/8 x 9 1/4. 0-486-47853-X

LINEAR ANALYSIS AND REPRESENTATION THEORY, Steven A. Gaal. Unified treatment covers topics from the theory of operators and operator algebras on Hilbert spaces; integration and representation theory for topological groups; and the theory of Lie algebras, Lie groups, and transform groups. 1973 edition. 704pp. 6 1/8 x 9 1/4. 0-486-47851-3